Food Proteins

With an awareness of the growing demand for inexpensive, nutritional, and functional proteins, this book provides an overview of various protein sources and their protein composition. It covers the behavior of proteins during processing and storage, as well as the potential applications of food proteins in different industries. It also features in-depth discussions of the chemical structure of food proteins, including how amino acid composition can determine nutritional qualities and functional properties, influencing the attributes of food products such as mouthfeel and flavor. Providing a comprehensive guide to the chemistry behind the proteins in food, this book is well suited for scientists and industry professionals seeking to develop their understanding of protein.

Food Proteins
Structure, Sources, and Applications

Edited by
Mudasir Ahmad Malik and Devinder Kaur

CRC Press
Taylor & Francis Group
Boca Raton London New York

CRC Press is an imprint of the
Taylor & Francis Group, an **informa** business

Designed cover image: Shutterstock

First edition published 2026
by CRC Press
2385 NW Executive Center Drive, Suite 320, Boca Raton FL 33431

and by CRC Press
4 Park Square, Milton Park, Abingdon, Oxon, OX14 4RN

CRC Press is an imprint of Taylor & Francis Group, LLC

ISBN: 9781041012436 (hbk)
ISBN: 9781041012429 (pbk)
ISBN: 9781003613909 (ebk)

DOI: 10.1201/9781003613909

Typeset in Times
by codeMantra

Contents

Preface

Proteins play an important role in building the blocks of body tissue and are the cornerstone of human nutrition. Proteins form the key functional ingredient in the food system. These are the most fascinating biomolecules to study due to their complex structure and wide range of applications in the food system.

This book, *Food Proteins: Structure, Sources, and Applications*, has been designed to provide comprehensive knowledge on proteins derived from various food sources including milk, cereals, pulses, oilseeds, fruits, vegetables, meat, fish, chicken, and eggs. Each chapter in this book is framed to offer a systematic understanding of protein composition, amino acid profiles, classification, and structural and physicochemical properties along with the effect of processing on the quality and functional attributes of proteins. Special focus is given to the industrial applications and the nutritional aspects of proteins from both plant and animal sources. This book is highly relevant to students, researchers, professionals, and academicians in the field of food science and technology. Each chapter in this book discusses the specific food protein source, exploring its detailed protein content and unique properties. This book brings together diverse information into one volume with the intention to bridge the gap between basic protein science and its industrial applications in different food systems.

I sincerely thank all my colleagues, mentors, and students, whose support and encouragement have been invaluable in shaping this book.

Dr. Mudasir Ahmad Malik
Dr. Devinder Kaur

About the Editors

Mudasir Ahmad Malik (Ph.D.) is an Assistant Professor in the Department of Food Engineering and Technology, Ghani Khan Choudhury Institute of Engineering & Technology (GKCIET), Malda, West Bengal, India. He has completed his B.Tech. in Food Technology from the Islamic University of Science and Technology and M.Tech. and Ph.D. from Sant Longowal Institute of Engineering and Technology Longowal. He has published a number of research articles in national and international journals with high impact factors. He has worked on the modification of buckwheat starch during his M.Tech. program and sunflower protein during his Ph.D. program. He has participated in a number of national and international conferences. He is currently working on waste utilization from the food industry for the development of value-added products. He has 7 years of teaching experience. He is an expert in food engineering, food chemistry, fruits, and vegetables.

Devinder Kaur (Ph.D.) is an Associate Professor at the Centre of Food Technology, University of Allahabad, Prayagraj. She obtained her degrees (B.Sc., M.Sc., and Ph.D.) in Food Technology from Guru Nanak Dev University, Amritsar. She has published more than 55 research papers in nationally and internationally reputed journals, 40 book chapters, three books, and one patent, and delivered invited, oral, and poster presentations. She is reviewer of various national and international journals. Dr. Kaur is actively involved in teaching undergraduate and postgraduate students. She has supervised several dissertations of undergraduate and postgraduate students, and guided doctorate students in the area of Food Technology. She has also completed research projects from UGC, DST, and UPCAR. She is a life member of the Association of Food Scientists and Technology and the National Academy of Science and a member of Vigyan Parishad Allahabad. Her area of research includes valorization of food waste, food pigments, and functional food.

Contributors

Tehreem Ashraf
National Institute of Food Science and Technology (NIFSAT)
University of Agriculture Faisalabad (UAF)
Faisalabad, Pakistan

Akshita Awasthi
Department of Plant Breeding and Genetics
Punjab Agricultural University
Ludhiana, India

Madhvi Awasthi
Department of Nutrition and Dietetics
School of Allied Health Sciences
Manav Rachna International Institute of Research and Studies
Faridabad, India

Syed Shafia Bashir
Department of Food Technology
School of Interdisciplinary Sciences and Technology, Jamia Hamdard
New Delhi, India

Swathika. E
Department of Food Science and Technology
Pondicherry University
Pondicherry, India

Rohit Ghatuary
Department of Horticulture
Lovely Professional University
Phagwara, India

Sundaramoorthy Haripriya
Department of Food Science and Technology
Pondicherry University
Pondicherry, India

Abida Jabeen
Division of Food Science & Technology
Sher-e-Kashmir University of Agricultural Sciences and Technology-Kashmir (SKUAST-K)
Srinagar, India

Jagriti Joshi
Department of Food Science and Technology
Pondicherry University
Pondicherry, India

Sweta Joshi
Department of Food Technology
School of Interdisciplinary Sciences and Technology, Jamia Hamdard
New Delhi, India

Pir Mohammad Junaid
Department of Post-Harvest Engineering and Technology
Aligarh Muslim University
Aligarh, India

Rupesh Kaushik
Department of Horticulture
Lovely Professional University
Phagwara, India

Tariq Khan
Department of Post-Harvest Engineering and Technology
Aligarh Muslim University
Aligarh, India

Amit Kotiyal
Department of Horticulture
Lovely Professional University
Phagwara, India

Mumtahin Ul Kousar
Division of Food Science & Technology
Sher-e-Kashmir University of Agricultural Sciences and Technology-Kashmir (SKUAST-K)
Srinagar, India

Noor Ul Ain Malhotrah
National Institute of Food Science and Technology (NIFSAT)
University of Agriculture
Faisalabad (UAF), Pakistan

Mudasir Ahmad Malik
Department of Food Engineering and Technology
Ghani Khan Choudhury Institute of Engineering and Technology
Malda, India

Sunita Mishra
Department of Food and Nutrition
Babasaheb Bhimrao Ambedkar University (A Central University)
Lucknow, India

Vikas Nanda
Department of Food Engineering and Technology
Sant Longowal Institute of Engineering and Technology
Longowal, India

Rabia Naseem
National Institute of Food Science and Technology (NIFSAT), University of Agriculture
Faisalabad (UAF), Pakistan

Gazia Nasir
Department of Bioengineering, Integral University
Lucknow, Uttar Pradesh, India

Zia Parveen
Department of Food and Nutrition, Babasaheb Bhimrao Ambedkar University (A Central University)
Lucknow, Uttar Pradesh, India

Aasima Rafiq
Division of Food Science & Technology
Sher-e-Kashmir University of Agricultural Sciences and Technology-Kashmir (SKUAST-K)
Srinagar, India

Bushra Shaida
Department of Food Technology
School of Interdisciplinary Sciences and Technology, Jamia Hamdard
New Delhi, India

Syed Shahid Reyaz Shah
Department of Food Technology
Islamic University of Science and Technology
Awantipora, India

Anamika Sharma
Department of Food Engineering and Technology
Sant Longowal Institute of Engineering and Technology
Longowal, India

Aditi Thakur
Department of Horticulture
Lovely Professional University
Phagwara, India

Kanchan Thakur
Department of Food Science Nutrition and Technology
CSK Himachal Pradesh Agricultural University
Palampur, India

Siddharth Thakur
Department of Food Science and Nutrition
Chaudhary Sarwan Kumar Himachal Pradesh Krishi Vishwavidyalaya
Palampur, India

Joshi Thodam
Department of Horticulture
Lovely Professional University
Phagwara, India

Ranjana Verma
Department of Food Science Nutrition and Technology
CSK Himachal Pradesh Agricultural University
Palampur, India

1 Milk Proteins

Sweta Joshi, Bushra Shaida,
Syed Shafia Bashir, and Gazia Nasir

INTRODUCTION

"The whole, fresh, clean, lacteal secretion obtained by the complete milking of one or more healthy milch animals, excluding that obtained within 15 days before or 5 days after calving" (De, 2008). The milk production in India in the year 2023–2024 was 239.3 million tonnes, in 2022–2023 was 230.6 million tonnes, and in 2021–2022 was 221.1 million tonnes, indicating an annual growth in milk production. Figure 1.1 shows the milk production in India since 1991 (National Dairy Development Board).

Since prehistoric times, humans have been known to ingest milk from both cattle (such as cows and buffalos) and non-cattle (such as goats and sheep) (Dunne et al., 2012, Evershed et al., 2008). Due to its extensive availability and high production quantities, cattle milk is the most consumed milk globally and is a convenient source of nutrition. People in poor nations and in regions where the climate is not conducive to dairy cattle's survival might benefit nutritionally from non-cattle milks. Goat milk also serves as an alternative for those suffering from cow milk intolerance (Faye &

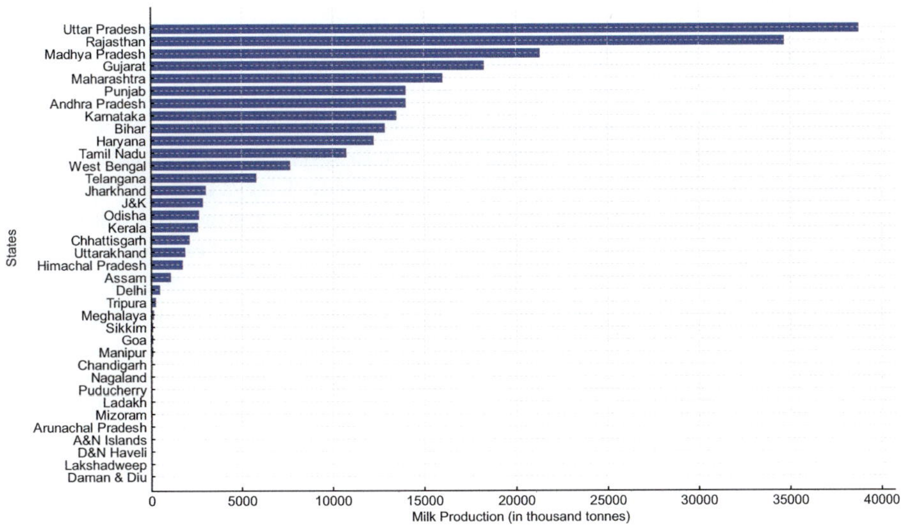

FIGURE 1.1 State-wise milk production.

DOI: 10.1201/9781003613909-1

Konuspayeva, 2012, Park, 2017). Examples are horse milk in central Asia's steppe regions; yak milk on the Tibetan plateau; reindeer milk in northern Scandinavia; musk ox milk in the Arctic; lamb milk in Europe and the Mediterranean basin (including the Middle East); goat milk in Africa and southern Asia; and camel milk, also known as "the white gold of the desert," in Africa. There are no global figures on the production of milk from other dairy animals, including llamas, yaks, horses, donkeys, deer, and musk ox. Due to the unknown quantities that are either sold directly to locals by farmers or consumed locally at their homes, a large portion of non-cattle milk production goes unrecorded, particularly in developing nations. In non-cattle agricultural systems, the addition of milk as a product increases value and aids farmers in managing the volatile costs of wool, meat, and hair. In many regions across the world, the milking of buffalo, goats, sheep, and camels is well established, growing in popularity, and proving to be a lucrative enterprise for those who have already adopted it (Roy et al., 2020).

Consuming dairy and milk are commonly cited as a crucial component of a nutritious, well-balanced diet. It is the first food that mammals eat and gives them all the energy and nutrients they need for healthy growth and development, which is especially important for the building of bone mass (Pereira, 2014). Numerous elements pertaining to the individual animal or the animal's surroundings affect the composition of bovine milk. Milk composition is influenced by several factors, many of which are not fully understood. These factors include diet (Larsen et al., 2010), breed (Palladino et al., 2010), individual animal genetics (Soyeurt et al., 2008), lactation stage (Stoop et al., 2009), management (Coppa et al., 2013), and season (Heck et al., 2009). Several mechanisms underlie these effects (Schwendel et al., 2015).

COMPOSITION OF MILK FROM DIFFERENT SOURCES

Milk primarily consists of water, fats, proteins, lactose (carbohydrates), and minerals (Gantner et al., 2015). The concentration of milk from different sources is summarized in Table 1.1. The physical characteristics and nutritional profile of milk are influenced by the concentrations of these components, which differ between species. The composition of milk varies significantly due to several factors, including breed, lactation stage, milking interval, feed type, and climate (Claeys et al., 2014).

The previous research has thoroughly examined the variations in milk composition across different species. Milk composition differs significantly among species, but all milks share four main components: protein, fat, lactose, and minerals. Non-ruminant milks, such as those from horses and donkeys, resemble human milk more closely in terms of protein, lactose, and ash content than do milks from dairy cattle and other ruminants. Ruminant milks generally contain higher levels of protein and fat compared to human and non-ruminant milks (Table 1.1). Human milk is notable for its much greater concentration of oligosaccharides derived from lactose, which far exceeds that found in other species. Goat milk is also recognized for its relatively high oligosaccharide content, with a composition considered like that of human milk (Roy et al., 2020).

TABLE 1.1
Milk Composition from Different Species

Species	Water (%)	Fat (%)	Protein (%)	Lactose (%)	Ash (%)	Total Casein (g/L)	Total Whey Proteins (g/L)	MUFA (%)	PUFA (%)	SFA (%)
					Ruminants					
Cow	86.30	4.9	3.4	4.1	0.7	24.6–28	5.5–7.0			
Buffalo	84.20	6.6	3.9	5.0	0.7	32–40	6.00	24.0–29.4	2.3–3.9	62.1–74
Goat	86.5	4.5	3.5	4.7	0.8	23.3–46.3	3.7–7.0	21.8–35.9	2.6–5.6	59.9–73.7
Sheep	83.70	6.0	5.4	4.9	0.8	41.8–52.6[a]	10.2–16.1	23–39.1	2.5–7.3	57.5–74.16
Camel	87.61	5.38	2.98	3.20	0.70	22.1–26.0	5.9–8.1	28.1–31.1	1.8–11.1	47–69.9
					Non-Ruminants					
Horse/Mare	90.8	1.3	2.1	6.6	0.5	9.4–13.6	7.4–9.1	18.9–36.2	12.8–51.3	37.5–55.8
Donkey	90.2	1.2	1.7	6.3	0.4	6.4–10.3	4.9–8.0	15.31–35	14.17–30.5	46.7–67.7
Reindeer	67.9	16.1	10.5	2.6	2.9	–	–	–	–	–
Moose	74.4	9.9	10.0	3.3	2.4	–	–	–	–	–
					Human					
Human	87.43	3.75	1.63	6.98	0.21	2.4–4.2	6.2–8.3	33.2–45.1	8.1–19.1	39.4–45

Source: Roy et al. (2020); Cheung and Mehta (2015); O'mahony and Fox (2012).

[a] Values (g/kg).

In terms of chemistry, milk is a complex mixture of fat that is emulsified as globules (2,000–6,000 nm), the main milk protein (casein, 50–300 nm), some mineral matter in the colloidal state, and lactose in the form of a true solution along with some minerals and soluble whey proteins (4–6 nm). The diameter of the lactose, salt, and other trace amounts of soluble substances is 0.5 nm. Under normal circumstances, the stability of milk is provided by the careful balance between the various elements in various states and phases. Normal milk has a pH between 6.5 and 6.7 (Cheung & Mehta, 2015). Buffalo milk contains higher levels of fat, lactose, protein, ash, calcium, and vitamins A and C, while having lower levels of vitamin E, riboflavin, and cholesterol. It lacks carotene but contains the blue-green pigment biliverdin, along with a bioactive Penta saccharide and gangliosides that are absent in cow milk. The fat globules in buffalo milk are larger but less stable, with a lower content of membrane materials compared to those in cow milk. Additionally, buffalo milk fat has a slightly higher proportion of saturated fatty acids and differs in triglyceride distribution and physical properties. The casein micelles in buffalo milk are larger and mineral-rich, and can withstand alkaline disruption at higher pH levels than those in cow milk (Abd El-Salam et al., 2011).

PROTEIN CONTENT AND AMINO ACID PROFILE OF MILK FROM DIFFERENT SOURCES

The diverse species have significantly diverse milk compositions (Ahmad et al., 2008). The primary class of milk proteins, caseins and whey proteins, are present in varying proportions in different milk species. The casein to whey protein ratio in human milk is 40:60; in animal milk, it is 50:50; and in the milk of cows, sheep, goats, and buffalo, it is 80:20. About 95% of the nitrogen constitutes protein, a crucial component of milk. The substantial differences occur between the various milk species in terms of protein fractions such as crude protein (CP), true protein (TP), casein, non-casein-nitrogen (NCN), whey proteins, and non-protein-nitrogen (NPN) levels. The most vital components of the human diet, proteins from milk have major nutritional, biological, and functional qualities. The amino acid profiles of whey and casein proteins hold a special place in human nutrition. They are considered high-quality proteins because of their high biological value, high digestibility (97%–98%), and quick body absorption and utilization. Because casein releases amino acids into the bloodstream gradually and continuously, it is a very effective source of nutrients (Schaafsma, 2000). Milk proteins include the largest concentration of branched-chain amino acids (BCAAs), which are crucial for tissue growth, healing, and preventing catabolic reactions during exercise. Similarly, the amino acid cysteine helps the body fight off several illnesses, has potent antioxidant qualities, and raises glutathione levels (Ha & Zemel, 2003). Furthermore, milk proteins are acknowledged for their use in medical nutritional formulae, baked goods, salad dressings, emulsifiers, infant formulas, and sports nutrition. The protein content of milk is significantly influenced by a number of parameters, including genetics, species/breed, lactation stages, diet type, udder health, and seasonal fluctuations (Pavić et al., 2002).

Milk is rich in key proteins such as casein and whey, which together provide a well-balanced mix of both essential and non-essential amino acids, each present in different amounts. The amino acid profile of milk from different sources is summarized in Table 1.2. Among the essential amino acids in casein, leucine was present in the greatest quantity, with lysine ranking as the second most abundant. The highest leucine levels in casein were observed in cow's milk (108 ± 2.30 mg/g), followed by camel (96 ± 2.20 mg/g), and buffalo (90 ± 2.40 mg/g). In contrast, whey proteins from various milk species showed only slight differences in their leucine content. Notably, leucine is especially important for protein metabolism and plays a specific role in initiating the process of muscle protein synthesis. The study identified a high concentration of the essential amino acid lysine in camel milk casein (67 ± 2.39 mg/g) and whey proteins (96 ± 2.20 mg/g). Significant amounts of valine, isoleucine, threonine, and phenylalanine were also present in both casein and whey fractions across milk species. The highest valine levels were detected in cow milk casein (54 ± 1.42 mg/g) and sheep milk whey proteins (53 ± 1.30 mg/g).

In goat milk, casein showed the greatest concentrations of isoleucine, phenylalanine, and histidine. Meanwhile, phenylalanine content was elevated in whey proteins from camel (57 ± 1.50 mg/g), sheep (51 ± 1.39 mg/g), and cow (44 ± 1.25 mg/g) milk. Among non-essential amino acids, glutamic acid was the most abundant in both casein and whey proteins. Notably, buffalo (367 ± 4.50 mg/g) and goat (359 ± 5.29 mg/g) whey proteins contained the highest glutamic acid levels compared to casein. Proline was notably present in goat casein (144 ± 3.29 mg/g) and camel whey proteins (129 ± 4.79 mg/g). Cysteine was detected in both casein and whey proteins, with small variations among the milk species. Additionally, buffalo milk whey proteins exhibited higher levels of asparagine and serine compared to the other species analyzed (Rafiq et al., 2015).

COMPOSITION OF MILK PRODUCTS

Considering that milk is a complex, nutrient-rich liquid, mostly composed of water (around 87%), it serves as an excellent medium for dissolving and suspending other important nutrients. The composition of different milk products has been summarized in Table 1.3. Its structural and functional qualities are influenced by the proteins (~3.3%) it contains, of which 80% are caseins and 20% are whey proteins (Mehta, 2015). Additionally, milk contains lactose, a type of carbohydrate (~4.8%) that provides energy. Essential fatty acids, fat-soluble vitamins, and energy are provided by the fat content, which varies from about 3.25% in whole milk to almost none in skimmed milk. Minerals essential for bone health and metabolic processes include calcium, phosphorus, magnesium, and potassium (~0.7%). Vitamins A, D, B2, and B12 are also present in milk, enhancing its nutritional profile. Different animal species, processing techniques, and nutritional variances all impact the milk composition and its derivative products (Pereira, 2014).

TABLE 1.2
Amino Acid Profile of Milk from Different Sources

Amino Acid	Buffalo	Cow	Goat	Sheep	Camel
Essential Amino Acid Content in Caseins of Different Milk Species (mg/g)					
Threonine	40	40	23	40	34
Valine	49	54	49	49	60
Methionine	17	21	19	17	23
Isoleucine	34	60	62	44	34
Leucine	90	108	69	79	96
Phenylalanine	49	49	56	45	45
Histidine	18	21	25	19	19
Lysine	49	63	61	55	67
Essential Amino Acid Content in Whey Proteins of Different Milk Species (mg/g)					
Threonine	38	41	31	30	25
Valine	47	49	47	53	33
Methionine	17	22	22	26	27
Isoleucine	39	34	35	25	29
Leucine	74	81	71	81	83
Phenylalanine	38	44	39	51	57
Histidine	15	16	16	18	23
Lysine	53	65	63	71	96
Non-Essential Amino Acid Content in Caseins of Different Milk Species (mg/g)					
Aspartate	66	62	38	69	61
Serine	50	47	39	50	44
Glutamate	254	218	241	338	303
Proline	57	96	144	55	73
Glycine	15	12	15	14	14
Alanine	22	21	28	21	25
Cysteine	8	12	10	8	13
Tyrosine	51	56	65	49	59
Arginine	112	40	36	32	28
Non-Essential Amino Acid Content in Whey Proteins of Different Milk Species (mg/g)					
Aspartate	74	65	70	45	36
Serine	52	47	51	45	36
Glutamate	367	350	359	316	257
Proline	77	59	57	102	129
Glycine	15	11	14	12	12
Alanine	22	20	24	35	39
Cysteine	9	12	12	13	16
Tyrosine	44	43	42	45	58
Arginine	8	10	7	5	5

Source: Rafiq et al. (2015).

TABLE 1.3
Composition of Milk and Milk Products

S. No.	Milk Product	Water (%)	Fat (%)	Protein (%)	Lactose (%)	Minerals (%)
1	Whole milk	87	3.27–4	3.3	4.8	0.7
2	Skimmed/ low-fat milk	90	Skimmed <0.5 low-fat 1–2	3.4	5	0.7
3	Cream	55–65	10–48	2–3	3–4	0.5–0.6
4	Butter	15–18	80–85	1	Trace amounts	0–2
5	Cheese	30–60	20–35	10–30	Trace to low	3–5
6	Yogurt	85	0.1–10	5	3–4	0.7–1
7	Powdered milk	<5	Whole 26–28 Skimmed <1.5	25–27	38–50	5–8
8	Condensed milk	25–30	8–10	8	12–16	–
9	Ghee	<0.5	99.5	Trace Amounts	None	–
10	Ice cream	55–65	10–20	3–5	5	–
11	Whey	94	0.1–0.5	0.8–1	4.5	0.5–0.7

Source: Indian Food Composition tables, National Institute of Nutrition, ICMR.

CLASSIFICATION AND STRUCTURE OF MILK PROTEIN

When milk acidification happens at pH 4.6 (isoelectric point) at normal temperature, almost 80% of cow and buffalo milk protein precipitates from the solution. That portion is called casein, while the protein that is still soluble in the solution is known as whey or serum protein (Fox et al., 2015). Small amounts of nitrogenous substances that are not proteins are also present in the solution. The classification of milk protein is illustrated in Figure 1.2. Each species has a different ratio between casein and whey protein. Human milk is 40:60, and in cow, buffalo, sheep, and goat milk, it is 80:20.

Casein

The total nitrogen in human milk is 40%, and in bovine milk, it is 80%, which makes casein an essential milk protein. The amount of casein in milk typically ranges from 24 to 29 g/L and depends on the type of breed and lactation stage. Caseins exhibit a unique structure (Figure 1.3) and physical, biological, and nutritional characteristics typical for milk proteins. It is found in milk as casein micelles, which are mostly 0.02–0.30 μm in diameter (Jianqin et al., 2015). Except cysteine, the protein of casein includes almost all necessary amino acids (Fox et al., 2015).

Additionally, these micelles include inorganic materials, primarily calcium phosphate, at a rate of roughly 8 g/100 g of casein. Heat does not affect caseins; since casein molecules naturally reside in an expanded and open condition (an increase in water entropy) at higher temperatures, heat has little to no effect. Casein is comprised of amino acids that form a polymeric chain through a peptide bond.

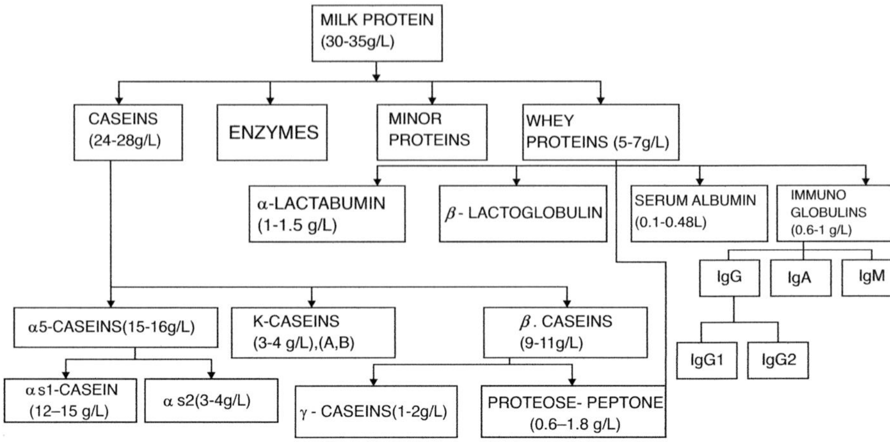

FIGURE 1.2 Classification of milk proteins.

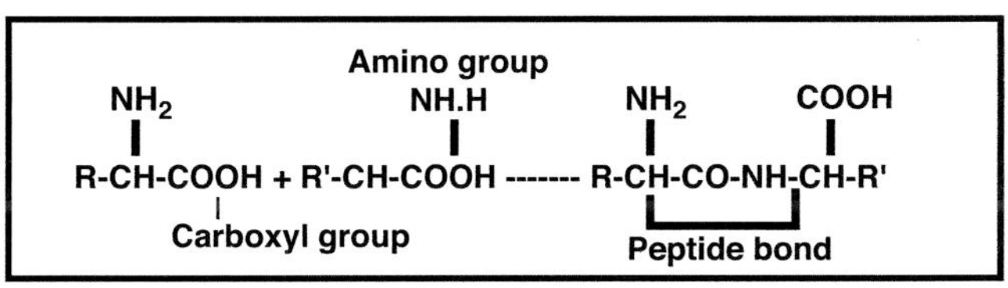

FIGURE 1.3 Casein structure.

Long-term heating of milk at high temperatures, however, may cause hydrolytic breakage of the phosphate and peptide bonds, compromising the stability of the complex and causing milk to coagulate. Compared to individual casein molecules, the voluminous casein micelles can hold more water. Each of the casein proteins—α-casein, β-casein, κ-casein, and γ-casein—has distinct functional characteristics. Caseins aid in muscular repair and satiety because of their gradual rate of digestion, which releases amino acids gradually (Fernandez & Marette, 2020).

At its isoelectric point, casein exhibits a zero net change at pH 4.6. It is positively charged below this pH and negatively charged above it. Because casein has a negative charge at milk's pH (6.6–6.8), it moves in an electric field in the direction of the anode.

Both α- and β-casein at their isoelectric points (pH 4.7 and 4.9) are insoluble, but they dissolve easily on either side of pH (Bhat et al., 2016). At pH 4.2 at 20°C, α-casein is less soluble, whilst β casein is more soluble, which leads to their separation. Additionally, casein aggregates when it binds calcium ions above its isoelectric point, decreasing its solubility. Because of its amino acids, casein also exhibits oxidation and reduction, optical rotation, and globular nature, and can react with formalin and formaldehyde. During this reaction with formalin, an equivalent number of protons (H^+ ions) are released, and they are subsequently titrated against an alkali. The quantity of protein present is equal to the amount of protons released (Luo et al., 2014).

Whey Proteins

Whey is the portion of milk that remains after casein has been separated and coagulated during the production of cheese and yoghurt. Whey proteins are a blend of globular proteins with different functions and characteristics (Kilara & Vaghela, 2018). β-lactoglobulin and α-lactalbumin are the main fractions; heat causes both to coagulate to varying degrees depending on the temperature, holding period, concentration of salt, and solution pH (e.g., pH 4.5) (Crowley et al., 2016). β-Lactoglobulin, α-lactalbumin, bovine serum albumin, immunoglobulins (mostly IgG1, IgG2, and IgM), lactoferrin, protease peptone, serum transferring, glycoprotein, and enzymes are all components of whey proteins (Table 1.4).

β-Lactoglobulin

The major whey protein present in buffalo and cow milk is β-lactoglobulin, wherein buffalo milk displays a molecular weight of about 38,500 Daltons, compared to 37,600 Daltons observed in the cow milk β-lactoglobulin. Indian cattle breed milk mainly consists of two genetic variants (A and B) of β-lactoglobulin, whereas in the western breeds, C and D variants have also been reported. Milk of Murrah buffalo shows the predominant presence of two genetic variants of β-lactoglobulin. Their electrophoretic mobility is similar to that of A and C variants of cow milk β-lactoglobulin. Buffalo milk also shows the presence of yet another variant of β-lactoglobulin, which has faster electrophoretic mobility than the A variant of cow milk β-lactoglobulin (Sawyer, 2012).

α-Lactalbumins

The characteristics of the α-lactalbumins found in cow and buffalo milk are nearly identical. It has been determined that the molecular weight of cow α-lactalbumins is 14,176 Daltons, while that of buffalo α-lactalbumins is 16,200 Daltons. Both cow and buffalo milks contain two genetic variants of α-lactalbumins, i.e., A and B. Their electrophoretic mobility is nearly the same (Yadav et al., 2015).

Immunoglobulins (Ig): These are antibodies that are produced in reaction to particular antigens stimulating the body. These provide a non-specific humoral reaction to respiratory and gastrointestinal microorganisms that are Gram-negative. Immunoglobulins fall into five main classes: IgA, IgD,

TABLE 1.4
Properties of Whey Protein

Property	β-Lactoglobulin	α-Lactalbumin	Bovine Serum Albumin	Lactoferrin
Molecular Weight	36,000	16,000	65,000	80,000
Isoelectric Points	5.18	4.2	4.72	8.2

IgE, IgG, and IgM. All Igs share a similar basic structure consisting of two identical heavy chains (53,000 Daltons) and two identical light chains (23,000 Daltons). IgG makes up almost 90% of the total colostral immunoglobulins in cow and buffalo milk, is the most abundant portion of the five immunoglobulin classes. When compared to cow milk, buffalo milk has a much greater average immunoglobulin concentration (76.2 µg/100 g whey proteins) in the first milking after delivery. On the sixth day after giving birth, it drops to 43.4 µg/100 g (Bell, 2000).

Bovine Serum Albumin

The milk-derived protein known as bovine serum albumin (BSA) has a molecular weight of 66.5 kDa and an isoelectric point of 4.7. Its polypeptide chain is 583 amino acids long. BSA accounts for 5%–10% of whey protein. It contains additional bioactive properties, reduces the absorption of cholesterol, and is high in essential AA. The biological characteristics of proteins are greatly influenced by the presence of methionine, isoleucine, leucine, tyrosine, valine, and proline in their structure. Tryptophan (with two residues, Trp-213 and Trp-134), phenylalanine, and tyrosine are the three natural fluorophores. The protein's outer surface had Trp-134, while its hydrophobic structure contained Trp-213. BSA is a non-glycosylated globular transport protein that is present in serum and is essential for the circulatory system (Solanki et al., 2021).

Lactoferrin: This glycoprotein's capacity to bind free ionic iron, which is fundamentally necessary for the growth of enteropathogenic microorganisms, gives it a potent inhibitory effect on Gram-negative enteropathogenic bacteria. In addition to its antibacterial properties, lactoferrin has been linked to a nutritional function in iron metabolism. Compared to cow milk, which has an average lactoferrin level of 0.15–0.32 mg/mL, buffalo milk has a comparatively higher lactoferrin content of 0.32–0.5 mg/mL (Huppertz et al., 2018).

Proteose-Peptone: Compared to cow milk, which has an average amount of 240 mg/100 ml, buffalo milk has a substantially greater level of this protein, at roughly 330.5 mg/100 mL (Murtaza et al., 2017).

PROPERTIES OF MILK PROTEINS

Physical and Chemical Properties of Casein

Solubility: One important functional characteristic of casein that affects its possible use is its solubility. Casein micelles are very stable at isoelectric pH, although cationic salts of casein are generated at higher pH (above 5.5) and very low pH (less than 3.5), and they are very soluble. Several variables, including the solvent's structure and characteristics, protein composition, temperature and pH, dielectric constant, and ionic strength, affect how soluble casein is. Sodium and ammonium caseinates are more soluble than calcium caseinates because of the high rate of ionization. Calcium caseinates produce stable clusters at pH > 5.5. Several studies have examined the effects of NaCl on different forms of casein at a pH of 2.5. They found that the addition of 0.1 M NaCl reduced casein's solubility by 98% and 90%, respectively, and dephosphorylated casein.

The hydrolysis of casein by enzymes increases its solubility. As a result, smaller charged groups are formed, which increases the hydrophilia of the protein. Augusta Rolim Biasutti et al. (2007) investigated how trypsin hydrolysis affected casein's functional properties and found that as reaction time increased, solubility improved as well.

Moisture Binding Properties: The micellar matrix structure of casein has a water-binding capacity of 2–4 g/g protein and allows for the mechanical trapping of a significant amount of water. A greater capacity to bind water is also made possible by the k-casein "hairs" that emerge from the micelle's surface. Casein's ability to bind moisture is greatly enhanced by enzymatic hydrolysis (Wang et al., 2022). Gani et al. (2015) investigated how pepsin affected casein's water-binding capacity and found that hydrolysis at 20°C raised it from 18.47 to 35.37 mL/g. The breakdown of casein into smaller molecules with numerous water-binding sites could be the cause of this. Similarly, compared to casein hydrolysates, unhydrolyzed sodium caseinate has a greater capacity to bind water (Augusta Rolim Biasutti et al., 2007).

Gelation: Gelation, sometimes referred to as coagulation, happens when milk undergoes limited proteolysis by acidification. Renneting primarily affects κ-casein, which hydrolyzes κ-casein that stabilizes micelles to produce para-κ-casein-containing micelles that agglomerate in milk serum and coagulate when Ca^{2+} ions are present. Acidification simultaneously impacts the casein micelle's core and hairy brush (Schorsch et al., 2001). Gelation was significantly impacted by the solution's pH. Transparent gel was created at the pH of maximal stability. But gels made at low pH have a high rate of syneresis and are opaque (Panouillé et al., 2004). Steric and electrostatic repulsion are not present at this pH.

Viscosity: Due to the random and open structure of casein molecules, casein solutions have a high viscosity. Casein viscosity grows logarithmically with protein concentration and is linearly correlated with it. There is a well-established relationship between temperature and casein viscosity, which decreases logarithmically with temperature (Carr & Golding, 2016). The viscosity of the casein solution is also significantly influenced by the pH. Casein's viscosity peaks at a pH of almost neutral and sharply increases at lower or higher pH values. According to Liu and Guo (2008), a high (20%) concentration of sodium and calcium caseinate shows an inverse viscosity relationship with pH.

Casein solution viscosity was impacted by the presence of cations, solution's pH, and is also greatly affected by the addition of salts. The protein solution's viscosity decreased when calcium was added at a pH of 5.5. Casein molecules aggregate due to calcium, which also makes the particles bigger. However, proteins become more viscous when calcium chloride is added at a pH of 6.5 (Konstance & Strange, 1991). Casein is broken down by proteinase enzymes, which lowers the solution's viscosity. In their study, Solorza-Feria et al. (2011) examined how several plant and animal enzymes affected the apparent viscosity of casein protein and found that enzyme treatment decreased casein's viscosity.

Emulsion: The capacity to stabilize oil or oil–water emulsions at low concentrations under predetermined circumstances is known as emulsifying power. As a result, the oil droplets' surface tension, interface adsorption, and cohesive film formation were all reduced. The peptide chain's several hydrophobic regions cause bonds to form between the protein and the oil. Casein interacts hydrophobically with milk fats and other lipids, making it a great emulsifier. Casein's solubility is a key factor in its emulsifying capabilities, and proteins with low solubility perform poorly in emulsification (Darewicz et al., 2006).

The emulsifying characteristics of casein may not be considerably impacted by the presence of contaminants such as γ-casein. Yet some fractions lower surface tension because they are surface-active (Atamer et al., 2017). Na-caseinate-based emulsions demonstrated improved stability during spray drying, heat stability, and freeze-thaw endurance. The emulsion system becomes unstable due to the significant flocculation propensity of calcium caseinate emulsions (Carr & Golding, 2016). Casein partial hydrolysis enhances emulsion characteristics but decreases stability. Furthermore, bacterial *Streptoverticillium mobaraense* transglutaminase and fungal *Trichoderma reesei* tyrosinase used in enzymatic cross-linking show improved emulsifying qualities (Monogioudi et al., 2011).

Foaming Properties: Generally speaking, foaming is linked to casein's surface activity at the air/aqueous interface, where surface tension drops. Factors such as temperature, ionic strength, pH, and protein concentration affect casein's capacity to foam (Silva et al., 2013). Of all the caseins, β-casein has a greater propensity to lower interfacial tension and a high capacity for foaming, followed by α- and κ-caseins. However, because of their tertiary structure and intermolecular interactions, foams made with α-casein are more stable (Ho et al., 2022). The foaming characteristics of milk protein dispersions made with different ratios of casein and whey proteins were investigated by Xiong et al. (2020). They found that dispersions with a high casein content (80:20) had a low capability for foaming.

Physical and Chemical Properties of Whey Protein

Solubility: One of the most important functional characteristics of proteins is their solubility, which greatly influences the functional characteristics of other proteins. Proteins must have more solubility to generate the proper foam, emulsion, whipping, and gelation. Surface hydrophilic (protein–solvent) and hydrophobic (protein–protein) interactions dictate a protein's solubility. One of whey protein's unique qualities for a variety of beverage applications is its solubility in water throughout a broad pH range (2–9). Whey proteins have a difficult time remaining soluble when heated.

Water-Holding Capacity: The quantity of water held in a protein matrix under circumstances is quantified by the term "water-holding capacity" (Zayas, 1997). Numerous theories regarding the fundamentals of protein–water interactions have been put forth in the past. Kinsella and Whitehead (1989) and Goff et al. (1989) identified six fundamental forms of water that are connected to proteins.

- Hydrophobically hydrated water surrounds polar residues in a cage-like structure.
- The first water to be absorbed and coupled to protein groups is monolayer water, which may be accessible for specific processes.
- Unfrozen water is any water that, depending on the polar side chain concentration and AA composition, does not freeze at a sharp transition temperature.
- Hydrodynamic hydration water that loosely envelops the protein.
- Capillary water which is mechanically held within the protein molecule by surface forces.
- Structural water that isn't suitable for use in chemical processes.

The type of the solution, dispersion, suspension, powder, or isolation is determined by the chemical makeup of whey protein. A three-dimensional network that absorbs and retains water is created by chemical and physical interactions. Additionally, this mixture exhibits swelling properties, meaning that water can be held, bound, and trapped by it (Kneifel et al., 1993).

Emulsification: The process of dispersion of one or more phases into a continuous phase creates heterogeneous systems known as emulsions. Amphiphilic surface-active substances that have affinities for both dispersing/continuous and dispersed phases can be added to stabilize them (Farooq & Boye, 2011). Reducing interfacial tension and facilitating easier diffusion of the dispersed phase are the main objectives of adding an emulsifier. In protein-stabilized emulsion systems, an oil droplet is surrounded by an interfacial membrane to prevent coalescence, flocculation, creaming, or oiling-off. To function as an effective emulsifier, proteins must be soluble and able to reduce the interfacial tension at the oil/water interface (Shimizu et al., 1981). Temperature, protein content, salts, and pH all affect how proteins adsorb onto the surface of fat globules, changing the whey proteins' capacity to emulsify.

Whey protein–coconut oil emulsions had higher protein adsorption and viscosity and less creaming stability at pH 5 (Yamauchi et al., 1980). This suggests that emulsion stability is significantly impacted by the electrostatic character of the protein.

Gelation: Whey proteins can create gels with various properties like hard, rubbery, viscous, and smooth, soft curds. They differ in terms of mouthfeel, stickiness, color, and cohesiveness (Kinsella & Whitehead, 1989). Coagulation, or curd-like gel formation, is caused by extreme thermal processing, while mild heating of whey protein can produce specific protein–protein interactions that result in gel formation. The proteins create a structural network that traps water during gelation, which reduces syneresis. The development of a gel's structure, elasticity, mechanical strength, and color is essential during culinary applications (Mulvihill & Kinsella, 1988). Whey proteins, when heated above their critical temperature, experience structural alterations and aggregation. On cooling, clear to opaque and soft to stiff viscous gels are formed by these aggregates.

Temperature, pH, concentration, calcium, and protein type influence gel formation. To form a three-dimensional network, a balance between the repulsive and attractive forces is necessary. In certain situations, due to limited protein unfolding during heating, hydrophobic residues are exposed and subsequently linked on cooling. A gel network is created based on the number of contact sites, molecular reactivity, and the balance and strength of repulsive interactions between various solute molecules. Therefore, greatly impacted by the ionic strength and pH. Roughly 1–2 molecules of the protein create a thick viscous network, depending on the degree of cross-linking in protein gels containing serum albumin (Minj & Anand, 2020).

EFFECT OF PROCESSING ON MILK PROTEINS

During processing, various chemical modifications occur in milk proteins, which alter the digestion and overall physiological impact that milk consumption has. Processing of milk majorly causes the following protein (amino acid) alterations (Van Lieshout et al., 2020):

- Protein Denaturation
- Protein Aggregation
- Chemical Modifications
- Oxidation
- Glycation
- Dephosphorylation

Protein Denaturation

Various factors during milk processing like heat, acid, or mechanical forces can change the normal shape of milk protein, which leads to protein denaturation. This process also affects the structural and functional aspects of milk proteins (Van Lieshout et al., 2020).

Heating whey proteins, such as α-lactalbumin (α-La) and β-lactoglobulin (β-Lg), above 70°C–75°C denatures and unfolds the tertiary structure of the proteins, exposing free thiol groups (Crowley et al., 2016).

Casein micelles do not change as much as whey proteins when heated. However, denatured whey proteins interact (Figure 1.4) with κ-casein to form disulfide bonds, which can affect the structure.

Amino Acid Aggregation: Heat treatment and pressure during milk processing aggregate the amino acids in milk protein, which in turn changes the protein quality (Table 1.5). This aggregation happens through disulfide or non-covalent bonds (heat-generated protein aggregation at the fat surface through disulfide bonds and pressure generated aggregation through non-covalent bonds). Aggregation impacts the overall quality and structure of protein, which can impact its aminoacidemia. It can also affect the structure of milk proteins, which can affect the quality of dairy products like cheese and yogurt (Horstman & Huppertz, 2023).

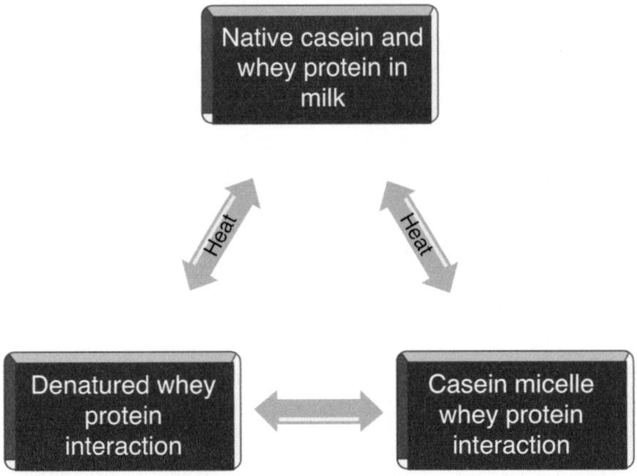

FIGURE 1.4 Protein denaturation.

TABLE 1.5
Factors Influencing Chemical Modifications

Factors	Impact	Reference
Temperature	High temperatures directly impact protein denaturation and aggregation.	Krishna et al. (2021)
pH	Changes in pH can affect protein stability and denaturation.	Law and Leaver (2000)
Processing Time	Longer processing times can increase the extent of chemical modifications.	van Lieshout et al. (2020)

Oxidation: Alterations in protein function occur due to exposure to oxygen during processing which leads to oxidation of amino acid side chains, potentially contributing to off-flavors.

Glycation: Lactose binds with protein amino group during heating, forming a Maillard reaction product which affects protein digestibility.

Dephosphorylation: The removal of phosphate groups from casein proteins can happen during processing, which impacts the calcium-binding ability.

APPLICATIONS OF MILK PROTEINS

Due to the diverse functional, nutritional, and bioactive properties, milk proteins have a wide range of applications in the food industry, healthcare, and beyond. Table 1.6 illustrates some of the most common applications:

- Food Industry
- Nutraceuticals and Dietary Supplements
- Biotechnology Uses
- Pharma Industry
- Cosmetic Industry

TABLE 1.6
Application of Milk Proteins in Food Industries

Milk Proteins	Industry	Application	Reference
Casein Protein Products	Bakery and confectionery	Used as stabilizers in creams and cakes because of its whipping properties.	Rollema and Muir (2009)
	Liquor industry	Used in fermentation of ethanolic beverages, as fining agent, to reduce color and astringency and to aid in clarification.	
Sodium Caseinate	Frozen food manufacturing industry	Acts as a stabilizer and used in ice-cream substitutes and frozen desserts to improve whipping properties, and texture.	Patel et al. (2014)
Whey Protein Products	Beverage industry	Whey protein concentrate is used as a supplement in ready to drink fruit beverages.	Patel et al. (2014)
Casein and Whey	Meat industry	Used as fillers, binders in cooked meat products to improve texture, water-binding capacity, and to improve nutritive value.	Pennings et al. (2011)
Milk Protein Concentrate	Milk replacement (non-industrial use)	Can be used as a replacer for whole milk powder and skimmed milk powder, helping to formulate products with higher protein and low lactose profile as milk.	Szigeti and Lupski (2009)

CONCLUSION

The study of milk proteins highlights their fundamental importance in both nutrition and dairy technology. Casein and whey proteins, the principal protein fractions in milk, are distinguished by their unique amino acid profiles, including a rich supply of essential and branched-chain amino acids such as leucine, isoleucine, and valine. These proteins contribute not only to the nutritional value of milk but also to its functional properties, influencing the texture, stability, and processing characteristics of dairy products. Comparative analysis across different milk species—such as cow, buffalo, goat, sheep, and camel—reveals notable differences in protein composition and amino acid distribution. Such diversity offers opportunities for the dairy industry to develop specialized, nutritionally enhanced products tailored to specific dietary needs, including those of vulnerable population groups. Moreover, the favorable balance of amino acids in both casein and whey proteins underscores the potential for utilizing non-bovine milk sources in functional food formulations. In summary, milk proteins are central to the nutritional quality and technological versatility of milk and dairy products. Continued research and innovation in this area will further enable the dairy sector to meet evolving consumer demands and address global nutritional challenges.

REFERENCES

Abd El-Salam, M. H., & El-Shibiny, S. (2011). A comprehensive review on the composition and properties of buffalo milk. *Dairy Science & Technology*, *91*, 663–699. https://doi.org/10.1007/s13594-011-0029-2

Ahmad, S., Gaucher, I., Rousseau, F., Beaucher, E., Piot, M., Grongnet, J. F., & Gaucheron, F. (2008). Effects of acidification on physico-chemical characteristics of buffalo milk: A comparison with cow's milk. *Food Chemistry*, *106*(1), 11–17. https://doi.org/10.1016/j.foodchem.2007.04.021

Atamer, Z., Post, A. E., Schubert, T., Holder, A., Boom, R. M., & Hinrichs, J. (2017). Bovine β-casein: Isolation, properties and functionality. A review. *International Dairy Journal*, *66*, 115–125. https://doi.org/10.1016/j.idairyj.2016.11.010

Augusta Rolim Biasutti, E., Regina Vieira, C., Capobiango, M., Dias Medeiros Silva, V., & Pinto Coelho Silvestre, M. (2007). Study of some functional properties of casein: Effect of pH and tryptic hydrolysis. *International Journal of Food Properties*, *10*(1), 173–183. https://doi.org/10.1080/10942910600853865

Bell, S. J. (2000). Whey protein concentrates with and without immunoglobulins: A review. *Journal of Medicinal Food*, *3*(1), 1–13. https://doi.org/10.1089/jmf.2000.3.1

Bhat, M. Y., Dar, T. A., & Singh, L. R. (2016). Casein proteins: Structural and functional aspects. In *Milk Proteins-From Structure to Biological Properties and Health Aspects* (vol. 10, pp. 64–187). https://doi.org/10.5772/64187

Carr, A., & Golding, M. (2016). Functional milk proteins production and utilization: Casein-based ingredients. In *Advanced Dairy Chemistry: Volume 1B: Proteins: Applied Aspects* (pp. 35–66). https://doi.org/10.1007/978-1-4939-2800-2_2

Cheung, P. C. K., & Mehta, B. M. (Eds.). (2015). *Handbook of Food Chemistry* (Vol. 11). Berlin, Heidelberg: Springer. https://doi.org/10.1007/978-3-642-36605-5

Claeys, W. L., Verraes, C., Cardoen, S., De Block, J., Huyghebaert, A., Raes, K., … & Herman, L. (2014). Consumption of raw or heated milk from different species: An evaluation of the nutritional and potential health benefits. *Food Control*, *42*, 188–201. doi.org/10.1016/j.foodcont.2014.01.045.

Coppa, M., Ferlay, A., Chassaing, C., Agabriel, C., Glasser, F., Chilliard, Y., … & Martin, B. (2013). Prediction of bulk milk fatty acid composition based on farming practices collected through on-farm surveys. *Journal of Dairy Science*, *96*(7), 4197–4211. https://doi.org/10.3168/jds.2012-6379

Crowley, S. V., Dowling, A. P., Caldeo, V., Kelly, A. L., & O'Mahony, J. A. (2016). Impact of α-lactalbumin: β-lactoglobulin ratio on the heat stability of model infant milk formula protein systems. *Food Chemistry*, *194*, 184–190. https://doi.org/10.1016/j.foodchem.2015.07.077

Darewicz, M., Dziuba, J., & Dziuba, M. (2006). Functional properties and biological activities of bovine casein proteins and peptides. *Polish Journal of Food and Nutrition Sciences*, *15*(1), 79.

De, S. (2008). *Outlines of Dairy Technology*, 2nd Edition (pp. 463–464). New Delhi: Oxford University Press.

Dunne, J., Evershed, R. P., Salque, M., Cramp, L., Bruni, S., Ryan, K., … & di Lernia, S. (2012). First dairying in green Saharan Africa in the fifth millennium BC. *Nature*, *486*(7403), 390–394. doi.org/10.1038/nature11186

Evershed, R. P., Payne, S., Sherratt, A. G., Copley, M. S., Coolidge, J., Urem-Kotsu, D., … & Burton, M. M. (2008). Earliest date for milk use in the Near East and southeastern Europe linked to cattle herding. *Nature*, *455*(7212), 528–531. doi.org/10.1038/nature07180

Farooq, Z., & Boye, J. I. (2011). Novel food and industrial applications of pulse flours and fractions. In B. K. Tiwari, A. Gowen, & B. McKenna (Eds.). *Pulse Foods: Processing, Quality and Nutraceutical Applications* (pp. 283–323). Academic Press.

Faye, B., & Konuspayeva, G. (2012). The sustainability challenge to the dairy sector–The growing importance of non-cattle milk production worldwide. *International Dairy Journal*, *24*(2), 50–56. doi.org/10.1016/j.idairyj.2011.12.011

Fernandez, M. A., & Marette, A. (2020). Dairy products and diabetes: Role of protein on glycaemic control. In *Milk and Dairy Foods* (pp. 173–203). Academic Press. https://doi.org/10.1016/B978-0-12-815603-2.00007-3

Fox, P. F., Uniacke-Lowe, T., McSweeney, P. L. H., & O'Mahony, J. A. (2015). Milk proteins. In *Dairy Chemistry and Biochemistry* (pp. 145–239). Springer International Publishing. https://doi.org/10.1007/978-3-319-14892-2_4

Gani, A., Broadway, A. A., Masoodi, F. A., Wani, A. A., Maqsood, S., Ashwar, B. A., … & Gani, A. (2015). Enzymatic hydrolysis of whey and casein protein-effect on functional, rheological, textural and sensory properties of breads. *Journal of Food Science and Technology*, *52*, 7697–7709. https://doi.org/10.1007/s13197-015-1840-1

Gantner, V., Mijić, P., Baban, M., Škrtić, Z., & Turalija, A. (2015). The overall and fat composition of milk of various species. *Mljekarstvo: časopis za unaprjeđenje proizvodnje i prerade mlijeka*, *65*(4), 223–231. doi.org/10.15567/mljekarstvo.2015.0401

Goff, H. D., Kinsella, J. E., & Jordan, W. K. (1989). Colloidal aspects of ice cream – A review. *Journal of Dairy Science*, *72*, 18–29.

Ha, E., & Zemel, M. B. (2003). Functional properties of whey, whey components, and essential amino acids: Mechanisms underlying health benefits for active people. *The Journal of Nutritional Biochemistry*, *14*(5), 251–258. https://doi.org/10.1016/S0955-2863(03)00030-5.

Heck, J. M. L., Van Valenberg, H. J. F., Dijkstra, J., & Van Hooijdonk, A. C. M. (2009). Seasonal variation in the Dutch bovine raw milk composition. *Journal of Dairy Science*, *92*(10), 4745–4755. https://doi.org/10.3168/jds.2009-2146

Ho, T. M., Bhandari, B. R., & Bansal, N. (2022). Functionality of bovine milk proteins and other factors in foaming properties of milk: A review. *Critical Reviews in Food Science and Nutrition*, *62*(17), 4800–4820. https://doi.org/10.1080/10408398.2021.1879002

Horstman, A. M., & Huppertz, T. (2023). Milk proteins: Processing, gastric coagulation, amino acid availability and muscle protein synthesis. *Critical Reviews in Food Science and Nutrition*, *63*(30), 10267–10282. https://doi.org/10.1080/10408398.2022.2078782

Huppertz, T., Fox, P. F., & Kelly, A. L. (2018). The caseins: Structure, stability, and functionality. In *Proteins in Food Processing* (pp. 49–92). Woodhead Publishing. https://doi.org/10.1016/B978-0-08-100722-8.00004-8

Jianqin, S., Leiming, X., Lu, X., Yelland, G. W., Ni, J., & Clarke, A. J. (2015). Effects of milk containing only A2 beta casein versus milk containing both A1 and A2 beta casein proteins on gastrointestinal physiology, symptoms of discomfort, and cognitive behavior of people with self-reported intolerance to traditional cows' milk. *Nutrition Journal*, *15*, 1–16. https://doi.org/10.1186/s12937-016-0147-z

Kilara, A., & Vaghela, M. N. (2018). Whey proteins. In *Proteins in Food Processing* (pp. 93–126). Woodhead Publishing. https://doi.org/10.1016/B978-0-08-100722-8.00005-X

Kinsella, J. E., & Whitehead, D. M. (1989). Proteins in whey: Chemical, physical, and functional properties. *Advances in Food and Nutrition Research*, *33*, 343–438. https://doi.org/10.1016/S1043-4526(08)60130-8

Kneifel, W., Seiler, A., & Seiler, A. (1993). Water-holding properties of milk protein products: A review. *Food Structure*, *12*(2), 297–308.

Konstance, R. P., & Strange, E. D. (1991). Solubility and viscous properties of casein and caseinates. *Journal of Food Science*, *56*(2), 556–559.

Krishna, T. C., Najda, A., Bains, A., Tosif, M. M., Papliński, R., Kapłan, M., & Chawla, P. (2021). Influence of ultra-heat treatment on properties of milk proteins. *Polymers*, *13*(18), 3164. https://doi.org/10.3390/polym13183164

Larsen, M. K., Nielsen, J. H., Butler, G., Leifert, C., Slots, T., Kristiansen, G. H., & Gustafsson, A. H. (2010). Milk quality as affected by feeding regimens in a country with climatic variation. *Journal of Dairy Science*, 93(7), 2863–2873. https://doi.org/10.3168/jds.2009-2953

Law, A. J., & Leaver, J. (2000). Effect of pH on the thermal denaturation of whey proteins in milk. *Journal of Agricultural and Food Chemistry*, *48*(3), 672–679. https://doi.org/10.1021/jf981302b

Liu, Y., & Guo, R. (2008). pH-dependent structures and properties of casein micelles. *Biophysical Chemistry*, *136*(2–3), 67–73. https://doi.org/10.1016/j.bpc.2008.03.012

Luo, Y., Pan, K., & Zhong, Q. (2014). Physical, chemical and biochemical properties of casein hydrolyzed by three proteases: Partial characterizations. *Food*. https://doi.org/10.1016/j.foodchem.2014.01.048

Mehta, B. M. (2015). Chemical composition of milk and milk products. In *Handbook of Food Chemistry* (pp. 511–553). https://doi.org/10.1007/978-3-642-36605-5_37

Minj, S., & Anand, S. (2020). Whey proteins and its derivatives: Bioactivity, functionality, and current applications. *Dairy*, *1*(3), 233–258. https://doi.org/10.3390/dairy1030016

Monogioudi, E., Faccio, G., Lille, M., Poutanen, K., Buchert, J., & Mattinen, M. L. (2011). Effect of enzymatic cross-linking of β-casein on proteolysis by pepsin. *Food Hydrocolloids*, *25*(1), 71–81. https://doi.org/10.1016/j.foodhyd.2010.05.007

Mulvihill, D. M., & Kinsella, J. E. (1988). Gelation of β-lactoglobulin: Effects of sodium chloride and calcium chloride on the rheological and structural properties of gels. *Journal of Food Science*, *53*(1), 231–236.

Murtaza, M. A., Pandya, A. J., & Khan, M. M. H. (2017). Buffalo milk utilization for dairy products. In Y. W. Park, G. F. W. Haenlein, & W. L. Wendorff (Eds.). *Handbook of Milk of Non-Bovine Mammals* (pp. 284–342). Wiley.

O'mahony, J. A., & Fox, P. F. (2012). Milk proteins: Introduction and historical aspects. In *Advanced Dairy Chemistry: Volume 1A: Proteins: Basic Aspects*, 4th Edition (pp. 43–85). Boston, MA: Springer US. https://doi.org/10.1007/978-1-4614-4714-6_2

Palladino, R. A., Buckley, F., Prendiville, R., Murphy, J. J., Callan, J., & Kenny, D. A. (2010). A comparison between Holstein-Friesian and Jersey dairy cows and their F1 hybrid on milk fatty acid composition under grazing conditions. *Journal of Dairy Science*, *93*(5), 2176–2184. https://doi.org/10.3168/jds.2009-2453

Panouillé, M., Nicolai, T., & Durand, D. (2004). Heat-induced aggregation and gelation of casein submicelles. *International Dairy Journal*, *14*(4), 297–303. https://doi.org/10.1016/j.idairyj.2003.09.003.

Park, Y. W. (2017). Goat Milk–Chemistry and Nutrition. *Handbook of Milk of Non-bovine Mammals*, 42–83.

Patel, H., Patel, S., & Agarwal, S. (2014). *Milk Protein Concentrates: Manufacturing and Applications* (pp. 3–4). Arlington, VA: US Dairy Export Council.

Pavić, V., Antunac, N., Mioč, B., Ivanković, A., & Havranek, J. L. (2002). Influence of stage of lactation on the chemical composition and physical properties of sheep milk. *Czech Journal of Animal Science*, 47(2), 80–84.

Pennings, B., Pellikaan, W. F., Senden, J. M. G., van Vuuren, A. M., Sikkema, J., & van Loon, L. J. C. (2011). The production of intrinsically labeled milk and meat protein is feasible and provides functional tools for human nutrition research. *Journal of Dairy Science, 94*(9), 4366–4373.

Pereira, P. C. (2014). Milk nutritional composition and its role in human health. *Nutrition*, *30*(6), 619–627. https://doi.org/10.1016/j.nut.2013.10.011

Rafiq, S., Huma, N., Pasha, I., Sameen, A., Mukhtar, O., & Khan, M. I. (2015). Chemical composition, nitrogen fractions and amino acids profile of milk from different animal species. *Asian-Australasian journal of animal sciences*, *29*(7), 1022. https://dx. doi.org/10.5713/ajas.15.0452

Rollema, H. S., & Muir, D. D. (2009). Casein and related products. In A. Y. Tamime (Ed.). *Dairy Powders and Concentrated Products* (pp. 235–254). Hoboken, NJ: Wiley-Blackwell.

Roy, D., Ye, A., Moughan, P. J., & Singh, H. (2020). Composition, structure, and digestive dynamics of milk from different species—A review. *Frontiers in Nutrition*, *7*, 577759. doi: 10.3389/fnut.2020.577759

Sawyer, L. (2012). β-Lactoglobulin. In *Advanced Dairy Chemistry: Volume 1A: Proteins: Basic Aspects*, 4th Edition (pp. 211–259). Boston, MA: Springer US. https://doi.org/10.1007/978-1-4614-4714-6_7.

Schaafsma, G. (2000). The protein digestibility–corrected amino acid score. *The Journal of Nutrition*, *130*(7), 1865S–1867S. https://doi.org/10.1093/jn/130.7.1865S

Schwendel, B. H., Wester, T. J., Morel, P. C. H., Tavendale, M. H., Deadman, C., Shadbolt, N. M., & Otter, D. E. (2015). Invited review: Organic and conventionally produced milk—An evaluation of factors influencing milk composition. *Journal of Dairy Science*, *98*(2), 721–746. https://doi.org/10.3168/jds.2014-8389

Shimizu, M., Kamiya, T., & Yamauchi, K. (1981). The adsorption of whey proteins on the surface of emulsified fat. *Agricultural and Biological Chemistry*, *45*(11), 2491–2496. https://doi.org/10.1080/00021369.1981.10864922

Silva, N. N., Piot, M., de Carvalho, A. F., Violleau, F., Fameau, A. L., & Gaucheron, F. (2013). pH-induced demineralization of casein micelles modifies their physico-chemical and foaming properties. *Food Hydrocolloids*, 32(2), 322–330. https://doi.org/10.1016/j.foodhyd.2013.01.004

Solanki, R., Rostamabadi, H., Patel, S., & Jafari, S. M. (2021). Anticancer nano-delivery systems based on bovine serum albumin nanoparticles: A critical review. *International Journal of Biological Macromolecules*, *193*, 528–540. https://doi.org/10.1016/j.ijbiomac.2021.10.040

Solorza-Feria, J., Méndez-Montealvo, G., González-Soto, R., Osorio-Díaz, P., & Jiménez-Pérez, A. (2011). Changes in the apparent viscosity profiles of casein suspensions as affected by plant enzymes. *LWT-Food Science and Technology*, *44*(2), 414–420. https://doi.org/10.1016/j.lwt.2010.08.025

Soyeurt, H., Dardenne, P., Dehareng, F., Bastin, C., & Gengler, N. (2008). Genetic parameters of saturated and monounsaturated fatty acid content and the ratio of saturated to unsaturated fatty acids in bovine milk. *Journal of Dairy Science*, *91*(9), 3611–3626. https://doi.org/10.3168/jds.2007-0971

Stoop, W. M., Bovenhuis, H., Heck, J. M. L., & Van Arendonk, J. A. M. (2009). Effect of lactation stage and energy status on milk fat composition of Holstein-Friesian cows. *Journal of dairy science*, *92*(4), 1469–1478. https://doi.org/10.3168/jds.2008-1468

Szigeti, K., & Lupski, J. R. (2009). Charcot–Marie–tooth disease. *European Journal of Human Genetics*, *17*(6), 703–710.

van Lieshout., G. A., Lambers, T. T., Bragt, M. C., & Hettinga, K. A. (2020). How processing may affect milk protein digestion and overall physiological outcomes: A systematic review. Critical reviews in food science and nutrition, 60(14), 2422–2445. https://doi.org/10.1080/10408398.2019.1646703

Wang, L., Shao, X., Cheng, M., Fan, X., Wang, C., Jiang, H., & Zhang, X. (2022). Mechanisms and applications of milk-derived bioactive peptides in food for special medical purposes. *International Journal of Food Science & Technology*, *57*(5), 2830–2839.

Xiong, X., Ho, M. T., Bhandari, B., & Bansal, N. (2020). Foaming properties of milk protein dispersions at different protein content and casein to whey protein ratios. *International Dairy Journal*, *109*, 104758. https://doi.org/10.1016/j.idairyj.2020.104758

Yadav, J. S. S., Yan, S., Pilli, S., Kumar, L., Tyagi, R. D., & Surampalli, R. Y. (2015). Cheese whey: A potential resource to transform into bioprotein, functional/nutritional proteins and bioactive peptides. *Biotechnology Advances*, *33*(6), 756–774. https://doi.org/10.1016/j.biotechadv.2015.07.002

Yamauchi, K., Shimizu, M., & Kamiya, T. (1980). Emulsifying properties of whey protein. *Journal of Food Science*, *45*(5), 1237–1242. https://doi.org/10.1111/j.1365-2621.1980.tb06529.x

Zayas, J. F. (1997). Water holding capacity of proteins. In *Functionality of Proteins in Food* (pp. 76–133). Springer. https://doi.org/10.1007/978-3-642-59116-7

2 Cereal Proteins

Swathika. E, Jagriti Joshi,
and Sundaramoorthy Haripriya

INTRODUCTION

Cereals typically contain carbohydrates (65%–75%) and proteins (7%–12%). Protein ranks the second-largest component in cereal grains, essential for its functionality and varying molecular weights of diverse amino acids. This composition emphasises the value of cereals as an important aspect of diet, calling for more in-depth investigation within the scope of thorough nutritional research. Cereal-derived protein products are vital for improving human health. Thomas Osborne conducted a comprehensive investigation into plant proteins, classifying them based on their solubility in various solvents. They are divided into four classes on their solubility: glutenins (soluble in acid or alkaline), globulins (soluble in salt), albumins (soluble in water), and prolamins (soluble in ethanol–water). Various cereal grains such as rice, maize, wheat, oats, and barley contain distinct types of proteins. Wheat contains gluten (80%–85%), which is divided into glutenins (insoluble in alcohol) and gliadins (soluble in alcohol). They have unique structures, differing in molecular weights and amino acid compositions. Rice protein is hypoallergenic and very well tolerated, with protein content that differs among its milling fractions including paddy (5.6%–7.7%), rice bran (11.3%–14.9%), hulled rice (2.0%–2.8%), brown rice (7.1%–8.3%), and milled rice (6.3%–7.1%). In comparison, maize contains approximately 10% protein, which can be classified into four types based on solubility: glutelins (34%), globulins (3%), albumins (3%), and prolamins (over 60%). Oats are cereal grains rich in protein, which is found throughout their structure: 12% in the endosperm, 18%–26% in the bran, and 29%–38% in the germ. They can be categorised into four types: glutelins, globulins, albumins and prolamins. Globulins make up 70%–80% of the protein content and serve as major storage proteins, being soluble in salt solutions. Consequently, plant-based proteins can more effectively replace animal-based proteins due to their considerable potential to provide a sustainable solution. Furthermore, selecting an appropriate protein source is paramount, as its structural composition and linkages significantly influence the effective binding of minerals. This interaction subsequently promotes the formation of protein-mineral complexes. Moreover, these innovative legumes can be foundational for developing novel uses of cereal-derived proteins, explicitly targeting the mitigation of toxicity concerns through their binding with inorganic minerals. Cereal-based protein products are not only resource-efficient but also environmentally sustainable, making them an increasingly popular choice in today's food industry. The proteins derived from cereals, including wheat, rice, and maize, offer a practical and strategic option for improving food security around the globe. With the world population steadily

 DOI: 10.1201/9781003613909-2

increasing, the need for innovative and sustainable food sources has never been more critical (Jindal et al., 2023).

By examining cereal-based protein products, we can better understand how they can contribute to addressing the pressing challenges posed by this rising global population and the growing demand for sustainable food production. These protein sources are produced through resource-efficient technologies that minimise waste and reduce environmental impact. As we strive to develop food systems that can support future generations, cereal-based proteins will play a vital role in creating a balanced and sustainable food landscape.

COMPOSITION OF DIFFERENT CEREALS

Rice

Rice, or *Oryza sativa*, often referred to as the "grain of life", is a staple crop that is widely cultivated and consumed by a large segment of the global population, particularly in Asia. Brown rice, a whole-grain product, is low in calories yet high in fibre and essential vitamins, especially vitamin B. Additionally, it is a high source of iron and manganese, provided being cholesterol-free. This whole-grain product, brown rice, is low in calories but high in fibre and vitamins, particularly vitamin B, iron, manganese, and cholesterol. The composition of rice is shown in Table 2.1. It has also been beneficial for cleansing the digestive tract and preventing blood clots. When applied topically, crushed rice powder has been shown to have a medicinal effect on skin issues. The pigment "proanthocyanidin" gives this rice its red colour. Rice is also recognised for its chemoreceptive and antioxidant qualities (Baniwal et al., 2021). According to Verma and Srivastav (2017), rice is a potential source of minerals (calcium, magnesium, phosphorus, iron, copper, zinc, and manganese), amino acids, B complex vitamins (thiamine, niacin, and riboflavin), and carbohydrates (amylose and amylopectin). However, rice is recognised as a gluten-free cereal, making it suitable for those who have problems with gluten intake. Rice is a favoured food for babies because it contains lysine. Cleaning, husking, separating, de-stoning, whitening, polishing, and sifting are all steps in the rice processing process that result in the byproducts of bran and husk (Verma & Srivastav, 2020). The grain chemical composition varies considerably based on variety, soil, and the environment in which it is grown. The values for protein, fat, and starch also differ depending on the analysis technique. Compared to brown rice, milled rice had more readily available carbohydrates, primarily starch (Zhou et al., 2002).

Wheat

Wheat (*Triticum*) was among the first grains cultivated by humans. Within this genus, several species, such as common wheat (*Triticum aestivum* L.), durum wheat (*Triticum durum* Desf.), and spelt (*Triticum spelta* L.) are grown worldwide. Common wheat is the most widely produced species globally and ranks as the second most extensively cultivated crop in the world (Biel et al., 2020). Wheat serves as a

TABLE 2.1
Nutritional Composition of Milled and Brown Rice per 100 g

Properties	Milled Rice	Brown Rice	Wheat	Maize	Oats	Barley	Rye	Sorghum	Quinoa	Buckwheat	Flaxseed
Energy (Kcal)	129	122	339	365	433.8	352	401.4	329	368	343	530
Protein (g)	7.3–8.3	6.5–10.0	138.52	8.84	13.7	9.9	8.2	4.40-21.10	13.51	13.27	20.3
Fat (g)	0.3–0.65	1.9–3.9	1.9	4.57	8.6	2.3	1.31	2.10–7.60	6.48	3.84	24
Fibre (g)	0.3–0.5	1.6–2.8	29.49	2.15	2.3	15.6	15.0–25.0	1.00–3.40	7	10	37.1
Carbohydrates (g)	91.07	85.2–88.9	72.6	71.88	74.3	77.7	89.2	72.09	64.16	82.14	28.9
Source	Mohidem et al. (2022)	Mohidem et al. (2022)	Baniwal et al. (2021); Kowieska et al. (2011)	Kazerooni, (2019)	Alemayehu et al. (2021)	Baniwal et al. (2021)	Cardoso et al. (2019)	Tasie and Gebreyes (2020)	Baniwal et al. (2021); Li et al. (2021)	Baniwal et al., (2021); Sofi et al. (2023); Tien et al. (2018)	Baniwal et al. (2021)

significant source of carbohydrates, containing around 13% protein, which is relatively high compared to other major cereals, although its protein quality, specifically in providing essential amino acids, is lower (Table 2.1). When consumed as whole grain, wheat also offers dietary fibre and various other nutrients. However, gluten, which constitutes the majority of wheat's protein, can trigger conditions such as dermatitis herpetiformis, non-celiac gluten sensitivity, gluten ataxia, and celiac disease in a small percentage of the population.

A staple crop grown all over the world, wheat is a member of the *Poaceae* family. It has a high potential for providing dietary fibres, antioxidants (black wheat), free and esterified phenolic acids, vitamins, and minerals (phosphorus, magnesium, and selenium) that are essential for maintaining and controlling several bodily metabolic processes. Wheat contains several bioactive ingredients, including flavonoids, phenolic acids (such as hydroxycinnamic and hydroxybenzoic acids), benzoxazinoids (BXs), carotenoids, alkylresorcinols, and other compounds. These substances have a variety of medicinal uses, particularly in the prevention and treatment of conditions such as type 2 diabetes, obesity, cardiovascular disease (CVD), and colorectal cancer (Baniwal et al., 2021).

Maize (Corn)

A significant annual cereal crop worldwide, maize, often known as maize (*Zea mays* L.), is a member of the Poaceae family. The Taino word may mean "life giver," while the ancient Greek term zea refers to "sustaining life." It is regarded as the "queen of cereals" throughout the world because of its maximum potential yield. America is the world's largest producer of maize, accounting for around 35% of global production. It is the engine of the US economy and is referred to as the "mother grain" of Americans (Rouf Shah et al., 2016). Corn has numerous pharmacological properties, including antibacterial, antioxidant, and anticancer properties, which are displayed by aromatic plants. Corn silk is used to reduce the blood creatinine level and cure kidney stones, urinary tract infections, and irritations. Corn contains low levels of fat and cholesterol. Corn serves as a great source of fibre, magnesium, vitamins, minerals, and complex carbohydrates. The most popular vegetable oil on the market is maize oil. It also has a high sugar and carbohydrate content that gives energy and calories. There are carbohydrates in the form of the so-called simple sugars, sucrose, glucose, and fructose. Corn oil is made from the starch, gluten, and germ of corn and is utilised in paints, coatings, soaps, and biodiesel industries. Along with saturated palmitic acid, the primary fatty acids in maize oil are linoleic and oleic acids; stearic and myristic acids are present in trace amounts (Kazerooni, 2019). The composition of maize is shown in Table 2.1.

Oats

Oats (*Avena sativa* L.) are unique among cereal crops considering their diverse range of nutrients beneficial for human consumption, animal feed, medical treatments, and cosmetic products. This cereal is rich in carbs, soluble fibre, proteins, lipids, various phenolic compounds, vitamins, and minerals (Table 2.1). With the developing recognition of healthier eating practices, oats have garnered increased interest from both businesses and researchers. Food companies are innovating by adding oats into products such as breakfast cereals, beverages, bread, and baby food. Oats beta-glucan (OBG) is a major component of the oat cereal. This viscous polymer consists of a linearly branched chain of D-glucose monosaccharides linked by a combination of (1 → 3) and (1 → 4) bonds. It is situated in the kernel's endosperm cell wall. It is thought to be the principal active ingredient in oats and has a number of nutritional and functional benefits, chief among them being the reduction of cholesterol and the prevention of diabetes. Oat groats have a protein level of 13%–20%. About 30% of the embryo is made up of proteins. Antioxidant substances such as tocols, phenolic compounds, and other trace amounts of oat sterols, which are likewise linked to positive health effects. Tocopherols and tocotrienols, which together form tocols, are the primary contributors to vitamin E activity (Paudel et al., 2021). The composition of oats is shown in Table 2.1.

Barley

Nowadays, one of the most efficient cereal crops cultivated worldwide is barley (*Hordeum vulgare* L.). Barley shares a tight evolutionary relationship with two other small-grain wheat, rye, and other cereal species. Barley grain is widely used in animal feeds and human nutrition as flakes, groats, and a raw material for making malt. Barley comes in a variety of morphological varieties now, including two-row and six-row varieties, as well as various colours (black, blue, purple, or yellow). Another way to identify different types of barley is by their genotypes: hulled. Together, cultivar-specific characteristics, meteorological factors, and farming practices alter the chemical makeup and quality of barley grain (Biel et al., 2020). Highland barley has a high capacity to scavenge free radicals and is high in phenolic compounds (2.42%–7.33%), in bound and free forms of phenolic acids, flavonoids, and anthocyanins. It is among the significant natural antioxidants that can prevent and treat metabolic illnesses, as well as control human lipid and glucose metabolism (Dang et al., 2022). Barley grains are identified as a source of improved dietary fibre (β-glucan) content, low energy, fat, and an inexpensive price, which is why they are recognised as a functional food. The composition of barley is shown in Table 2.1.

Rye

One cereal that may be used to produce wholesome, high-fibre dishes is rye (Secale cereale). Since rye is high in carbohydrates, protein, dietary fibre, vitamins, and minerals—particularly magnesium and zinc—it has long been used for baking and as an ideal source of energy and nutrients (Table 2.1). In addition to possible nutritional and health advantages, increased rye consumption can alleviate the incidence of some non-communicable diseases, namely diabetes, heart disease, hyperlipidaemia, and obesity, and improve digestion, protective action against a variety of hormone-based cancer types (El-Mahis et al., 2023). Rye's protein composition can vary greatly (8%–15%) based primarily on growth conditions. Rye's primary proteins, albumins, make up between 29% and 40% of its overall protein concentration. Globulins comprise

8%–11% of the overall protein content, whereas prolamins and glutelins, the rye storage proteins known as secalins, comprise 17%–19% and 9%–15% of the entire protein composition, respectively. According to Osborne fractionation, between 21% and 30% of the proteins cannot be extracted. Rye proteins possess a better amino acid profile than wheat because they include more lysine, but they are deficient in tryptophan and isoleucine (Németh & Tömösközi, 2021).

Sorghum

An indigenous crop in Africa, sorghum [*Sorghum bicolour* (L.) Moench] continues to be a staple grain in many rural areas. With an annual yield at 25°C, sorghum is primarily grown in dry regions, particularly on shallow and heavy clay soils. Except for a reduced oil content, the composition of sorghum grain and its constituent parts is mostly comparable to that of maize. About 8%–12% of the grain is protein, 65%–76% is starch, and about 2% is fibre. The germ has large amounts of protein (19%) and ash (10%) in addition to being a strong source of oil (28% of the germ). Significant amounts of starch are deposited in the mesocarp tissue of this fraction, even though cellulose and hemicellulose make up most of the bran. Wax, not oil, makes up the majority of bran lipid. Numerous factors, such as the type of hybrid, the soil and climate, and the crop management practices, can affect the composition of sorghum grain from various sources (Abah et al., 2020). About 85% of the crops are directly consumed as human food in northern China, India, and southern Russia, where it is primarily consumed (Tasie & Gebreyes, 2020). The composition of sorghum is shown in Table 2.1.

Quinoa

The grain-like crop known as quinoa (*Chenopodium quinoa* Willd.) was first cultivated in South American Andean nations. Although quinoa is a part of the Amaranthaceae family rather than the Poaceae (Gramineae), it is known as a pseudocereal since its seeds are mostly ground into flour and utilised as a cereal grain. Quinoa seeds can be eaten in a variety of ways, such as powdered into flour for use in baked goods like breads, pancakes, and tortillas, boiled and used similarly to rice, or dried and added to salads and mueslis (Li et al., 2021). The main edible part of quinoa is its grains, which are gluten-free and packed with protein, essential amino acids, and important vitamins and minerals. Due to these nutritional benefits and health advantages, quinoa is often referred to as a "superfood". Additionally, quinoa greens—which include quinoa leaves, sprouts, and microgreens—are also highly nutritious and offer health benefits such as antimicrobial, anticancer, antidiabetic, anti-obesity, antioxidant, and heart-protective properties, similar to quinoa grains (Pathan & Siddiqui, 2022). The composition of quinoa is shown in Table 2.1.

Buckwheat

Consumers' diets have included more functional foods with bioactive substances in recent years. These functional foods provide users with both health and nutritional advantages (Sofi et al., 2023). Buckwheat (Fagopyrum esculentum), a member of the Polygonaceae family, is classified as a pseudocereal due to its similarities with regular cereals in terms of production, use, and chemical composition. It is also recognised as a functional food, particularly when used in doughs supplemented with buckwheat flour. This is because buckwheat inhibits the formation of nitrosamines in the digestive tract and offers several health benefits, including low calorie content, a high amount of unsaturated fatty acids, an essential amino acid profile, and potential antioxidant properties (Tien et al., 2018). Buckwheat is rich in various bioactive components that have strong effects in preventing chronic illnesses. These include proteins, dietary fibre, vitamins, flavonoids, fagopyrins, d-fagomine, and phenolic acids. The seeds of buckwheat contain different types of proteins such as albumins, globulins, prolamins, and glutelins, offering a well-balanced amino acid profile. Additionally, because buckwheat proteins are gluten-free, they are more suitable for individuals with celiac disease. The composition of buckwheat is shown in Table 2.1.

Flaxseed

Flaxseed (*Linum usitatissimum* L.), also known as linseed, is an annual crop primarily cultivated for its oil, fibre, food, and feed properties. Due to its exceptional nutritional value, the significance of this crop has increased greatly in recent years. Flaxseed exhibits high biological activity, which has led to its incorporation into various products, including cosmetics, health supplements, and functional foods (Mueed et al., 2022). Flaxseed is known for its abundance of alpha-linolenic acid (ALA, 18:3n-3), and flaxseed is regarded as the best source of ω-3. The amount of polyunsaturated fatty acids (PUFAs) is high compared to marine algae, soybean, maize, or fish oil. Docosapentaenoic acid (DPA, 22:5n-3), eicosapentaenoic acid (EPA, 20:5n-3) and docosahexaenoic acid (DHA, 22:6n-3) are examples of long-chain polyunsaturated fatty acids that are thought to have a number of advantageous physiological and functional qualities, such as promoting the development of cranial nerves, improving memory, boosting immunity, increasing the body's metabolism of saturated fatty acids, preventing cerebral thrombosis, and preventing myocardial infarction, among other health benefits (Yang et al., 2021). The composition of flaxseed is shown in Table 2.1.

Millet

Millets are classified as small cereals in the Poaceae family of grasses. Millets are small-seeded grain crops that may grow in a variety of tropical and desert environments. They can even survive in less fertile soil. Major and minor millets are the two primary categories into which millets fall. While kodo, barnyard, proso, small, and foxtail millet are minor millets, sorghum, pearl, and finger millet, called ragi, are important millets (Bhatt et al., 2022). Currently, millets are not regarded as a single significant commodity in the food baskets of North America and Europe, but their significance as a component of gluten-free and multigrain cereal products has been emphasised. Furthermore, millets are a nutritious diet due to their high fibre content, gluten-free proteins, and abundance of bioactive substances (Mishra et al., 2022). The nutritional composition of different types of millets is shown in Table 2.2.

TABLE 2.2
Nutritional Composition of Millets

Properties	Pearl Millet	Finger Millet	Foxtail Millet	Proso Millet	Kodo Millet
Energy (Kcal)	361–363	328–336	352–391	352.5	349.5–353
Protein (g)	11.4–11.8	7.3–7.7	11.2–15	10.0–13.0	8.3–10.0
Fat (g)	4.87–5	1.3–1.5	3.3–4.3	3.1	3.03–3.6
Fibre (g)	1.2–2.3	11.5	6.7–8.23	8.47	5.2
Carbohydrate (g)	67.0–69.10	71.52–72	60.9–67.3	67.09	63.82–66.6

Source: Ali and Bhattacharjee (2023); Mishra et al. (2022).

PROTEIN AND AMINO ACID PROFILES OF DIFFERENT CEREALS

Approximately 40% of the annual production of cereal grains is utilised as animal feed, as grains are often used in a refined form in food products. The primary cereal crops produced globally are wheat and maize, followed by rice, barley, sorghum, and oats. (Poutanen et al., 2022). Unlike monocotyledonous cereals, pseudocereals—such as buckwheat, quinoa, and amaranth—are dicotyledonous plants, which is how they vary from actual cereals from a botanical standpoint. Due to customer desires for a variety of protein sources that are nutritionally balanced, cereal and pseudocereal proteins have recently attracted increasing attention (Sá & House, 2024). Protein Content of Different Cereals is shown in Table 2.3. The largest source of plant protein globally is cereal foods, which typically provide more protein in the diet than meat products in Central America, Africa, Asia, and Europe. Since 50–70 g of protein per day is the recommended daily intake, excessive protein consumption (and food intake) is widespread. When considered in this light, reducing the amount of animal protein consumed without substituting it would improve environmental sustainability while also increasing the proportion of grain protein in the diet (Poutanen et al., 2022). The human body requires protein, a macronutrient composed of amino acids, for healthy growth and functioning. The purest form of protein is protein isolate, which serves as an excellent nutritional supplement and can aid in weight management and enhance physical performance (Adenekan et al., 2018). As the global population continues to grow and high-protein diets become more popular, the demand for protein has risen. Additionally, as individuals seek healthier dietary options, interest in plant-based protein products is consistently increasing (Sá & House, 2024).

Grain proteins are dispersed unevenly throughout the grain, and the species, variations, and development conditions of the plants all affect the total protein content of cereals. Cereal proteins are typically regarded as low quality despite their significant role in human nutrition because of their usually inadequate amino acid profile, as well as problems with allergenicity and antinutritional consequences. However, cereal proteins play a crucial role in the final uses of the grain and its derivative products in addition to nourishment (Avelar et al., 2024). Comparative estimates for the protein contents of various cereals are challenging to acquire due to their diverse production processes and amounts of fertiliser application. The nutritional significance

TABLE 2.3
Protein Content of Different Cereals per 100 g

Cereal	Protein (g)	References
Rice	7.3–8.3	Mohidem et al. (2022)
Wheat	138.52	Kowieska et al. (2011)
Maize	8.84	Rouf Shah et al. (2016)
Oats	13.7	Alemayehu et al. (2021)
Barley	9.9	Baniwal et al. (2021)
Rye	8.2	Cardoso et al. (2019)
Sorghum	4.40–21.10	Tasie and Gebreyes (2020)
Millet	9.9	Baniwal et al. (2021)
Quinoa	13.51	Li et al. (2021)
Buckwheat	13	Baniwal et al. (2021)
Flaxseed	20.3	Baniwal et al. (2021)

of protein is determined by the balance of essential amino acids, which must be included in the diet since animals cannot produce them. If any of these amino acids are deficient, the body will break down the others, resulting in inadequate growth for both livestock and humans, as well as a loss of nitrogen from the diet. There are ten amino acids that are classified as essential: lysine, leucine, isoleucine, phenylalanine, tyrosine, threonine, tryptophan, valine, histidine, and methionine. Cysteine is often regarded as essential since it can only be synthesised from methionine, which is itself essential (Shewry, 2007). Figure 2.1 displays the classification of amino acids.

The physical state of the plant, particularly its nitrogen status, has a significant impact on the number of proteins produced in the seed. The quantity and timing of nitrogen fertiliser applications determine this status in and of itself. Effects on dry matter yield and protein content of that dry matter combine to produce the yield of protein from cereals; generally speaking, an increase in fertiliser nitrogen increases both factors, but the effect on dry matter yield is diminished and the effect on protein content increases as nitrogen application time is postponed (Kowieska et al., 2011).

Although lysine is considered the limiting amino acid in all cereals, the amount varies by species. Oats contain the highest amount, while rice and maize have the lowest. But the lysine found in rice protein is thought to be hypoallergenic, making it suitable for human consumption (Helm & Burks, 1996). Oats have the most advantageous amino acid composition of any grain because of their high protein and lysine content, the latter of which is scarce in the majority of cereals. 83.5 mg of total necessary amino acids per kilogram of body weight should be consumed daily. Considering this, the reconsideration of one hundred to one hundred grammes of oats could provide a recommended dose of essential amino acids (Sterna et al., 2016). Originally thought to be necessary exclusively for infants, research has since shown that this amino acid is also important for adults. Experts currently recommend that adults consume 8–12 mg/kg of body weight per day of histidine (Tomičić et al., 2022). Small amounts of phenylalanine can be consumed by young children, as it is necessary for growth and development (Kowieska et al., 2011).

ESSENTIAL AMINO ACIDS	CONDITIONALLY NON-ESSENTIAL	NON-ESSENTIAL
• Histidine	• Arginine	• Alanine
• Isoleucine	• Cystine	• Asparagine
• Leucine	• Glutamine	• Aspartate
• Lysine	• Glycine	• Glutamate
• Methionine	• Proline	• Serine
• Phenylalanine	• Tyrosine	
• Threonine		
• Tryptophan		
• Valine		

FIGURE 2.1 Categorisation of amino acids.

Aspartic acid, glutamic acid, serine, proline, alanine, tyrosine, glycine, cysteine, and arginine are the non-essential amino acids (NEAA) of cereals, which make up 62.24%–69.42% of the total amino acids. Both essential amino acids (EAA) and NEAA varied significantly ($p < 0.05$) between samples.

Oats and wheat had the highest overall amount of EAAl and NEAA, whereas the level of essential amino acids was lower than the level of NEAA in the studied samples. Serine, glutamic acid, proline, glycine, arginine, and aspartic acid were among the NEAA that demonstrated notable differences. While alanine, cysteine, and tyrosine differed non-significantly ($p > 0.05$), there were differences ($p < 0.05$).

The most abundant amino acid was discovered to be glutamic acid, which had values ranging from 1.39 to 2.94 g/100 g of material. Glutamic acid was found in the highest concentration in wheat and the lowest in rice (Tomičić et al., 2022).

The body produces glutamic acid, which is essential for the metabolism of amino acids and for preserving the nitrogen balance in the body, which is crucial in times of stress and illness. In the central nervous system, glutamic acid is also a well-established excitatory neurotransmitter (Kowieska et al., 2011). According to NHANES III (1988–1994), 15 g of glutamic acid should be consumed daily on average (Zafar et al., 2014). In addition to having the highest concentration of total non-essential amino acids, the wheat sample had the highest concentrations of serine, glutamic acid, proline, isoleucine, and phenylalanine (Tomičić et al., 2022). In actuality, histidine is a semi-essential amino acid, meaning that the human body only partially synthesises it. In this work, it is grouped alongside the necessary amino acids for simplicity's sake (Wiedemair et al., 2020). The amino acid profile of different cereals is shown in Table 2.4.

CATEGORISATION AND CHEMISTRY OF CEREAL PROTEINS

Cereal proteins encompass various proteins categorised into various types based on their solubilities, functionalities, biochemical attributes and genetic properties. The earliest classification that appeared based on the extractability and solubilisation of

TABLE 2.4
Amino Acid Profile of Different Cereals

Amino Acid	Rice (g/100 g)	Wheat (g/100 g)	Maize (g/100 g)	Oats (g/100 g)	Barley (g/100 g)
Threonine	0.24	0.37	0.35	0.42	0.33
Valine	0.40	0.59	0.48	0.61	0.44
Methionine	0.16	0.22	0.20	0.28	0.16
Isoleucine	0.29	0.53	0.34	0.45	0.31
Leucine	0.55	0.93	1.16	0.94	0.72
Phenylalanine	0.35	0.68	0.46	0.65	0.52
Histidine	0.16	0.32	0.29	0.33	0.23
Lysine	0.24	0.30	0.27	0.44	0.33
Aspartic acid/ asparagine	0.62	0.62	0.66	0.87	0.60
Serine	0.35	0.67	0.45	0.53	0.40
Glutamic acid/ glutamine	1.29	4.74	1.77	2.57	2.84
Proline	0.31	1.46	0.82	0.77	1.17
Glycine	0.30	0.50	0.39	0.58	0.46
Alanine	0.38	0.43	0.71	0.52	0.37
Tyrosine	0.22	0.36	0.38	0.36	0.21
Arginine	0.55	0.48	0.47	0.97	0.58

Source: Sharma and Saini (2022); Tomičić et al. (2022); Wiedemair et al. (2020); Zhang and Xu (2017).

cereal proteins in different solvents is "Osborne fractionation". Based on this classification, cereal proteins are categorised into four types: water-soluble albumins, saline-soluble globulins, prolamins that dissolve in alcohol/water mixtures, and glutenins that dissolve in dilute acids or alkalis (Sudheesh et al., 2022; Tatham & Shewry, 2012). Even though 'Osborne fractionation' remains a popular method, it is now usual to categorise seed proteins into three classes: storage proteins, structural and metabolic proteins, and protective proteins (Shewry & Halford, 2002). The endosperm in the cereals serves as a reservoir for storing starch and proteins. The storage proteins in cereals constitute about 50% of their total nitrogen content in the cereal. They have been labelled with familiar names, such as for wheat prolamins, the term is gliadin; for glutelin, glutenins; for maize, zeins; for rye, secalins; for barley, hordeins; for oats, avenin; for rice, oryzins; and for millet and sorghum, karfirin. The principal storage protein for most cereals is prolamin, while rice and oats contain glutenins and globulins as their primary protein source (Scherf et al., 2016; Sudheesh et al., 2022). Figure 2.2 represents significant cereal classifications of grain proteins.

Storage Prolamins

The term "prolamin" refers to a group of storage proteins that are primarily composed of the amino acids glutamine and proline, which together make up 30–70 mol% of the

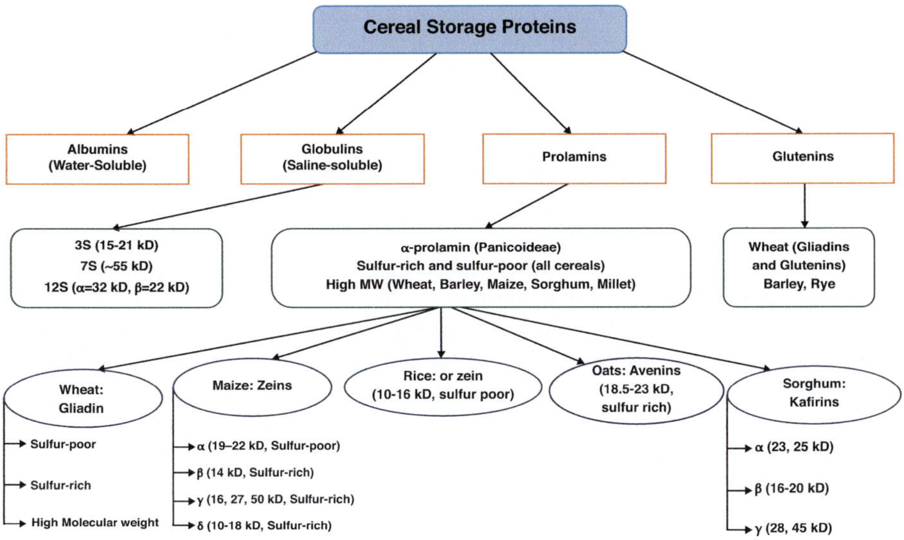

FIGURE 2.2 Classification of cereal storage proteins.

total amino acids in prolamin proteins. Prolamins serve as the main storage proteins found in cereals such as wheat, sorghum and corn. They are primarily produced during seed development within the endoplasmic reticulum and accumulate in vesicles surrounded by protein bodies in the endosperm. Prolamins are heterogeneous polypeptides that can be classified into monomers and polymers, with molecular masses that range from 10 kDa to 100 kDa, based on the type of cereal (Tapia-Hernández et al., 2019). All individual prolamin peptides are soluble in alcohol, while some polymers occur to be insoluble. Prolamins are categorised into three groups depending on amino acid sequence similarity and phylogenetic analysis. They are (1) α-prolamin, (2) sulphur-rich and sulphur-poor prolamins, and (3) high-molecular-weight (HMW) prolamins. Group 1 is found exclusively in Panicoideae, whereas Group 2 prolamins are present across all cereals. Group 3 is identified in wheat and barley but can also be found in maize, sorghum, and millet. Although the major groups of prolamins (*Triticeae* and *Panicoideae*) had different evolutionary origins, they share a common structural characteristic: the formation of repetitive blocks of one or more domains based on the short peptide motifs containing ten amino acid units. These motifs are abundant in glutamine and other amino acids such as proline, glycine and phenylalanine (Labuschagne, 2018; Shewry & Halford, 2002).

Prolamins from Triticeae: The prolamin superfamily includes prolamins found in wheat, barley, and rye. These prolamins share structural similarities and consist of a complex mixture of polymorphic components, with 20 to 30 components in barley and rye and about 50 components in wheat. Sulphur-rich prolamins are abundant, and their N-terminal end is linked to repetitive sequences of amino acids, with proline and glutamine being the most frequently occurring. The peptide pattern is the tetrapeptide

ProGlnGlnPro combined with a second peptide sequence. Meanwhile, the C-terminal end contains fewer proline and glutamine residues but has a high concentration of cysteine, which is rich in sulphur. In sulphur-poor prolamins, the entire sequence has repetitive peptides rich in prolines and glutamines that lack cysteines; hence, they cannot form disulphide bonds. The s-poor prolamins consist of octapeptides arranged in sequential order (ProGlnGlnProPheProGlnGln). Some examples of s-poor prolamins with octapeptides are C-hordeins, ω-secaline, and ω-gliadin. It showcased the evolutionary connection involving repetitions of ProGlnGlnPro found in the sulphur-rich prolamins (Sudheesh et al., 2022).

HMW prolamins contain 627–827 amino acid residues, with a central domain comprising 481–681 residues and non-repetitive side domains comprising 81–104 amino acid residues. Their C- and N-terminals possess 42 amino acid residues. Unlike sulphur-rich and sulphur-poor prolamins that have tetrapeptides and octapeptides, respectively, HMW prolamin subunits have hexapeptides (ProGlyGlnGlyGlnGln), nano peptides (GlyTyrTyrProThrSerPro or LeuGlnGln), and tripeptides (GlyGlnGln). The cysteine residues are confined to non-repetitive domains, similar to sulphur-rich prolamins (Sudheesh et al., 2022).

Wheat contains two primary prolamins, gliadins and glutenins, which combine to form wheat gluten. This complex protein network comprises disulphide bonds that aid in the aggregation of the proteins and significantly influence the dough's rheology. Gliadins are monomeric proteins with a molecular weight ranging from 30 to 80 kDa. They are classified into three groups: α + β-gliadins, ω-gliadins, and γ-gliadins, based on their amino acid content and molecular weight. The intrachain disulphide bridges found in gliadins promote the viscous properties of wheat gluten.

On the other hand, glutenins are polymeric proteins with molecular weights varying from 100 to 10,000 kDa. They consist of Low-Molecular-Weight Glutenin Subunits (LMW-GS) and High-Molecular-Weight Glutenin Subunits (HMW-GS). Glutenins contain both intrachain and interchain disulphide bonds, allowing them to form dense and rigid structures that enhance the elasticity of gluten (Tapia-Hernández et al., 2019).

Prolamins from Maize: Zein, the maize prolamin, is categorised into α, β, γ, and δ-zein, among which α-zeins are predominantly (75%–85%) present in the maize grains, while others contribute minor percentages. α-Zeins have molecular weights of 19 and 22 kD and are either oligomers or monomers. They possess one or two cysteine residues and very limited methionine. γ-Zeins (molecular weights of 16, 27, and 50 kD) are also rich in cysteine and represent less than 10% of the total zeins. While α-zeins are soluble in 95% ethanol, γ-zeins are soluble in 0%–80% (v/v) 2-propanol with a reducing agent. β-Zein accounts for 10%–15% of the total zein content, is soluble in 30%–85% ethanol, and has a molecular weight of 14 kD. δ-Zein makes up 1%–5% of total zein and is related to the prolamins of the *Triticeae* family. They exhibit solubility comparable to α-zein, with molecular weights ranging from 10 to 18 kD. Both β- and δ-zein are rich in methionine, making them insoluble in water (Sudheesh et al., 2022; Tapia-Hernández et al., 2019).

Prolamins from Rice and Oats: Rice and oats contain fewer prolamins than other cereals, with levels below 10%. Prolamins in rice have a smaller molecular size with no repetitive units and are distinct from other cereals. The relative molecular mass of rice prolamins is 10,000, 13,000, and 16,000. They are soluble in propanol (55%) with an extraction percentage of 20%–25% and aqueous ethanol (60%–70%) (Jayaprakash et al., 2022; Sudheesh et al., 2022).

The prolamins in oats are called avenins, and they are closely related to the prolamins of the Triticeae family due to their similar repeated sequences. They belong to S-rich prolamins with a relative molecular mass ranging from 18,500 to 23,000. Oat prolamins are present in the globular matrix of the endosperm but are entirely absent in the aleurone layer. While oat prolamins are high in glutamic acid and glutamine, they have lower levels of proline than prolamins from the Triticeae family and lack tryptophan. The oat prolamins can be isolated using 70% (v/v) ethanol without affecting the disulphide bonds (Sudheesh et al., 2022).

Prolamins from Sorghum: Kafirins are prolamins found in sorghum grain that are alcohol-soluble and mainly located in the endosperm, making up 70–80% of the grain's total protein content. They are divided into three categories: α-, β-, and γ-kafirins, based on their molecular mass, solubility, and structure. The most prevalent type is α-kafirin (23 and 25 kD), whereas β-kafirin (16–20 kD) and γ-kafirin (28 and 45 kD) are present in smaller amounts, creating both intra- and interchain linkages. Kafirins originate in the rough endoplasmic reticulum and create protein bodies. Their high proline and glutamine levels influence their structural characteristics, accounting for 30% of the total amino acids. Additionally, their hydrophobic properties make them ideal for coating applications (Labuschagne, 2018; Tapia-Hernández et al., 2019).

Storage Glutenins and Globulins

Glutelin, a storage protein, is a large polymer majorly present in wheat, followed by barley and rye. Glutelins contain hydrophobic amino acids, namely phenylalanine, valine, proline, tyrosine, and leucine. Therefore, they are insoluble in water, salt solutions, and aqueous alcohols while mildly soluble in reducing agents, dilute acids and alkalis. The wheat glutelins are called glutens (further divided into gliadins and glutenins), vastly influencing the rheological properties of baked products by enhancing the dough's elasticity and gas-holding properties. They are also responsible for cross-linking with other proteins by inter- and intra-disulphide bonds (Kumar et al., 2019). A combination of aqueous alcohol, a reducing reagent, an increased temperature of 50°C–60°C, and an addition of a disaggregating agent aids in the cleaving of disulphide bonds, breaking them into monomeric subunits, therefore encouraging a complete extraction of glutenins (Scherf et al., 2016).

Globulin storage proteins occupy the embryo and aleurone layer of cereal, including maize, wheat, barley, oats, and rice. These proteins are readily soluble in dilute salt solutions and have been thoroughly studied in maize. While similar proteins have been discovered in other cereals, their exact relationships are still not fully understood. The 7S globulins, which are stored in protein bodies, function as storage proteins but are not critical for normal seed development and germination, as shown by maize mutants that lack these proteins. Despite their high concentration in the embryo and aleurone layer, they possess less impact on the end-use characteristics of the grain (Shewry & Halford, 2002). Oat globulins have higher heterogeneity, and 12S globulins are the predominant components, with two subunits called α- and β-subunits with molecular weights of 32,000 and 22,000. α-Subunit polypeptides are acidic, while β-subunit polypeptides are basic. They combine to form a dimer of molecular weight of 54,000 with disulphide bonds in the native globulin. Globulin from the cell wall dissolves in water much more easily than the insoluble globulin found in the vacuolar protein bodies within the endosperm, which only dissolves in a 1 mol/L NaCl solution. The formation of insoluble protein aggregates in the protein bodies is likely due to interactions between carbohydrates and globulin, as well as potential protein–protein interactions among the globulin molecules. Oats have lesser quantities of 3S and 7S globulins. The 7S fraction comprises polypeptides with a molecular weight of around 55,000, while the 3S fraction contains components with molecular weights of approximately 15,000 and 21,000. Compared to 12S globulins, the 3S and 7S globulins are higher in glycine and lower in glutamine and glutamic acid. In oats, these proteins are predominantly found in the endosperm, whereas lysine-rich globulins and albumins are located near the embryo. Similar proteins can also be found in other cereals, including barley, rye, rice, and corn, suggesting a shared evolutionary lineage, despite variations in the composition of seed storage proteins (Klose & Arendt, 2012).

Other Minor Proteins

Albumins (contain eight polypeptides) are water-soluble, can coagulate, and are present in lower concentrations in seeds. They are readily soluble in water due to adequate net charge and the absence of significant disulphide cross-linking. The plants can easily break down albumins to provide nitrogen and sulphur for sprouting and further development. They are heat-sensitive and denature around 75.7°C and 73°C. The lesser disulphide bonds make them readily digestible and absorbable. The proteins in this fraction mainly span from 10 to 200 kDa, and glycoproteins are most prevalent in 16 and 60 kDa (Jayaprakash et al., 2022; Kumar et al., 2019).

Hydrolytic enzymes are proteins with specific metabolic roles. They are synthesised in the scutellum and aleurone layer and are responsible for the breakdown of carbohydrates, proteins, and fats during germination. Barley contains β-amylase, α-amylase, Z-protein, and limit dextrinase. Among them, β-amylase and limit dextrinase are essential for seed development and are activated during germination. Cereals also produce low-molecular-weight protein inhibitors that are significant and impactful in regulating the activity of proteases and carbohydrates.

Key inhibitors include BASI (barley α-amylase subtilisin inhibitor), WASI (wheat), and RASI (rice), each approximately 20,000 in molecular weight, which control α-amylase activity and enhance the quality of wheat flour. Furthermore, barley has chymotrypsin inhibitors (CI-1 and CI-2) that are rich in lysine but deficient in cysteine. These inhibitors are integral for maintaining grain quality and managing preharvest sprouting (Sudheesh et al., 2022).

STRUCTURAL AND FUNCTIONAL CHARACTERISTICS OF CEREAL PROTEINS

Structural Studies

Based on the structure and classification, cereal proteins are divided into the superfamilies, including prolamins, cupins, and plant pathogen-related proteins (BetV1) and the families denoting legumins, vicilins, nsLTPS, and albumins. Specific proteins, such as profilins, expansins, and chlorophyll-binding proteins, have not been fully classified yet. The prolamins contain eight conserved cysteine residues that form disulphide bonds, along with unique CC and CXC motifs that help identify members, such as 2S albumins and non-specific lipid transfer proteins (nsLTPs). nsLTPs function as lipid transporters and are plant allergens that trigger intense immune responses. 2S albumins are smaller protein units composed of disulphide-linked polypeptides and are often categorised as allergens due to their compact structure and stability. The Cupin superfamily is mainly divided into 7S vicilin and 11S legumin, which are highly immunogenic because of their ability to withstand heat and proteolysis. 7S vicilins serve various functions, including protection against oxidative stress, antimicrobial activity, and sugar-binding properties. In contrast, 11S legumins primarily form hexamers, which resist crystallisation; nonetheless, they play a major role in seed storage and nutrition. The Bet V1 (Pathogenesis-Related) Superfamily acts as a defence mechanism against pathogens. This family is vital for the plant's immunity and environmental stress response. They have two proteins, namely PR-1 protein with a unique $\alpha+\beta$ structure with hydrophobic core regions and PR-5, which has structural stability due to multiple disulphide-rich loops present in it (Jain, 2023).

Solubility: Solubility is primarily influenced by three key factors: (1) various functional characteristics like gelation, emulsification, and foaming; (2) the ratio of hydrophilic and hydrophobic residues in the protein and on its surface; and (3) the protein's pH level. Proteins that have a higher presence of hydrophilic amino acids on their surfaces usually exhibit greater water solubility because they can create hydrogen bonds with water. In contrast, hydrophobic proteins, such as prolamins and glutelins, are generally insoluble due to the presence of amino acids like leucine, valine, and isoleucine, which repel water. Additionally, the solubility of these proteins tends to increase as the pH deviates from the isoelectric point. Additionally, the solubility that depends on pH can also vary according to the protein source (Avelar et al., 2024). To demonstrate, Zhao et al. (2020) presented the result of wheat protein displaying a perfect 'U' shape in the pH/protein solubility graph, indicating that wheat protein is soluble in both acidic and basic conditions.

Binding Properties: The capacity of proteins to absorb and hold fats, along with their ability to engage with them in emulsions and various food systems, is crucial in food formulations (Haque et al., 2016). Proteins are amphiphilic in nature, which makes them capable of binding with both water and oil. This is a vital property of the proteins since it impacts the flavour, retention, texture, mouthfeel, and tenderness of the end product (Avelar et al., 2024). The water-binding capacity (WBC) and oil-binding capacity (OBC) measure the amount of water and oil absorbed per gram of protein. These capacities indicate how much water or oil a protein structure can hold under specific conditions while preventing it from being released from the matrix. A low WBC can negatively affect the texture of the food, leading to dryness and shrinkage, which may impact its taste. Incorporating hydrophilic compounds such as polar proteins or amino acids in the protein matrix can improve the WBC, while adding nonpolar proteins and amino acids can elevate the OBC (Hadidi et al., 2024).

Emulsifying Properties: Protein molecules have excellent emulsifying properties as they possess amphiphilic and interfacial film-forming abilities. When introduced into oil–water emulsions, these substances migrate to the oil–water interface. They align so that the hydrophobic units penetrate the oil droplets while the hydrophilic units enter the aqueous phase. This alignment leads to the formation of robust viscoelastic films surrounding the oil droplets. The formed films prevent emulsion destabilisation and act as a strong mechanical barrier. Meanwhile, modifying the protein structure, causing partial denaturation, can enhance the emulsifying capacity (EC) due to exposing the hidden hydrophobic amino acids to the surface (Haque et al., 2016).

Foaming Properties: Protein films' development, creating entrapment of air, is very essential for various food products, including soufflés, mousse, leavened bread, whipped creams, and meringue. In addition, these air cells ensure proper dispersion of flavours and aid in attaining precise texture and mouthfeel (Stone et al., 2019). The protein's capability to form foam is typically examined by two main factors: foaming capacity (FC) and foam stability (FS). FC is measured by determining the percentage of air incorporated into the foam after whipping. In contrast, FS refers to the maintenance of the foam's volume over a specific period (Avelar et al., 2024). Amphiphilic proteins facilitate the creation of stable foams by adhering to the air–water boundary and producing a cohesive layer around gas bubbles. These proteins reduce interfacial tension, making it easier to generate small gas bubbles during foaming. Additionally, they stabilise the bubbles by creating a viscoelastic layer that generates repulsive forces, helping to prevent the larger bubbles from merging or collapsing (Hadidi et al., 2024).

Gelation Properties: Proteins serve a crucial role in the food industry by forming gels, which influence the texture and sensory qualities of products like jellies, puddings, yoghurts, cheeses, eggs, and meats. Protein gelation takes place when the attractive interactions between protein molecules exceed the repulsive forces, resulting in aggregation and the formation of a network. A variety of interactions—including van der Waals forces, hydrophobic interactions, hydrogen bonding, electrostatic attractions, and covalent bonds—play a role in this process and are influenced by the types of proteins and the specific conditions present. Gelation can be triggered by techniques such as changing the pH, adding salt, applying heat, cooling, using high pressure, or employing enzymes. When proteins are heated beyond their denaturation temperature, they begin to partially unfold, revealing nonpolar amino acids, which promotes aggregation. Therefore, controlling factors like pH, ionic strength, and temperature is vital for desired gel formation (Hadidi et al., 2024).

EFFECTS OF PROCESSING ON CEREAL PROTEINS

Cereals are mostly consumed as processed foods, including bread, pasta, noodles, breakfast cereals, etc. Cereal processing is a vast sector with great technologies to improve shelf life, flavours, nutritional values, etc. They also aid in increasing the grain yield, preserving the nutrients, and enhancing the food's texture, sensory attributes, and functional properties. Emerging food processing techniques, including extrusion, cooking, germination, fermentation, drying, and protein isolation, aim to enhance protein content and digestibility (M. Li & Niu, 2023; Mollakhalili-Meybodi et al., 2022; Sá & House, 2024).

Thermal Processing Technologies

Thermal methods primarily focus on microbial control, food preservation, and shelf-life extension. Thermal processing provides a massive number of benefits, including the following: (1) enhanced digestibility, (2) removing allergens and toxins, (3) deactivation of antinutrients such as α-galactosides, protease inhibitors, tannins and lectins and improving bioavailability, and (4) delivering the preferred aroma, colour, flavour (Avelar et al., 2024; Mollakhalili-Meybodi et al., 2022). However, they can also lead to unfavourable conditions, such as producing undesired components, reducing nutrients, and forming disulphide bonds (Mollakhalili-Meybodi et al., 2022). The impact of various thermal processing technologies on cereal proteins is provided in Table 2.5.

Boiling and Microwave altered the amino acid structure by denaturation, unfolding, and exposing the peptide bonds to the digestive enzymes. These changes improved the protein digestibility of the treated buckwheat proteins (Deng et al., 2015). Microwave resulted in improved solubility of the cereal proteins with a short-span treatment, but a decrease in the solubility was noted after a prolonged treatment time. Microwave treatment also improved wheat proteins' OBC, FC and EC (Saeed, 2012). Extrusion increased the in vitro digestibility of different cereals by deactivating the antinutrients and linkage breakdown. Explosion-puffing aids in the structural opening of the puffed grains, therefore promoting the availability of the proteins for peptidase activity (Joye, 2019). The improved surface hydrophobicity

TABLE 2.5
The Impact of Diverse Thermal Processing Techniques on the Various Attributes of Cereal Proteins

Treatment	Cereal	Processing Condition	Implications	References
Cooking	Rice	Cooking time: 40 min Stewing time: 20 min	Decreased digestibility due to the unfolding and aggregation of proteins. Decreased solubility of proteins due to heat-induced denaturation and hydrophobicity	Liu et al. (2019)
Extrusion cooking	Rice	Extruder screw speed: 100–250 rpm Extruder temperature: 90°C–150°C Moisture: 25%–40%	Increased water-holding capacity (WHC), FC, and emulsion activity index. Decreased sulfhydryl, sulphide bonds and hydrophobicity Disruption in the protein's 3-D structure and a change in the microstructure of the protein	Gao et al. (2022)
Explosion puffing	Millet, barley, black rice, rice, glutinous rice, and wheat	Moisture: 13%–14% Pressure: 0.8–1.5 MPa Time: 70–80 s	Improved protein digestibility Decreased protein solubility due to protein aggregation and increased surface hydrophobicity	Huang et al. (2018)
Roasting	Oats	Roasting temperature: 150°C Roasting time: 45 min	Occurrence of glycation Increased surface hydrophobicity, thermal stability, and α-helix Decreased emulsifying properties, solubility and β-strands	Wang et al. (2022)
Baking	Purple rice Wheat	Baking temperature: 170°C Time: 9 min	A decrease in the protein digestibility of gluten proteins due to protein denaturation	Klunklin and Savage (2018)
Microwave oven	Wheat kernels	Humidity: 18%–20% Power: 1,000 W Temperature: 110°C–120°C Time: 2 min	Resulted in the polymerisation of proteins Improved water solubility of gluten Decreased gluten antigenicity	Lamacchia et al. (2016)

of the explosion-puffing treated cereals makes them less soluble in water (Huang et al., 2018). Extrusion cooking disrupted the overall protein structure, reducing sulfhydryl, sulphide bonds, and hydrophobicity. This enhanced water-holding capacity (WHC), FC, and emulsion activity index (Gao et al., 2022).

Nonthermal Processing Technologies

Nonthermal methods allow for enhanced food safety and extended shelf life while maintaining the nutritional and functional properties of the food with minimal alterations. Various nonthermal techniques that have become popular in cereal processing include cold plasma (CP), high hydrostatic pressure (HHP), ionising irradiation, pulsed electric field (PEF), pulsed light (PL), and ultrasonic (US). These methods also alter the nutritional and functional characteristics of cereal proteins by changing their structure. Findings from the processing of cereals using nonthermal methods demonstrated a decrease in the allergenicity of wheat proteins, an increase in the digestibility of corn flour, enhanced bioactivity of brown rice proteins, improved dough strength, greater solubility of rice proteins, and better emulsion stability of wheat flour (Mollakhalili-Meybodi et al., 2022). The influence of various nonthermal processing on different aspects of cereal proteins is presented in Table 2.6.

Irradiation is one of the most efficient methods of altering protein functionality. When rice protein was subjected to a gamma irradiation dose of 2KGy, it resulted in exceptional modifications of functional properties, including enhanced water absorption capacity (WAC), oil absorption capacity (OAC), water solubility, and highest emulsion activity and stability compared to the untreated proteins. The overall binding properties of the treated samples are at their best outcome since gamma irradiation uncovers the protein structure, exposing the internal hydrophilic groups and making them more favourable to bind with water molecules. Meanwhile, the hydrophobic side chains interact and stabilise the oil droplets via steric effects. On the contrary, Bhat et al., 2016 reported decreased OAC and WAC due to denaturation and aggregation of protein molecules. Cold plasma facilitates the oxidation of protein sulfhydryl groups and disulphide bond formation among the cysteine units, altering the proteins' overall physical and chemical attributes. The ion bombardment of the plasma during this modification is beneficial over other traditional modification methods as it disrupts protein aggregates, increasing the protein's contact surface and elevating the protein oxidation rate. These changes can affect the protein's solubility, surface hydrophobicity, molecular weight, etc. When wheat protein germ protein isolate was treated with cold atmospheric plasma, solubility increased with extended processing time. Exposing the protein isolates to cold plasma for 5 min enhanced both the EC and FC. The functional properties of the isolates were efficiently improved at a set parameter of 25 kV for 5 min while preventing oxidative damage (Abarghoei et al., 2023). Regarding the Pulse electric field, they scatter energy, which results in the breakage of the electrostatic interactions of protein chains, triggering cleavage or new combinations of amino acids. They sometimes increase the solubility of proteins, increasing the peptides' antioxidant activities (Mollakhalili-Meybodi et al., 2022). Ultrasonication involves using periodic mechanical movements of a probe to impart energy into a liquid medium, resulting in significant pressure changes that

TABLE 2.6
The Efficacy of Various Nonthermal Processing Techniques on Different Aspects of Cereal Proteins

Treatment	Cereal	Processing Condition	Implications	References
Gamma irradiation	Wheat	Gamma irradiation Doses: 0, 2.5, and 5 kGy	Increased FC and FS resulting from improved diffusion of unfolded and fragmented proteins in the air/water	Bhat et al. (2016)
Cold plasma	Wheat	Plasma source: Dielectric barrier discharge Voltage: 0.0, 9.0, 11.0, 13.0, 15.0, and 17.0 kV Gas flow: 1.5 L/min Time: 4 min Gas type: Dry air	Increased soluble protein content with increased voltage Increased seed germination and seedling growth Improved water uptake by the seed due to softening and cracking of the seed's surface	Guo et al. (2018)
Pulse Electric Field	Wheat gluten	Electric Field Strength: 1.65 kV/cm Pulse width (5 μs) Frequency (400 Hz) Number of pulses variations: 20,000 or 60,000.	Decreased protein solubility due to the unfolding of proteins and the formation of protein aggregates Increased WHC due to increased hydrophilicity on the gluten surface	Melchior et al. (2020)
Ultrasonication	barley	Treatment time: 2.5 min, cooling and another 2.5 min of treatment. Temperature: <35°C	Decreased particle size Improved protein solubility and colloidal stability of barley-derived proteins at alkaline pH	Wang et al. (2022)
Pulsed light	Wheat gluten	Lamp type: xenon Range: 200–1,000 nm (200–400 nm: 41%; 400–700 nm: 51%; 700–1,000 nm: 8%) Pulsed light: 1.75–26.25 J/cm^2	Modification of gluten structure due to disulphide exchange Destruction of gliadin epitopes Decreased allergenicity of wheat gluten	Panozzo et al. (2016)

cause the creation of tiny, swiftly expanding bubbles. This occurrence, referred to as "cavitation," can break down a protein's tightly packed structure, and the resultant partially unfolded protein frequently shows enhanced physicochemical characteristics and functionalities. High-intensity ultrasonication increased the exposure of hydrophobic and charged groups. It increased the solubility of oat protein isolates while decreasing the particle size. In addition, the ultrasonication-induced cavitation increased the enzymatic efficiency and in vitro digestibility of oat protein isolates (Li & Xiong, 2021).

Other Processing Methods

Fermentation and germination processes annihilate the antinutrients by producing/releasing the hydrolytic enzymes. These processes also aid in the fragmentation of the proteins, hence improving the availability and digestibility of the millet proteins (Annor et al., 2017). The lactic acid bacteria fermentation of the sorghum flour resulted in enhanced protein digestibility. This was achieved by the hydrolytic enzymes fractioning proteins into soluble and straightforward units. Additionally, a lower pH during fermentation boosts the activity of peptidases and increases protein digestibility (Ogodo et al., 2019).

APPLICATION OF CEREAL PROTEINS

Cereal Protein-Based Vegan Meat

Cereal protein-based meat substitutes are gaining significant traction due to a variety of compelling reasons. The expense involved in animal protein sources significantly influences the development of economically viable plant-based protein alternatives. In developing nations, the availability of high-quality animal proteins is often constrained, which has led to a growing interest in plant-based meat substitutes. Research has examined the health advantages of these alternatives compared to conventional meat products. Key benefits identified include reductions in blood lipid levels, enhanced satiety, improved cardiovascular health, and a decreased risk of certain cancers, particularly concerning soy-based alternatives (Senarathna et al., 2024).

Consumers are increasingly aware of health concerns associated with conventional meat consumption, while environmental problems stemming from large-scale meat production cannot be ignored. Additionally, many individuals choose these alternatives because of religious beliefs, concerns about animal diseases, and a strong commitment to animal welfare. The economic challenges of meat production further bolster the argument for adopting plant-based options. With all these factors at play, it is clear that shifting towards plant-based meat is not just a trend but a responsible choice for our health, the planet, and ethical considerations (Senarathna et al., 2024).

The main concepts for meat substitutes focus on structuring techniques, primarily on digestibility and bioavailability, aiming to enhance sensory properties, namely flavour, colour, and protein nutritional quality (Senarathna et al., 2024).

Seitan, a widely recognised plant-based protein source, originates from 6th-century China. Unlike other meat alternatives such as tofu and tempeh, seitan is composed primarily of gluten, a significant protein component derived from wheat grains. Its unique texture and properties make it a popular choice for mimicking the mouthfeel of meat in various culinary applications.

Currently, Soy protein remains as the primary ingredient in the plant-based meat alternative market. This leadership is due to its remarkable functional properties, enabling it to replicate conventional meat products' texture and flavour effectively. Additionally, soy protein is more affordable than many other plant-based protein options, making it a cost-effective choice for producers and consumers. As more people pursue sustainable and health-focused meat alternatives, the significant benefits of soy protein are expected to ensure its continued prominence in this expanding industry (Senarathna et al., 2024).

Consumers' acceptance of meat alternatives is influenced by several critical factors, including the protein source they provide, their taste, overall flavour, and nutritional value. These characteristics significantly shape consumer preferences. However, beyond these determinants, additional reasons have contributed to the growing popularity of meat alternatives in recent years. Notably, these products are becoming increasingly affordable, making them accessible to a broader audience. In addition to their cost-effectiveness, many meat alternatives boast an appealing texture that closely mimics traditional meat, which enhances the eating experience for many consumers. Furthermore, the health benefits of these products, particularly their cholesterol-free nature, appeal to health-conscious individuals and those aiming to improve their diet. As a result, the market for meat alternatives is witnessing substantial growth, reflecting a shift in consumer attitudes towards these products (Senarathna et al., 2024).

The primary objective of developing meat substitutes is to offer textures and sensory experiences comparable to those found in animal protein meat. Moisture, the predominant component in meat alternatives, works in conjunction with cereal protein, which serves as the principal ingredient that enhances the required texture and palatability of the product.

ALTERNATIVES TO DAIRY PRODUCTS

Consumer interest in replacing cow's milk with plant-based alternatives has recently grown. This shift is promoted by various factors such as allergies, lactose intolerance, and environmental concerns like sustainability, personal lifestyle choices, and ethical issues. Plant-based milk alternatives may offer various health benefits, such as decreasing cholesterol and enhancing blood sugar regulation. Renowned for its health benefits, soy milk stands out as an affordable, invigorating, and nutrient-dense drink. Approximately two millennia ago, soy milk originated in China, marking the first recorded practice of plant-based milk for nutritional purposes. In addition to its historic importance, soy milk is a versatile and advantageous plant-based drink option. Isoflavones represent the functionally active components associated with various health benefits, including protection against osteoporosis, cardiovascular diseases and cancer.

Research shows that oat milk undergoes little change during digestion, which has significant implications for developing new plant-based drinks. The structural integrity and physicochemical characteristics maintained during digestion facilitate nutrient absorption and enhance consumer bioavailability. In addition to oat milk and soy milk, various forms of cereal-based milk, such as rice milk, have been subject to investigation (Senarathna et al., 2024).

Recent developments in food technology have concentrated on improving the meltability and stretchability of plant-based cheeses, a significant area of interest in the growing plant-based food market. These improvements have been mainly achieved by utilising zein, a unique plant protein derived from corn. Zein is significant for its enhanced proline content, an amino acid that plays a crucial role in the structural properties of protein. The loose packing of zein molecules allows for the formation of a stable and functional network in water. This non-covalently linked network is vital because it contributes to the desired texture and behaviour of heated plant-based cheeses, enabling them to melt and stretch similarly to their dairy counterparts. The development of these characteristics not only enhances the culinary applications of plant-based cheeses but also makes them more appealing to consumers seeking alternatives to traditional cheese products (Senarathna et al., 2024).

A cheese substitute of 30% zein showed similar rheological, melt-stretch and textural behaviour compared to cheddar cheese. Alongside rice cheese, quinoa cheese, and oat-based cheese, plant-based cheese options have made notable progress.

Nevertheless, developing effective plant-based milk alternatives involves navigating many technological challenges. These challenges are particularly pertinent when replicating the critical aspects of cow's milk, including its appearance and taste, its flavour profile, stability during storage, and overall nutritional value. Researchers and innovators must address these hurdles with creativity and resourcefulness to achieve these goals. By tackling these complex issues head-on, there is a strong potential to create plant-based milk alternatives that taste and look similar to traditional dairy milk and provide comparable sensory experiences and nutritional benefits. Such advancements could significantly enhance consumer acceptance and broaden the market for these alternatives, ultimately contributing to a more sustainable food system (Senarathna et al., 2024).

Recently, significant progress in producing plant-based yoghurt has been observed, driven by the rising recognition of yoghurt as a vital source of probiotics. This change has sparked greater interest in non-dairy probiotic products, as consumers increasingly prioritise their health and wellness. The growing awareness of the benefits of probiotics, which can aid digestion and enhance gut health, has led to a diverse range of options in the market today. Additionally, the trend towards plant-based diets has encouraged manufacturers to innovate and improve their products, creating yoghurts that cater to various dietary preferences. This shift meets the needs of vegan and lactose-intolerant consumers and appeals to a broader audience interested in incorporating healthier food alternatives into their diets. As a result, the market for plant-based yoghurt is expected to continue expanding, reflecting both consumer demand and advancements in food technology (Senarathna et al., 2024).

A diverse range of experimental plant-based yoghurts has been investigated, employing non-dairy substitutes such as pea, lupin, oat and soy. Kokwar et al. (2022)

engineered a non-dairy probiotic beverage utilising multigrain ingredients such as red rice, oats, buckwheat, and barley as its primary components that meet nutritional needs and cater to the growing demand for diverse, health-focused foods. Additionally, dairy probiotics are often recognised for their potential health benefits, but they are also associated with certain disadvantages, including lactose intolerance and elevated cholesterol levels. This has led to a growing consumer preference for vegan alternatives (Senarathna et al., 2024).

Substitutes for Eggs

With the ongoing exploration of innovative solutions within the food industry to accommodate diverse dietary requirements, cereal protein-egg substitutes are a versatile and promising alternative. Eggs are recognised for their high protein content, which is crucial in various culinary applications, including enhancing aroma and colour, foaming, regulating sugar crystallisation, and coagulation and emulsification. Dietary preferences and individual needs, influenced by beliefs and religions, impact the acceptance of plant-based egg substitutes. Increasing consumer demand and the evolving needs of food producers are propelling the expansion of the egg substitute market. This evolution presents an exciting opportunity for businesses to tap into a lucrative segment that is reshaping the food industry. By embracing these changes, stakeholders can meet consumer preferences while contributing to a more sustainable food ecosystem. The demand for alternatives like plant-based eggs is on the rise, fuelled by various factors, including health considerations such as cardiovascular diseases, inherited disorders like phenylketonuria, and food allergies, particularly egg allergies. Using cereals as an egg alternative addresses nutritional needs and meets the rising demand for plant-based options. Cereal-based egg substitutes have become popular as sustainable options in numerous food applications. Substitutes often utilise the functional characteristics of cereal proteins, such as binding, emulsifying, and enhancing structural integrity, to achieve their desired effects. These alternatives improve texture, leavening, and moisture retention in final products, making them ideal for baking. Peas, flaxseeds, lupin, chia seeds, and lentils are common plant-based ingredients that serve as egg substitutes (Senarathna et al., 2024).

Protein-Enhanced Products

Recent advancements in the food industry include high-protein products like cereal protein isolates. Wheat, maize, peas, lupin, and faba beans are acknowledged as valuable high-protein ingredients. Cereal protein isolates greatly enhance the protein level and nutritional benefits of offerings such as plant-based meat substitutes, protein-enriched snacks, and energy bars, thus attracting health-conscious consumers. Protein isolates derived from grains like oats, wheat, and rice have received considerable attention due to their nutritional advantages and technological characteristics. These isolates, rich in essential amino acids, are important for developing high-protein food products (Senarathna et al., 2024).

CONCLUSION

Food security has emerged as a vital concern as the global population grows. Cereal-based proteins, recognised for their valuable agronomic qualities and environmental sustainability, provide a significant solution. Plant-based proteins present a sustainable approach to addressing the growing global food demand, primarily emphasising pulses as a source. Nevertheless, the functional characteristics of essential proteins, including 7S (β-conglycinin) and 11S (glycinin), are constrained by their propensity to aggregate, which arises from disulphide bonds and various interactions.

To adapt to varying consumer preferences focused on health and sustainability, cereal food manufacturers are experiencing major transformations. Consumers increasingly seek nutritious and suitable options, particularly gluten-free and cereal protein-based products that reflect these emerging trends. Cereal protein-based dairy and meat alternatives possess significant potential for diet diversification. Nonetheless, challenges such as a lack of familiarity and sensory attraction must be tackled to promote broader acceptance. Improving these aspects can help consumers embrace these innovative products and revolutionise their eating habits. Additionally, for cereal-based proteins to be recognised as functional protein ingredients, they must meet the requirements for solubility, water-binding capacities, fat stabilisation, gelling properties, and emulsification.

Innovative methodologies, including physical processing and enzymatic treatment, enhance the functional properties of proteins through structural modifications. These proteins are especially beneficial in 3D food printing, non-dairy alternatives, and meat analogues. They provide additional advantages such as antibacterial and antioxidant attributes, significantly contributing to their increasing prevalence and technological progress. Nevertheless, significant challenges persist in accurately replicating the sensory attributes of animal-derived products. Consumers' acceptance of plant-based alternatives is heavily influenced by characteristics such as flavour, texture, and appearance. Cost, flavour, and convenience greatly influence consumers' choices regarding cereal-based products. However, these same factors can also create significant barriers that hinder individuals from accepting a vegan lifestyle. Increasing shelf life results in greater consumer acceptance through innovative preservation techniques that maintain product quality over time.

The challenges associated with replicating the sensory attributes of meat and dairy products, along with consumer apprehensions regarding additives and the inherent off-flavours of pulses, underscore the importance of targeted research aimed at overcoming these barriers and enhancing the appeal and acceptance of plant-based proteins. Additionally, their adaptability in various food applications accentuates their competence to meet the increasing consumer demand for cereal-based protein options, which aligns with progressing alimentary trends and health-conscious lifestyles.

REFERENCES

Abah, C. R., Ishiwu, C. N., Obiegbuna, J. E., & Oladejo, A. A. (2020). Nutritional composition, functional properties and food applications of millet grains. *Asian Food Science Journal*, 9–19. https://doi.org/10.9734/afsj/2020/v14i230124

Abarghoei, M., Goli, M., & Shahi, S. (2023). Investigation of cold atmospheric plasma effects on functional and physicochemical properties of wheat germ protein isolate. *LWT*, 177, 114585. https://doi.org/10.1016/j.lwt.2023.114585

Adenekan, M. K., Fadimu, G. J., Odunmbaku, L. A., & Oke, E. K. (2018). Effect of isolation techniques on the characteristics of pigeon pea (*Cajanus cajan*) protein isolates. *Food Science & Nutrition*, 6(1), 146–152. https://doi.org/10.1002/fsn3.539

Alemayehu, G. F., Forsido, S. F., Tola, Y. B., Teshager, M. A., Assegie, A. A., & Amare, E. (2021). Proximate, mineral and anti-nutrient compositions of oat grains (*Avena sativa*) cultivated in Ethiopia: Implications for nutrition and mineral bioavailability. *Heliyon*, 7(8), e07722. https://doi.org/10.1016/j.heliyon.2021.e07722

Ali, A., & Bhattacharjee, B. (2023). Nutrition security, constraints, and agro-diversification strategies of neglected and underutilized crops to fight global hidden hunger. *Frontiers in Nutrition*, 10, 1144439. https://doi.org/10.3389/fnut.2023.1144439

Annor, G. A., Tyl, C., Marcone, M., Ragaee, S., & Marti, A. (2017). Why do millets have slower starch and protein digestibility than other cereals? *Trends in Food Science & Technology*, 66, 73–83. https://doi.org/10.1016/j.tifs.2017.05.012

Avelar, Z., Pereira, R. N., Vicente, A. A., & Rodrigues, R. M. (2024). Protein quality of cereals: Technological and functional perspectives. *Journal of Cereal Science*, 117, 103922. https://doi.org/10.1016/j.jcs.2024.103922

Baniwal, P., Mehra, R., Kumar, N., Sharma, S., & Kumar, S. (2021). Cereals: Functional constituents and its health benefits. *The Pharma Innovation*, 10(3), 1–7. https://doi.org/10.22271/tpi.2021.v10.i3a.5681

Bhat, N. A., Wani, I. A., Hamdani, A. M., Gani, A., & Masoodi, F. A. (2016). Physicochemical properties of whole wheat flour as affected by gamma irradiation. *LWT - Food Science and Technology*, 71, 175–183. https://doi.org/10.1016/j.lwt.2016.03.024

Bhatt, D., Fairos, M., & Mazumdar, A. (2022). *Millets: Nutritional composition, production and significance: A review. The Pharma Innovation*, 11, 1577–1582.

Biel, W., Kazimierska, K., & Bashutska, U. (2020). Nutritional value of wheat, triticale, barley and oat grains. *Acta Scientiarum Polonorum Zootechnica*, 19(2), 19–28. https://doi.org/10.21005/asp.2020.19.2.03

Cardoso, R. V. C., Fernandes, Â., Heleno, S. A., Rodrigues, P., Gonzaléz-Paramás, A. M., Barros, L., & Ferreira, I. C. F. R. (2019). Physicochemical characterization and microbiology of wheat and rye flours. *Food Chemistry*, 280, 123–129. https://doi.org/10.1016/j.foodchem.2018.12.063

Dang, B., Zhang, W.-G., Zhang, J., Yang, X.-J., & Xu, H.-D. (2022). Evaluation of nutritional components, phenolic composition, and antioxidant capacity of Highland Barley with different grain colors on the Qinghai Tibet Plateau. *Foods*, 11(14), 2025. https://doi.org/10.3390/foods11142025

Deng, Y., Padilla-Zakour, O., Zhao, Y., & Tao, S. (2015). Influences of high hydrostatic pressure, microwave heating, and boiling on chemical compositions, antinutritional factors, fatty acids, in vitro protein digestibility, and microstructure of buckwheat. *Food and Bioprocess Technology,* 8(11), 2235–2245. https://doi.org/10.1007/s11947-015-1578-9

El-Mahis, A., Baky, M. H., & Farag, M. A. (2023). How does rye compare to other cereals? A comprehensive review of its potential nutritional value and better opportunities for its processing as a food-based cereal. *Food Reviews International*, 39(7), 4288–4311. https://doi.org/10.1080/87559129.2021.2023817

Gao, Y., Sun, Y., Zhang, Y., Sun, Y., & Jin, T. (2022). Extrusion modification: Effect of extrusion on the functional properties and structure of rice protein. *Processes*, 10(9), 1871. https://doi.org/10.3390/pr10091871

Guo, Q., Meng, Y., Qu, G., Wang, T., Yang, F., Liang, D., & Hu, S. (2018). Improvement of wheat seed vitality by dielectric barrier discharge plasma treatment. *Bioelectromagnetics*, 39(2), 120–131. https://doi.org/10.1002/bem.22088

Hadidi, M., Garcia, S. R., Ziogkas, D., McClements, D. J., & Moreno, A. (2024). Cereal bran proteins: Recent advances in extraction, properties, and applications. *Critical Reviews in Food Science and Nutrition*, 64(29), 10583–10607. https://doi.org/10.1080/10408398.2023.2226730

Haque, M. A., Timilsena, Y. P., & Adhikari, B. (2016). Food proteins, structure, and function. In *Reference Module in Food Science* (p. B9780081005965030572). Elsevier. https://doi.org/10.1016/B978-0-08-100596-5.03057-2

Helm, R. M., & Burks, A. W. (1996). Hypoallergenicity of rice protein. *Cereal Foods World*, 41, 839–843.

Huang, R., Pan, X., Lv, J., Zhong, W., Yan, F., Duan, F., & Jia, L. (2018). Effects of explosion puffing on the nutritional composition and digestibility of grains. *International Journal of Food Properties*, 21(1), 2193–2204. https://doi.org/10.1080/10942912.2018.1514634

Jain, A. (2023). Seed storage protein, functional diversity and association with allergy. *Allergies*, 3(1), 25–38. https://doi.org/10.3390/allergies3010003

Jayaprakash, G., Bains, A., Chawla, P., Fogarasi, M., & Fogarasi, S. (2022). A narrative review on rice proteins: Current scenario and food industrial application. *Polymers*, 14(15), 3003. https://doi.org/10.3390/polym14153003

Jindal, A., Patil, N. D., Bains, A., Sridhar, K., Inbaraj, B. S, Tripathi, M., Chawla, P., & Sharma, M. (2023). Application of cereal- and legume-derived protein-mineral complexes. *Encyclopedia*. https://encyclopedia.pub/entry/51295

Joye, I. (2019). Protein digestibility of cereal products. *Foods*, 8(6), 6. https://doi.org/10.3390/foods8060199

Kazerooni, E. G. (2019). Maize (Corn)-A useful source of human nutrition and health: A critical review. *International Journal of Chemical and Biochemical Sciences*, 15, 35–41.

Klose, C., & Arendt, E. K. (2012). Proteins in oats; their synthesis and changes during germination: A review. *Critical Reviews in Food Science and Nutrition*, 52(7), 629–639. https://doi.org/10.1080/10408398.2010.504902

Klunklin, W., & Savage, G. (2018). Effect of substituting purple rice flour for wheat flour on physicochemical characteristics, *in vitro* digestibility, and sensory evaluation of biscuits. *Journal of Food Quality*, 2018, 8052847. https://doi.org/10.1155/2018/8052847

Kokwar, M. A., Arya, S. S., & Bhat, M. S. (2022). A cereal-based nondairy probiotic functional beverage: An insight into the improvement in quality characteristics, sensory profile, and shelf-life. *Journal of Food Processing and Preservation*, 46(1), e16147. https://doi.org/10.1111/jfpp.16147

Kowieska, A., Lubowicki, R., & Jaskowska, I. (2011). Chemical composition and nutritional characteristics of several cereal grain.

Kumar, A., Agarwal, D. K., Kumar, S., Reddy, Y. M., Chintagunta, A. D., Saritha, K. V., Pal, G., & Kumar, S. P. J. (2019). Nutraceuticals derived from seed storage proteins: Implications for health wellness. *Biocatalysis and Agricultural Biotechnology*, 17, 710–719. https://doi.org/10.1016/j.bcab.2019.01.044

Labuschagne, M. T. (2018). A review of cereal grain proteomics and its potential for sorghum improvement. *Journal of Cereal Science*, 84, 151–158. https://doi.org/10.1016/j.jcs.2018.10.010

Lamacchia, C., Landriscina, L., & D'Agnello, P. (2016). Changes in wheat kernel proteins induced by microwave treatment. *Food Chemistry*, 197, 634–640. https://doi.org/10.1016/j.foodchem.2015.11.016

Li, L., Lietz, G., & Seal, C. J. (2021). Phenolic, apparent antioxidant and nutritional composition of quinoa (*Chenopodium quinoa* Willd.) seeds. *International Journal of Food Science & Technology*, 56(7), 3245–3254. https://doi.org/10.1111/ijfs.14962

Li, M., & Niu, M. (2023). New technologies in cereal processing and their impact on the physical properties of cereal foods. *Foods*, 12(21), 4008. https://doi.org/10.3390/foods12214008

Li, R., & Xiong, Y. L. (2021). Ultrasound-induced structural modification and thermal properties of oat protein. *LWT*, 149, 111861. https://doi.org/10.1016/j.lwt.2021.111861

Liu, K., Zheng, J., & Chen, F. (2019). Effect of domestic cooking on rice protein digestibility. *Food Science & Nutrition*, 7(2), 608–616. https://doi.org/10.1002/fsn3.884

Melchior, S., Calligaris, S., Bisson, G., & Manzocco, L. (2020). Understanding the impact of moderate-intensity pulsed electric fields (MIPEF) on structural and functional characteristics of pea, rice and gluten concentrates. *Food and Bioprocess Technology*, 13(12), 2145–2155. https://doi.org/10.1007/s11947-020-02554-2

Mishra, A., Pattnaik, B., Dutta, T., & Baitharu, I. (2022). Nutritional values and potential health benefits of millets- a review. *Journal of Nutrients*, 8(1), 9–26. https://doi.org/10.18488/87.v8i1.3176

Mohidem, N. A., Hashim, N., Shamsudin, R., & Che Man, H. (2022). Rice for food security: Revisiting its production, diversity, rice milling process and nutrient content. *Agriculture*, 12(6), 741. https://doi.org/10.3390/agriculture12060741

Mollakhalili-Meybodi, N., Nejati, R., Sayadi, M., & Nematollahi, A. (2022). Novel nonthermal food processing practices: Their influences on nutritional and technological characteristics of cereal proteins. *Food Science & Nutrition*, 10(6), 1725–1744. https://doi.org/10.1002/fsn3.2792

Mueed, A., Shibli, S., Korma, S. A., Madjirebaye, P., Esatbeyoglu, T., & Deng, Z. (2022). Flaxseed bioactive compounds: Chemical composition, functional properties, food applications and health benefits-related gut microbes. *Foods*, 11(20), 3307. https://doi.org/10.3390/foods11203307

Németh, R., & Tömösközi, S. (2021). Rye: Current state and future trends in research and applications. *Acta Alimentaria*, 50(4), 620–640. https://doi.org/10.1556/066.2021.00162

Ogodo, A. C., Ugbogu, O. C., Onyeagba, R. A., & Okereke, H. C. (2019). Microbiological quality, proximate composition and in vitro starch/protein digestibility of *Sorghum bicolor* flour fermented with lactic acid bacteria consortia. *Chemical and Biological Technologies in Agriculture*, 6(1), 7. https://doi.org/10.1186/s40538-019-0145-4

Panozzo, A., Manzocco, L., Lippe, G., & Nicoli, M. C. (2016). Effect of pulsed light on structure and immunoreactivity of gluten. *Food Chemistry*, 194, 366–372. https://doi.org/10.1016/j.foodchem.2015.08.042

Pathan, S., & Siddiqui, R. A. (2022). Nutritional composition and bioactive components in quinoa (*Chenopodium quinoa* Willd.) *Greens: A Review. Nutrients*, 14(3), 558. https://doi.org/10.3390/nu14030558

Paudel, D., Dhungana, B., Caffe, M., & Krishnan, P. (2021). A review of health-beneficial properties of oats. *Foods*, 10(11), 2591. https://doi.org/10.3390/foods10112591

Poutanen, K. S., Kårlund, A. O., Gómez-Gallego, C., Johansson, D. P., Scheers, N. M., Marklinder, I. M., Eriksen, A. K., Silventoinen, P. C., Nordlund, E., Sozer, N., Hanhineva, K. J., Kolehmainen, M., & Landberg, R. (2022). Grains – A major source of sustainable protein for health. *Nutrition Reviews*, 80(6), 1648–1663. https://doi.org/10.1093/nutrit/nuab084

Rouf Shah, T., Prasad, K., & Kumar, P. (2016). Maize- A potential source of human nutrition and health: A review. *Cogent Food & Agriculture*, 2(1). https://doi.org/10.1080/23311932.2016.1166995

Sá, A. G. A., & House, J. D. (2024). Protein quality of cereals: Digestibility determination and processing impacts. *Journal of Cereal Science*, 117, 103892. https://doi.org/10.1016/j.jcs.2024.103892

Saeed, S. A. (2012). 45 Publications 453 citations see profile. *Int. J. Agric. Biol.*, 14(3).
Scherf, K. A., Koehler, P., & Wieser, H. (2016). Gluten and wheat sensitivities – An overview. *Journal of Cereal Science*, 67, 2–11. https://doi.org/10.1016/j.jcs.2015.07.008
Senarathna, S., Mel, R., & Malalgoda, M. (2024). Utilization of cereal-based protein ingredients in food applications. *Journal of Cereal Science*, 116, 103867. https://doi.org/10.1016/j.jcs.2024.103867
Sharma, M., & Saini, C. S. (2022). Amino acid composition, nutritional profiling, mineral content and physicochemical properties of protein isolate from flaxseeds (*Linum usitatissimum*). *Journal of Food Measurement and Characterization*, 16(1), 829–839. https://doi.org/10.1007/s11694-021-01221-0
Shewry, P. R. (2007). Improving the protein content and composition of cereal grain. *Journal of Cereal Science*, 46(3), 239–250. https://doi.org/10.1016/j.jcs.2007.06.006
Shewry, P. R., & Halford, N. G. (2002). Cereal seed storage proteins: Structures, properties and role in grain utilization. *Journal of Experimental Botany*, 53(370), 947–958. https://doi.org/10.1093/jexbot/53.370.947
Sofi, S. A., Ahmed, N., Farooq, A., Rafiq, S., Zargar, S. M., Kamran, F., Dar, T. A., Mir, S. A., Dar, B. N., & Mousavi Khaneghah, A. (2023). Nutritional and bioactive characteristics of buckwheat, and its potential for developing gluten-free products: An updated overview. *Food Science & Nutrition*, 11(5), 2256–2276. https://doi.org/10.1002/fsn3.3166
Sterna, V., Zute, S., & Brunava, L. (2016). Oat grain composition and its nutrition benefice. *Agriculture and Agricultural Science Procedia*, 8, 252–256. https://doi.org/10.1016/j.aaspro.2016.02.100
Stone, A. K., Nosworthy, M. G., Chiremba, C., House, J. D., & Nickerson, M. T. (2019). A comparative study of the functionality and protein quality of a variety of legume and cereal flours. *Cereal Chemistry*, 96(6), 1159–1169. https://doi.org/10.1002/cche.10226
Sudheesh, C., Bhat, Z. R., Aaliya, B., & Sunooj, K. V. (2022). Cereal proteins. In *Nutraceuticals and Health Care* (pp. 29–60). Elsevier. https://doi.org/10.1016/B978-0-323-89779-2.00010-7
Tapia-Hernández, J. A., Del-Toro-Sánchez, C. L., Cinco-Moroyoqui, F. J., Juárez-Onofre, J. E., Ruiz-Cruz, S., Carvajal-Millan, E., López-Ahumada, G. A., Castro-Enriquez, D. D., Barreras-Urbina, C. G., & Rodríguez-Felix, F. (2019). Prolamins from cereal by-products: Classification, extraction, characterization and its applications in micro- and nanofabrication. *Trends in Food Science & Technology*, 90, 111–132. https://doi.org/10.1016/j.tifs.2019.06.005
Tasie, M. M., & Gebreyes, B. G. (2020). Characterization of nutritional, antinutritional, and mineral contents of thirty-five sorghum varieties grown in Ethiopia. *International Journal of Food Science*, 2020, 8243617. https://doi.org/10.1155/2020/8243617
Tatham, A. S., & Shewry, P. R. (2012). The S-poor prolamins of wheat, barley and rye: Revisited. *Journal of Cereal Science*, 55(2), 79–99. https://doi.org/10.1016/j.jcs.2011.10.013
Tien, N. N. T., Trinh, L. N. D., Inoue, N., Morita, N., & Hung, P. V. (2018). Nutritional composition, bioactive compounds, and diabetic enzyme inhibition capacity of three varieties of buckwheat in Japan. *Cereal Chemistry*, 95(5), 615–624. https://doi.org/10.1002/cche.10069
Tomičić, Z., Pezo, L., Spasevski, N., Lazarević, J., Čabarkapa, I., & Tomičić, R. (2022). Diversity of amino acids composition in cereals. *Food and Feed Research*, 49(2), 11–22. https://doi.org/10.5937/ffr49-34322
Verma, D. K., & Srivastav, P. P. (2017). Proximate composition, mineral content and fatty acids analyses of aromatic and non-aromatic Indian rice. *Rice Science*, 24(1), 21–31. https://doi.org/10.1016/j.rsci.2016.05.005
Verma, D. K., & Srivastav, P. P. (2020). Exploring the physicochemical and cooking properties of some Indian aromatic and non-aromatic rice (*Oryza sativa* L.) cultivars. *Oryza-An International Journal on Rice*, 57(2), 146–161. https://doi.org/10.35709/ory.2020.57.2.9

Wang, H., Xiang, L., Rao, P., Ke, L., Wu, B., Chen, S., Wang, S., Shi, Y., & Su, P. (2022). Effects of pretreatments on structural and functional changes of oat protein isolate. *Cereal Chemistry*, 99(1), 90–99. https://doi.org/10.1002/cche.10480

Wiedemair, V., Scholl-Bürgi, S., Karall, D., & Huck, C. W. (2020). Amino acid profiles and compositions of different cultivars of *Panicum miliaceum* L. *Chromatographia*, 83(7), 829–837. https://doi.org/10.1007/s10337-020-03899-8

Yang, J., Wen, C., Duan, Y., Deng, Q., Peng, D., Zhang, H., & Ma, H. (2021). The composition, extraction, analysis, bioactivities, bioavailability and applications in food system of flaxseed (*Linum usitatissimum* L.) oil: A review. *Trends in Food Science & Technology*, 118, 252–260. https://doi.org/10.1016/j.tifs.2021.09.025

Zafar, S., Naz, N., Nazir, S., Abbas, M., & Khan, A. M. (2014). Analysis of selected amino acids in different varieties of wheat available in Punjab, Pakistan. *Chromatography Research International*, 2014, 867070. https://doi.org/10.1155/2014/867070

Zhang, Q., & Xu, J.-G. (2017). Determining the geographical origin of common buckwheat from China by multivariate analysis based on mineral elements, amino acids and vitamins. *Scientific Reports*, 7(1), 9696. https://doi.org/10.1038/s41598-017-08808-y

Zhao, H., Shen, C., Wu, Z., Zhang, Z., & Xu, C. (2020). Comparison of wheat, soybean, rice, and pea protein properties for effective applications in food products. *Journal of Food Biochemistry*, 44(4). https://doi.org/10.1111/jfbc.13157

Zhou, Z., Robards, K., Helliwell, S., & Blanchard, C. (2002). Composition and functional properties of rice. *International Journal of Food Science and Technology*, 37(8), 849–868. https://doi.org/10.1046/j.1365-2621.2002.00625.x

3 Pulse Proteins

Ranjana Verma, Akshita Awasthi, Madhvi Awasthi, and Kanchan Thakur

INTRODUCTION

Pulses are important alternatives for proteins in regions where animal proteins are either not available or are scarce due to geographical cultural, or religious reasons. These foods are rich in energy, dietary fibres, proteins, vitamins, and minerals that aid human health. New research suggests that if pulses are consumed regularly, they may be useful in the treatment of various diseases, such as a decrease in the likelihood of developing cardiovascular diseases, diabetes, or cancers, as well as osteoporosis and hypertension, disorders of the gastrointestinal tract and adrenal glands, and even reduced levels of LDL cholesterol. All the major pulses of the legume group are chickpea, *Cicer arietinum*; pea, *Pisum sativum*; lentil, *Lens culinaris*; and common bean, all of which are good protein sources for both human and animal alike (Mudryj et al., 2014). Pulses contain a high content of amino acids like leucine, lysine, aspartic acid, glutamic acid, and arginine. Pulses, if mixed with cereals and other foods that include sulphur-containing amino acids and tryptophan, are a perfect amino acid supplement (Boye et al., 2010). In the majority of the pulses, 17%–30% of the protein is of dry weight type.

The functional properties of pulse proteins, in addition to their nutritional properties, render them indispensable during the development and processing of food. Solubility, fat and water binding, emulsification, gelation, and foaming are considered key functional properties. These pulse proteins find applications in dairy alternatives, gluten-free products, and substitutes for plant-based meats. In addition, they also help in enhancing the texture and extending the shelf life of the foods due to the water- and oil-holding capacity of the ingredients (Boye et al., 2010). The presence of these pulse proteins in enriched protein foods—energy bars, shakes, food products for infants, and diet foods—proves to be beneficial due to their nutritional value and easy digestibility. Pulses help the cause of sustainable agriculture in an environmentally friendly way. They enhance soil health by fixing atmospheric nitrogen, which decreases the dependency on synthetic fertilisers, as well as require lesser water and energy compared to animal-based protein sources (FAO, 2019). Pulses are therefore considered a crucial element in the production of sustainable plant-based food products that help achieve international environmental goals in an era when consumer demand for clean-label and environmentally friendly food basically keeps growing. Pulse proteins are said to resemble the functional properties of very well-known proteins such as whey or soy. Pulse proteins have been utilised in the manufacture of wide-ranging food products: baked goods, soups, extruded snacks, and ready-to-eat meals. Processing, like air classification, isoelectric precipitation, and ultrafiltration,

DOI: 10.1201/9781003613909-3

has had significant improvements that have enabled the production of high-quality pulse protein isolates and concentrates and thereby enhance their use in food and animal feed. Increasing awareness of pulses' health benefits has stimulated innovations around their application, particularly in addressing global challenges like protein deficiency and food security. Combining with cereal proteins, pulses provide a viable solution to form functional and nutritionally balanced protein sources, thus helping meet the ever-growing demand for sustainable and affordable nutrition worldwide

The functional properties of pulse proteins, which include emulsification, gelling, and water-binding capacity are mainly associated with their globulins and albumins; hence, they are useful in food processing applications (Jarpa-Parra, 2018). Bioactive peptides derived from pulse proteins have shown implications in preventative nutrition by their antioxidant, antihypertensive, and immunomodulatory potential health benefits (Dahl et al., 2012). Pulses are not only rich in macronutrients but also provide essential micronutrients like iron, zinc, and folate. Consuming pulses helps to counteract micronutrient deficiencies that are common throughout most developing areas and, as a result, lead to health problems such as anaemia and a weakened immune system (Mudryj et al., 2014). There exist, however, certain limitations of pulse proteins, as they might not be very digestible and sometimes contain anti-nutritional factors. Sophisticated processing methods have been created in order to enhance the functionality and bioavailability of pulse proteins to try to cross the barriers (Boye et al., 2010).

Increased demand for protein pulses has enhanced because of their ability to provide functionality and nutritional value and help solve food security problems. Protein sources and substitutes have become more important with a growing population of the world. Science involving the biochemical character, functional behaviour, and food quality of pulse proteins is extremely important to provide them with feasibility for application within various food processes and to increase sustainability as well as health, as pulse proteins play a key role in fighting malnutrition because they provide quality protein and important minerals. They are also good for the environment because they can be grown using eco-friendly and sustainable farming methods. Therefore, in the realm of functional properties, they are of very high value for improving food composition. Encouraging the intake of pulses would have tremendous potential beneficial impacts on the health of the world in malnourished areas.

COMPOSITION OF PULSES

Pulses are a heterogeneous group of leguminous crops that form part of the global diet (FAO, 2021). Lentils, chickpeas, peas, black gram, green gram, pigeon peas, kidney beans, and other beans are some of them. Being a low-cost source of nutrition, pulses have a varied range of nutrients, hence a staple food in most parts of the world (Kumar et al., 2017). Their dietetic composition consists mostly of carbohydrates, proteins, dietary fibre, vital vitamins, minerals, and traces of fat (Table 3.1). They have high protein, which is one of the unique characteristics of pulses and is an excellent vegetable protein for vegans and vegetarians. Their protein content varies depending on pulses. For example, lentils contain about 24%–26% protein and are rich in lysine but deficient in methionine (USDA, 2020). Chickpeas contain 19%–22% protein,

TABLE 3.1
Proximate Composition of Pulses per 100 g

Legume Crop	Protein (g)	Carbohydrates (g)	Total Lipid/Fat (g)	Dietary Fibre (g)
Cowpeas	23.85	59.64	2.07	10.70
Mung beans	23.86	62.62	1.15	16.30
Black beans	21.6	62.36	1.42	15.5
Faba broad beans	26.12	58.29	1.53	25.00
Lupin	36.17	40.37	9.74	18.90
Adzuki bean	19.90	62.90	0.53	12.70
Cranberry bean	23.00	60.00	1.23	24.70
Great northern bean	21.90	62.40	1.14	21.90
Lima bean, large	21.50	63.40	0.69	19.00
Lima bean, baby	20.60	62.80	0.93	20.60
Moth bean	22.94	61.52	1.61	–
Navy bean	22.33	60.75	1.50	15.30
Pink bean	21.00	64.20	1.13	12.70
Pinto bean	21.42	62.55	1.23	15.50
White bean	23.40	60.30	0.85	15.20

Source: USDA (2022).

which is high in lysine but low in sulphur amino acids (Mudryj et al., 2014). Kidney beans contain 21%–23% protein, which contains arginine and leucine. Black gram offers 25%–27% protein, making it one of the most protein-dense pulses (Jukanti et al., 2012). Green gram, with 23%–24% protein, is highly digestible and commonly consumed in sprouted form. Pigeon peas provide 20%–22% protein and are widely consumed in tropical regions. While pulses are high in protein, they are incomplete proteins, lacking certain amino acids, but combining them with cereals such as rice or wheat ensures a complete protein profile (Shewry, 2007).

Carbohydrates form the primary macronutrient in pulses, delivering sustained energy. Their carbohydrate content varies. Lentils have 50%–60% carbohydrates, primarily as complex starches (USDA, 2020). Chickpeas contain 55%–60% carbohydrates and feature a low glycaemic index, ideal for diabetic diets (Jenkins et al., 1981). Green gram provides 50%–55% carbohydrates and is high in resistant starch, promoting gut health (Hoover & Zhou, 2003). Pigeon peas contain 60% carbohydrates, while kidney beans have 55%–60%, including substantial dietary fibre (Mudryj et al., 2014). Black gram also offers 55%–60% carbohydrates with notable fibre content. The low glycaemic index of many pulses supports blood sugar regulation (Jenkins et al., 1981).

Dietary fibre in pulses supports digestion and gut health. Lentils offer 7%–9% fibre, while chickpeas contain 8%–9% dietary fibre (Mudryj et al., 2014). Kidney beans have 6%–8% fibre, and black gram contributes 8%–9%, aiding cholesterol management (Anderson et al., 2009). Green gram gives 7%–8% fibre, and pigeon peas give 6%–7%. Fibres that are soluble and insoluble in pulses are needed for the cardiovascular system and for avoiding constipation (Slavin, 2005). Pulses are low in fat, and therefore, they are appropriate for low-fat diets. Pulses contain 1–5% fat that is mostly unsaturated and heart-friendly. Lentils contain 1%, chickpeas have 3%–6% (mostly unsaturated), kidney beans and black gram have 1.5%–2%, and green gram has 1%–2%, while pigeon peas have 2%–3% (USDA, 2020).

Pulses are very good sources of B-vitamins such as thiamine, riboflavin, niacin, folate, and vitamin B6 for energy metabolism, brain function, and red blood cell formation. Folate content is high in lentils, chickpea, and black gram with 200–350 μg for every 100 g (USDA, 2020) required for the synthesis of DNA (Bailey, 2009). Vitamin B6 is predominately present in kidney beans, chickpea, and green gram. Thiamine is supplied through the metabolism of carbohydrates, and niacin is abundantly found in chickpeas, lentils, and pigeon peas (Jukanti et al., 2012). Iron, calcium, potassium, magnesium, phosphorus, and zinc are found in abundance in pulses. Lentils have 7–8 mg/100 g of iron, and black gram has 8–10 mg/100 g. Non-haeme iron in pulses is better absorbed with foods that have high vitamin C content (Hallberg & Hulthén, 2000). Black gram contains 120–150 mg/100 g calcium, and chickpeas contain 50–60 mg/100 g (USDA, 2020). Lentils, chickpeas, and kidney beans contain potassium, an essential element that is used for blood pressure control (Whelton et al., 1997). Magnesium in green gram and chickpeas contributes to bone health and enzyme activity. Zinc in black gram, chickpeas, and lentils supports immunity (Prasad, 2013). Pulses also include bioactive compounds like polyphenols, flavonoids, and phytates with health-promoting effects. Polyphenols in lentils, kidney beans, and black gram act as antioxidants (Xu & Chang, 2008). Flavonoids in chickpeas and green gram reduce oxidative stress (Anderson et al., 2009). Though phytates may limit mineral bioavailability, they also provide antioxidant benefits (Reddy, 2002). Dried pulses have low water content (10%–12%), suitable for storage, but this increases after soaking or cooking. As nutrient-dense foods, pulses combine macronutrients, micronutrients, and bioactive compounds, making them a cornerstone of a healthy diet. While their composition varies slightly, all pulses are beneficial for combating protein-energy malnutrition, ensuring food security, and fostering sustainable agriculture. Regularly including pulses in diets promotes overall health and wellness (FAO, 2021).

Pulses, including pea, chickpea, lentil, and bean, are notable for their high protein content, which generally ranges between 17% and 30% on a dry weight basis (Sathe et al., 1984). The primary proteins in pulses are globulins and albumins. Albumins, which are water-soluble, consist of various enzymatic proteins, protease inhibitors, amylase inhibitors, and lectins, with molecular masses spanning from 5,000 to 80,000 Da. Conversely, globulins, which are salt-soluble, primarily include legumin (11S) and vicilin (7S), which are storage proteins encoded by multigene families (Schwenke, 2001). The legumins typically have hexameric quaternary structures, with acidic subunits (~40,000 Da) and basic subunits (~20,000 Da). Vicilins are trimeric in character with molecular masses of 175,000–180,000 Da. Convicilin is another unique storage protein of the pea seed along with legumin and vicilin. Convicilin is characterised by unusual amino acid composition, trace amounts of carbohydrate material, and subunits with molecular weights ranging from 71,000 Da.

The protein is characterised by the presence of sulphur amino acids, in contrast to a lack of these amino acids in vicilin (Croy et al., 1980). The small pulse proteins are prolamins and glutelins as well. Prolamins are alcohol-soluble and consist of high proportions of proline and glutamine, whereas glutelins are diluted acid- or base-soluble with higher proportions of methionine and cysteine and thus are more nutritious (Osborne, 1924). Scientists have suggested that developing pulse varieties with a higher proportion of glutelin proteins could improve overall protein quality of pulses (Singh & Jambunathan, 1982).

The amino acid profile of pulse proteins is also well balanced with excessive levels of lysine, leucine, aspartic acid, glutamic acid, and arginine. They lack proper amounts of sulphur-containing amino acids such as methionine and cysteine, and tryptophan (Swanson, 1990). Nutritional composition is genetically variable depending on environment and method of analysis used (Sathe et al., 1984; Maheri-Sis et al., 2008).

Protein Fraction in Pulses

Pulse crop shows immense variation in the structure of protein fraction of pulses. Albumins contain more sulphur amino acids, while vicilins possess a larger fraction of aromatic and branched-chain amino acids. Albumin/globulin and legumin/vicilin ratios vary across pulse crops and influence their functional and nutritional attributes (Gupta & Dhillon, 1993). For example, the albumin:globulin ratio is 1:3 in lentil and 1:6.3 in black gram, and legumin:vicilin ratio is 10.5:1 in lentil and 1:6–9 in French bean.

Pea (*Pisum sativum*): Peas contain 23.1%–30.9% protein, in which albumins and globulins account for 15%–25% and 50%–60% of protein, respectively. Legumin and vicilin are major proteins in which convicilin accounts for 3%–7% of protein. Pea albumins have high contents of essential amino acids such as lysine and threonine, while globulins have high amounts of arginine, phenylalanine, leucine, and isoleucine. The protein digestibility of peas ranges between 60% and 74%, and cooking or autoclaving improves digestibility significantly (Gueguen & Barbot, 1988).

Chickpea (*Cicer arietinum*): Chickpeas are divided into kabuli and desi types, with protein content ranging from 20.9% to 25.27%. Globulins constitute the largest protein fraction (53.44%–60.29%), followed by albumins (8.39%–12.31%), glutelins (19.38%–24.40%), and prolamins (3.12%–6.89%). Kabuli varieties are larger and lighter in colour, while desi varieties are smaller, darker, and richer in protein. Chickpeas are also high in glutamic acid, aspartic acid, and arginine but lack sulphur-containing amino acids like methionine (Dhawan et al., 1991).

Lentil (*Lens culinaris*): Lentils exhibit a protein content of 25.9%–31.1%, with albumins (16.8%), legumins (44.8%), vicilins (4.2%), glutelins (11.2%), and prolamins (3.5%) making up the major fractions. Lentils are rich in lysine and leucine but are deficient in tryptophan and sulphur-containing amino acids. Their protein digestibility improves with cooking, ranging from 73.5% to 99.88% depending on processing techniques (Gupta & Dhillon, 1993).

Bean (*Phaseolus* spp.): Beans, including kidney, lima, and mung beans, vary in protein content from 16.89% to 26.4%. Globulins account for up to 70% of the total protein, followed by albumins (10%–30%). Beans are rich in lysine and leucine but are limited in methionine. Digestibility ranges between 38% and 79%, depending on the variety and processing methods.

PROTEIN CONTENT AND AMINO ACID PROFILE OF DIFFERENT PULSES

The comprehensive overview of the total protein content in various pulses cultivated and consumed in South Asia is presented in Table 3.2. According to the Indian Council of Medical Research (ICMR) in 2024, white soybean emerges as the undisputed protein food among these pulses, boasting an impressive 37.8 g of

TABLE 3.2
Protein Content of Different Pulses in Descending Order

S. No.	Name of Pulse	Protein Content (g/100 g)
1	Soy bean, white (*Glycine max*)	37.8
2	Soy bean, brown (*Glycine max*)	35.58
3	Lentil dal (*Lens culinaris*)	24.35
4	Green gram, dal (*Phaseolus aureus*)	23.88
5	Black gram, dal (*Phaseolus mungo*)	23.06
6	Lentil whole, yellowish (*Lens culinaris*)	22.87
7	Green gram, whole (*Phaseolus aureus*)	22.53
8	Lentil whole, brown (*Lens culinaris*)	22.49
9	Black gram, whole (*Phaseolus mungo*)	21.97
10	Horse gram, whole (*Dolicus biflorus*)	21.73
11	Red gram, dal (*Cajanus cajan*)	21.7
12	Bengal gram, dal (*Cicer arietinum*)	21.55
13	Cowpea, white (*Vigna catjang*)	21.25
14	Red gram, whole (*Cajanus cajan*)	20.47
15	Peas, dry (*Pisum sativum*)	20.43
16	Cowpea, brown (*Vigna catjang*)	20.36
17	Rice bean (*Vigna umbellata*)	19.97
18	Field bean, black (*Phaseolus vulgaris*)	19.93
19	Kidney bean, red (*Phaseolus vulgaris*)	19.91
20	Field bean, brown (*Phaseolus spp.*)	19.9
21	Field bean, white (*Phaseolus spp.*)	19.84
22	Moth bean (*Vigna aconitifolia*)	19.75
23	Kidney bean, brown (*Phaseolus vulgaris*)	19.5
24	Kidney bean, black (*Phaseolus vulgaris*)	19.01
25	Bengal gram, whole (*Cicer arietinum*)	18.77

Source: Longvah et al. (2017).

protein/100 g of edible portion. In contrast, Bengal gram, while still a valuable protein source, contains significantly less protein, approximately 18 g/100 g. This comparative analysis underscores the nutritional diversity within the South Asian pulses, highlighting the potential of white soybean as a potent dietary protein source (Longvah et al., 2017).

Chemically, pulse proteins are primarily composed of globulins, which dissolve in salt solutions, and albumins, which dissolve in water. Prolamins (alcohol-soluble) and glutelins (dilute acid/base-soluble) are minor proteins that contribute to less than 5% of the total protein content. The albumin-to-globulin ratio can vary from 1:3 to 1:6.3 in different types of pulses (Gupta and Dhillon, 1993). Albumins, constituting 10%–20% of seed proteins, primarily consist of metabolic proteins, including enzymes and non-enzymatic proteins. They have a low molecular weight (5–80 kDa) but have high levels of cysteine and methionine as compared to globulins (Boye et al., 2010). While albumins are nutritionally superior in terms of amino acid composition, they may contain anti-nutritional factors like haemagglutinins/lectins, amylase inhibitors, and trypsin and chymotrypsin inhibitors (Boulter and Croy, 1997; Boye et al., 2010).

The major protein in legumes/pulses is globulins, which are around 70%–80% of seed proteins. Their major function is as storage proteins. Globulin further can be of two types: Legumins and vicilins. As per the sedimentation coefficients, they are also called 7S globulins and 11–12S globulins, respectively. The ratio of legumins to vicilins varies across different pulses: 10.5:1 in lentils, 1:6–9 in French beans, 1:9 in cowpeas, 1–3:1 in peas, and 4–6:1 in chickpeas. Here, legumins are important because of their high levels of methionine and cysteine (sulphur-containing amino acids) as compared to vicilins (Gupta and Dhillon, 1993). When compared structurally, the vicilins and legumins are complex proteins found in various legumes. Legumin, a hexamer with a molecular weight of 300–400 kDa, consists of larger acidic subunits (40 kDa) and smaller basic subunits (20 kDa) linked by disulphide bonds. The basic subunits form the hydrophobic core and acidic subunits are primarily surface-exposed (Gueguen and Cerleti, 1994).

The other type of legume globulin, vicilin is a trimer of 50–70 kDa monomers that are held together by non-covalent interactions (Boulter and Croy 1997). It has a molecular weight of 145–190 kDa and lacks cysteine residues, thus preventing disulphide bond formation. Studies have found that it is the major storage protein in cowpea, red bean, mung bean, kidney bean, and urad bean, constituting up to 88% of total globulins (Meng and Ma, 2001; Tang et al., 2009; Tang and Sun, 2010; Shevkani et al., 2015). Legumins are predominant in chickpeas, while both vicilins and legumins are present in fava/faba bean and field pea (Kimura et al., 2008; Shevkani et al., 2015).

On the other hand, vicilins are present in different pulses in different molecular weights along with variations in composition. The molecular weight of vicilins in cowpea, kidney beans, mung bean, red bean, peas, and fava/faba beans can vary from 133–140 kDa to 163 kDa (Tang et al., 2009; Shevkani et al., 2015; Vioque et al., 2012; Parmar et al., 2017). It is important to note that vicilins from lentils, cowpeas, and French beans are glycosylated, while in faba bean, they lack carbohydrates. Vicilins from lentils have been found to contain approximately 2.8% carbohydrates (Gupta and Dhillon, 1993; Kimura et al., 2008).

Besides vicilins and legumins, peas also contain small amounts of convicilin, a 7S globulin with a molecular weight of around 70 kDa. Convicilin can exist as a trimer of three convicilin molecules or as heteromeric trimers with vicilins (Lam et al. 2018). Unlike vicilins, convicilin contains sulphur-containing amino acids (McLean et al., 1974; Boye et al., 2010). The pulse protein composition is influenced by both intrinsic and extrinsic factors. Intrinsic factors, such as the specific pulse variety, play a significant role in determining the protein profile. Studies have reported that the ratio of legumin to vicilin in pea and fava beans is high during seed development, while legumins accumulate predominantly in the later growth stages (Chandler et al., 1984).

If extrinsic factors are taken into consideration, then it is evident from studies that agronomic practices and environmental conditions affect the composition of pulse proteins. In a study by Mertens et al. (2012), it was observed that sulphur was crucial for the synthesis of legumins in pulses. Under sulphur-deficient conditions, the synthesis of legumins was absent, while the proportion of vicilins remained relatively stable throughout the stages of seed development.

Table 3.3 summarises the information on the protein quality and protein digestibility of several foods in particular legumes. The Atomic Absorption Spectroscopy (AAS) was used to indicate the level of sufficiency of the essential

TABLE 3.3
Comparison of PDCAAS Values of Different Pulses with Selected Food Grains

Name of Food Item	Amino Acid Score[a]	True Protein Digestibility (%)[b]	PDCAAS[c]
Pea yellow spilt	0.73	87.9	0.64
Peas green Split	0.59	85.2	0.50
Lentil green whole	0.71	87.9	0.63
Lentil red split	0.59	90.6	0.54
Chickpeas	0.61	85.0	0.52
Pinto beans	0.77	76.2	0.59
Kidney beans	0.70	78.6	0.5
Black beans	0.76	70.0	0.53
Navy beans	0.83	80.0	0.67
Soy flour	0.92	92.3	0.77
Wheat flour	0.47	92.0	0.43
Rice flour	0.54	91.0	0.50
Casein	1.04	96.6	1.00

Source: Pulse Canada (2011).

[a] Amino acid score is limiting the amino acid with the lowest ratio relative to the established amino acid requirement values for humans, aged 2–5 years.

[b] AOAC Method 991.29 ($n = 10$).

[c] PDCAAS = Amino Acid Score × % True Protein Digestibility.

amino acids in a protein source, while the Protein Digestibility Corrected Amino Acid Score (PDCAAS) included both levels of digestibility and the composition of the amino acid content. Casein, one of the proteins in milk, has a PDCAAS value of 1.00 which is regarded as the gold standard. In legumes, however, soy flour has the highest PDCAAS value of 0.77, which means its digestibility and proteomic absorption is great. This is more than that of wheat flour (0.43) and rice flour (0.50), which makes soy flour a good source of protein. Other legumes such as lentils, chickpeas, and beans also have good values of PDCAAS that are mostly higher than that of wheat and rice (Pulse Canada, 2011a). But it should be borne in mind that even though most pulses are a reasonably good source of protein, they are not as well rounded as animal proteins, casein, for instance. For a well-rounded diet, it is, however, possible to mix plant sources of protein pulses with other non-legumes to obtain all essential amino acids.

While pulses are good sources of protein, they lack the potential to be called complete proteins. Pulses provide a well-balanced profile of amino acids, particularly those essential for human nutrition, making them an indispensable part of plant-based diets. One of the notable strengths of pulses is their high lysine content, an essential amino acid that is often limited in cereals. Lysine plays a crucial role in growth, tissue repair, enzyme production, and immune system function. This property makes pulses a complementary protein source when consumed with lysine-deficient cereals such as rice and wheat. Pulses are also rich in arginine, glutamic acid, and aspartic acid, amino acids that contribute to vital physiological processes, including protein synthesis, nitrogen metabolism, and immune response.

It is well recognised that they tend to have low levels of some important amino acids, particularly methionine, cysteine and tryptophan (Tiwari and Singh, 2012). Methionine is essential for protein synthesis and other metabolic functions, while cysteine supports antioxidant defence mechanisms within the body. Despite this, beans are said to be rich in lysine, which is one of the amino acids that cereals usually lack. If the beans are eaten together with the cereals, then a higher and better protein profile is achieved. This ensures that all required amino acids are consumed. Such a complementary protein system is a common feature of many traditional diets where beans and grains are consumed. Table 3.4 presents the essential amino acid content of pulses of the South Asian region. The numbers in parentheses denote the rank of the pulses in terms of their richness in essential amino acids. Understanding the different profiles of amino acids amongst different beans is also vital. In broad terms, beans are deficient in methionine, cysteine, and tryptophan (Longvah et al., 2017). They, however, are rich in lysine, leucine, aspartic acid, glutamic acid, and arginine. This variation or lack of amino acids accentuates the dire need for a diverse diet, as using different types of beans with contrasting amino acids would allow an easy balance of required amino acids. For instance, blending beans with grains that are high in methionine and cysteine to obtain a complete protein source.

Essentially, as soon as these concepts are assimilated, it becomes possible for individuals to make judicious decisions that aim at maximising the quantity of protein consumed from plant-based sources (Shobana et al., 2013). The protein composition of pulses is a complex trait influenced by both genetic and environmental factors. Understanding these factors is essential for optimising the nutritional quality

TABLE 3.4
Essential Amino Acids Contents (g/100 g) of Pulses

	Histidine	Isoleucine	Leucine	Lysine	Methionine	Phenylalanine	Threonine	Tryptophan	Valine
Bengal gram, dal (*Cicer arietinum*)	2.39 ± 0.34	4.25 ± 0.26	6.91 ± 0.65	6.06 ± 0.52	1.12 ± 0.31	5.97 ± 1.06	3.24 ± 0.34	1.09 ± 0.25	4.09 ± 0.41
Bengal gram, whole (*Cicer arietinum*)	2.51 ± 0.18	4.34 ± 0.23	7.40 ± 0.31	6.59 ± 0.25	1.16 ± 0.16	6.26 ± 0.70	3.55 ± 0.31	0.95 ± 0.07	4.58 ± 0.51
Black gram, dal (*Phaseolus mungo*)	2.64 ± 0.11	3.75 ± 0.92	7.93 ± 0.45	6.22 ± 0.21	1.31 ± 0.35	5.68 ± 0.91	2.99 ± 0.19	1.07 ± 0.12	4.61 ± 1.06
Black gram, whole (*Phaseolus mungo*)	2.85 ± 0.08	3.42 ± 0.82	7.58 ± 0.62	6.35 ± 0.53	1.17 ± 0.38	5.90 ± 0.38	2.87 ± 0.63	0.98 ± 0.14	4.11 ± 0.88
Cowpea, brown (*Vigna catjang*)	2.93 ± 0.42	4.10 ± 0.58	7.49 ± 1.23	6.67 ± 1.26	1.38 ± 0.16	5.47 ± 0.49	3.80 ± 0.61	1.05 ± 0.03	4.87 ± 0.59
Cowpea, white (*Vigna catjang*)	3.25	4.40	7.96	7.14	1.53	5.63	4.10	0.92	5.31
Field bean, black (*Phaseolus vulgaris*)	3.21	4.57	8.91	6.79	1.36	5.88	4.12	0.73	5.24
Field bean, brown (*Phaseolus spp.*)	3.23	4.59	8.88	6.75	1.38	5.72	3.97	0.78	5.16
Field bean, white (*Phaseolus spp.*)	2.82 ± 0.76	4.41 ± 0.36	8.48 ± 0.66	6.13 ± 0.68	1.40 ± 0.18	5.76 ± 0.32	3.97 ± 0.31	0.89 ± 0.07	4.96 ± 0.36
Green gram, dal (*Phaseolus aureus*)	2.55 ± 0.26	4.07 ± 1.11	7.90 ± 0.99	6.09 ± 1.06	1.05 ± 0.23	6.20 ± 0.61	3.36 ± 0.60	1.24 ± 0.42	5.21 ± 1.28
Green gram, whole (*Phaseolus aureus*)	2.85 ± 0.26	4.17 ± 0.37	7.45 ± 0.50	6.60 ± 0.33	1.21 ± 0.20	6.21 ± 0.39	3.20 ± 0.18	1.02 ± 0.14	5.45 ± 0.37
Horse gram, whole (*Dolicus biflorus*)	2.70 ± 0.28	3.72 ± 0.83	6.79 ± 0.81	6.24 ± 0.57	0.73 ± 0.30	8.08 ± 0.60	3.32 ± 0.10	1.08 ± 0.25	4.16 ± 0.96
Lentil dal (*Lens culinaris*)	1.93 ± 0.32	3.74 ± 0.91	7.10 ± 0.86	6.12 ± 0.71	0.54 ± 0.19	5.10 ± 0.73	3.31 ± 0.24	0.81 ± 0.07	5.02 ± 0.39
Lentil whole, brown (*Lens culinaris*)	2.07 ± 0.14	4.17 ± 0.05	7.36 ± 0.34	6.78 ± 0.51	0.84 ± 0.03	4.61 ± 0.68	3.35 ± 0.05	0.76 ± 0.04	4.85 ± 0.06
Lentil whole, yellowish (*Lens culinaris*)	2.26	4.12	7.47	6.43	0.84	4.67	3.49	0.75	4.75
Moth bean (*Vigna aconitifolia*)	2.80 ± 0.29	4.17 ± 0.57	7.85 ± 0.73	6.63 ± 0.60	0.97 ± 0.20	5.85 ± 0.70	3.63 ± 0.11	0.92 ± 0.04	4.90 ± 0.47

(Continued)

TABLE 3.4 (*Continued*)
Essential Amino Acids Contents (g/100 g) of Pulses

	Histidine	Isoleucine	Leucine	Lysine	Methionine	Phenylalanine	Threonine	Tryptophan	Valine
Peas, dry (*Pisum sativum*)	2.34 ± 0.09	3.87 ± 0.46	7.02 ± 0.38	7.12 ± 0.51	0.68 ± 0.19	4.76 ± 0.23	3.65 ± 0.15	0.86 ± 0.19	4.67 ± 0.66
Kidney bean, black (*Phaseolus vulgaris*)	2.44	3.91	6.84	6.01	0.79	5.32	3.91	1.22	4.83
Kidney bean, brown (*Phaseolus vulgaris*)	2.71 ± 0.22	4.27 ± 0.50	7.93 ± 0.43	6.67 ± 0.18	0.97 ± 0.26	5.91 ± 0.56	4.35 ± 0.21	1.04 ± 0.13	5.21 ± 0.69
Kidney bean, red (*Phaseolus vulgaris*)	2.70 ± 0.30	4.23 ± 0.38	7.78 ± 0.71	6.71 ± 0.81	0.88 ± 0.40	5.90 ± 0.56	4.18 ± 0.65	1.05 ± 0.27	5.07 ± 0.71
Red gram, dal (*Cajanus cajan*)	3.16 ± 0.27	3.42 ± 0.43	6.73 ± 0.56	6.16 ± 0.28	0.87 ± 0.22	8.76 ± 1.02	3.26 ± 0.17	0.71 ± 0.11	4.38 ± 0.30
Red gram, whole (*Cajanus cajan*)	3.28 ± 0.13	3.38 ± 0.49	6.88 ± 0.48	7.07 ± 0.37	0.92 ± 0.32	8.71 ± 0.82	3.38 ± 0.21	0.75 ± 0.17	4.17 ± 0.65
Rice bean (*Vigna umbellata*)	3.23	3.80	7.86	6.75	1.12	7.56	3.52	0.82	5.23
Soya bean, brown (*Glycine max*)	2.40 ± 0.28	4.57 ± 0.33	8.27 ± 0.17	7.67 ± 0.11	1.46 ± 0.11	4.73 ± 0.77	3.47 ± 0.17	1.59 ± 0.26	5.04 ± 0.59

Source: Longvah et al. (2017).

TABLE 3.5
Comparison of Amino Acid Content across Pulses

Pulses	Lysine Content (%)	Methionine Content (%)	Essential Amino Acid (%)
Moth bean (*Vigna aconitifolia*)	7–8	0.23	53.8
Pigeon pea (*Cajanus cajan*)	6–7	1.03	51.7
Chickpea (*Cicer arietinum*)	6	0.12	50
Lentil (*Lens culinaris*)	6–8	1.18	49–51

Source: Boye et al. (2010).

of pulse-based foods and developing innovative products that meet the needs of consumers (Calles et al., 2019).

Research highlights notable variations in the amino acid composition among different pulse varieties (Table 3.5). For instance, *Vigna aconitifolia* (moth bean) contains approximately 53.8% essential amino acids, the highest among commonly studied pulses. Moth bean is particularly valued for its resilience to harsh environmental conditions and its high lysine content, making it a promising protein source in arid regions. Similarly, *Cajanus cajan* (pigeon pea) provides about 51.7% essential amino acids, positioning it as another nutritionally rich pulse. As a staple crop in many tropical regions, pigeon pea significantly contributes to dietary protein intake. Additionally, pulses like *Lens culinaris* (lentil) and *Cicer arietinum* (chickpea) exhibit high proportions of lysine and glutamic acid. Lentils, in particular, are known for their superior protein digestibility and relatively low anti-nutritional factors compared to other pulses, further enhancing their nutritional value.

Overall, the amino acid composition of pulses underscores their role as an affordable and sustainable protein source, especially in regions where protein malnutrition is prevalent. Combining pulses with cereals not only overcomes their amino acid limitations but also ensures a nutritionally balanced diet.

COMPOSITION OF DIFFERENT PULSE-BASED FOOD PRODUCTS

Pulses, including lentils, chickpeas, and beans, have long been a dietary staple in numerous cultures, particularly in the South Asian region. Renowned for their nutritional value and versatility, pulses offer a wide range of food products (Table 3.6).

Nutritional Composition of Pulse-Based Food Products

In the South Asian region, pulses have been traditionally incorporated into diverse food formulations. Dal, a staple dish made from split chickpeas or pigeon peas, is a prime example. Besan, a flour derived from chickpeas, is used to create a variety of dishes, including pakoras, kadhi, and rotis. Other innovative uses of pulses encompass *phutna* (roasted grains) and *satu* (a blend of roasted chickpea flour and cereal flours) (Kaur and Prasad, 2021).

TABLE 3.6
Composition of Different Pulse-Based Food Products

Food Product	Pulse Used	Country	Nutritional Value	Reference
			Fermented Foods	
Sepu bari	Black gram	India	Proteins, essential amino acids and fats	Sharma et al. (2013)
Kinema	Soybean	India, Nepal, and Bhutan	Antioxidant activity, increased protein digestibility, essential amino acids, B complex vitamins	Tamang (2010)
Natto	Soybean	Japan	Protein, fibre, and vitamins and minerals and probiotic micro organisms	Nagai et al. (2000)
Thua nao	Soybean	Thailand	Protein, fibre, and vitamins and minerals and probiotic microorganisms	Okada (2008)
Chungkokjang	Soybean	Korea	Protein, fibre, and vitamins and minerals and probiotic micro organisms	Lee (2008)
Pepok	Soybean	Myanmar	Protein, fibre, and vitamins and minerals and probiotic micro organisms	Tanaka (2008)
Sieng	Soybean	Cambodia	Protein, fibre, and vitamins and minerals and probiotic micro organisms	Tanaka (2008)
Tempe	Soybean	Indonesia	Protein, fibre, and vitamins and minerals and probiotic micro organisms	Nagai and Tamang (2010)
Thua nao	Soybean	Thailand	Protein, fibre, and vitamins and minerals and probiotic micro organisms	Chunhachart et al. (2006)
Douchi	Soybean	China	Protein, fibre, and vitamins and minerals and probiotic micro organisms	Kang (2001)
Sufu	Soybean	China and Taiwan	Protein, fibre, and vitamins and minerals and probiotic micro organisms	Han et al. (2001)
Miso	Soybean	Japan	Protein, fibre, and vitamins and minerals and probiotic micro organisms	Sugawara (2010)
Doenjang	Soybean	Korea	Protein, fibre, and vitamins and minerals and probiotic micro organisms	Back et al. (2010)
Shoyu	Soybean	Asian countries, except in the Indian continent and in middle east Asia	Protein, fibre, and vitamins and minerals and probiotic micro organisms	Sasaki and Nunomura (2003)
Tauco	Soybean	Indonesia	Protein, fibre, and vitamins and minerals and probiotic micro organisms	Winarno et al. (1986)

(Continued)

TABLE 3.6 (*Continued*)
Composition of Different Pulse-Based Food Products

Food Product	Pulse Used	Country	Nutritional Value	Reference
Maseura	Green gram/black gram	India and Nepal	Protein, essential amino acids, B complex vitamins	Tamang (2010)
Wari	Black gram/green gram	India and Pakistan	Protein, essential amino acids, B complex vitamins, postbiotics	Sharma et al. (2013)
Dhokla and khaman	Bengal gram	India	Protein, essential amino acids, B complex vitamins, postbiotics	Joshi et al. (1989)
Dosa	Black gram	India and Sri Lanka	Protein, essential amino acids, B complex vitamins, postbiotics	Nout et al. (2003)
Idli	Black gram	India and Sri Lanka	Protein, essential amino acids, B complex vitamins, postbiotics	Nout et al. (2003)
			Non-Fermented Foods	
Papad	Bengal gram/black gram/lentil red gram/ green gram	India	Protein, carbohydrate, vitamin B, contains calcium, magnesium, potassium, iron, niacin, thiamine, riboflavin, dietary fibre	Roy et al. (2007)
Teliye mah	Black gram	India	Proteins, amino acids, fats	Thakur and Bhalla (2004)
Plant-based meat alternatives	Any pulse	–	Lower in Saturated Fat and Cholesterol, Higher in Fibre	Andreani et al. (2023)
Extruded pulse products	Any pulse	–	Calcium, Riboflavons, Pantothenic acids Total Carbohydrates, Potassium, Zinc Thiamin, Vitamin B-6 Copper, Sodium, Folate, Iron Protein &Dietary fibre	Roy et al. (2007)
Nutrition bars	Dehydrated flour of lentils and yellow pea flour	–	High fibre, oligosaccharides, isoflavones, selenium, zinc and resistant starch, non-gluten	Rehab et al. (2024)

Soybeans, in particular, have been extensively utilised in the development of value-added products. Tempeh, tofu, and soy nuggets are popular meat alternatives produced from soybeans. These products offer a plant-based protein source and are frequently used in vegetarian and vegan diets (Jambunathan & Singh, 1990).

Pulses in Contemporary Food Innovation

In today's food industry, pulses are gaining increasing recognition as a versatile and sustainable ingredient. Their ease of processing into flours, concentrates, and isolates makes them suitable for a wide range of applications (Lascialfari et al., 2019). These include:

- **Enriched Foods:** Pulse-based ingredients can be used to fortify snacks, breads, cakes, cereals, and pasta with protein and fibre.
- **Gluten-Free and Allergy-Friendly Products:** Pulses are naturally gluten-free and can be used to create products suitable for individuals with celiac disease or gluten sensitivity.
- **Plant-Based Protein Sources:** Pulse-based protein ingredients can be used to develop plant-based meat alternatives, dairy alternatives, and other protein-rich products.
- **Functional Foods:** Pulses can be used to create functional foods with specific health benefits, such as reduced cholesterol levels, improved digestion, and enhanced immune function (Escobedo & Mojica, 2021).

Pulse-Based Product Development: Challenges and Solutions

While pulses offer numerous benefits, there are challenges associated with their use in modern food products. One significant challenge is the extended cooking time required for many pulse-based dishes. To address this, innovative processing techniques, such as extrusion and fermentation, can be employed to reduce cooking time and enhance the texture and flavour of pulse-based products.

Another challenge is the perception of pulses as a traditional and less trendy ingredient. To overcome this, food manufacturers can develop innovative and appealing products that incorporate pulses in unexpected ways. For instance, pulse-based snacks, such as chips and crackers, can be marketed to a younger demographic (Cotacallapa-Sucapuca et al., 2021).

As consumers become increasingly health-conscious and seek out sustainable and nutritious food options, the demand for pulse-based products is likely to grow. By understanding the unique properties of different pulses and exploring innovative processing techniques, food manufacturers can develop a diverse range of delicious and nutritious pulse-based products that cater to diverse consumer preferences (Guruswamy et al., 2022).

Fermented Pulse-Based Foods: A Nutritional Powerhouse

Fermented foods, derived from various pulses, particularly soybeans, have been integral to human diets for centuries. These foods undergo a process of microbial fermentation, which enhances their nutritional value and imparts unique flavours.

They are abundant in protein, fibre, vitamins, and minerals and have live probiotic microorganisms that promote gut health. Beyond that, they can also enhance bioavailability of certain nutrients, such as iron and zinc. Some fermented foods, such as Kinema, exhibit strong antioxidant activity, which helps combat oxidative stress and reduce the risk of chronic diseases. Thus, they promote a balanced gut microbiome, aiding digestion, nutrient absorption, and immune function (Frias et al., 2016).

The diversity of fermented foods is vast, reflecting regional culinary traditions and local microbial ecosystems. Examples include:

- **Asian Fermented Foods:** Natto, tempeh, miso, and doenjang are popular soybean-based fermented foods in Asia.
- **Indian Fermented Foods:** Idli, dosa, and dhokla are fermented rice and lentil-based foods commonly consumed in India.
- **African Fermented Foods:** Ogi and kenkey are fermented cereal-based foods popular in Africa (Nout, 2003).

Non-Fermented Pulse-Based Foods: A Nutritional Profile

Non-fermented pulse-based foods, such as lentils, chickpeas, and beans, are versatile and nutritious. They are also excellent protein sources, rich in fibre, low in fat and cholesterol, and are gluten-free. Different non-fermented food products are available depending on geographical region. Pulses are also being used for innovative non-fermented food products such as meat alternatives, extruded products, and use of use powder in nutritional bars (Escobedo and Mojica, 2021).

CLASSIFICATION OF PULSE PROTEINS

Pulse proteins are essential dietary components with significant nutritional and functional properties. Based on their solubility in different solvents, they are classified using Osborne's Protein Solubility Classification into albumins, globulins, prolamins, and glutelins (Figure 3.1). Among these, albumins and globulins dominate the protein content of pulses, while prolamins and glutelins are present in smaller proportions. Each category has unique solubility characteristics, functions, and nutritional significance, making them crucial for human nutrition and food industry applications.

Albumins

Albumins are water-soluble proteins, representing approximately 10%–20% of the total protein content in pulses. Albumins include enzymes, protease inhibitors, and lectins, which play critical roles in metabolism, seed germination, and defence mechanisms. These proteins are involved in enzymatic regulation, transport, and protection against pests and pathogens. Their physiological roles make albumins indispensable in the plant system and significant for human nutrition. Albumins are rich in essential amino acids, particularly lysine, which is vital for tissue repair, enzyme synthesis, and growth. The high lysine content of albumins enhances the

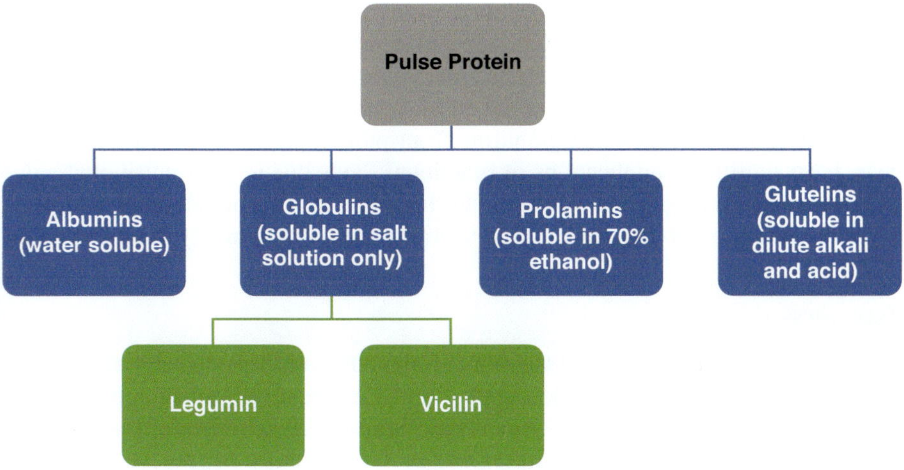

FIGURE 3.1 Classification of legume/pulse proteins. (Gupta and Dhillon, 1993.)

nutritional quality of pulse proteins, especially when combined with cereals deficient in lysine. For example, lentil and chickpea albumins are particularly valued for their functional properties, including solubility and bioavailability.

Research highlights that lentil albumins possess high digestibility, making them an excellent protein source for human consumption. Studies also indicate that albumins extracted from lentils and chickpeas exhibit bioactive properties, such as antioxidant and antimicrobial activities. These functional characteristics make albumins valuable not only as a nutrient source but also for potential applications in health-promoting foods.

Globulins

Globulins are soluble in salt solutions and represent the major protein fraction in pulses, constituting approximately 50%–70% of the total protein content. They serve as the primary storage proteins in pulse seeds, supplying amino acids and nitrogen during germination. They play a significant role in determining the nutritional quality and functional properties of pulses, including emulsification, water-holding capacity, and gelling.

Types of Globulins

- **Legumin (11S Globulin):** Found in peas and lentils, legumin is rich in glutamic acid and lysine. It forms the bulk of the protein storage fraction and contributes to the high digestibility of pulse proteins.
- **Vicilin (7S Globulin):** Present in chickpeas, lentils, and peas, vicilin exhibits emulsifying and foaming properties, making it useful in food processing applications.

Globulins are abundant in lysine, arginine, and glutamic acid, which are essential for protein synthesis and overall metabolic functions. However, they are deficient in sulphur-containing amino acids like methionine and cysteine, limiting their stand-alone nutritional completeness. Chickpea globulins are known for their excellent functional properties, such as emulsification, foaming, and gelling, which make them highly suitable for use in protein-enriched food products.

Research demonstrates that globulins from pulses, particularly peas and chickpeas, have significant applications in functional foods due to their nutritional value and technological properties. These studies highlight the role of globulins in improving emulsification stability, texture, and overall protein functionality in food systems.

Prolamins

Prolamins are soluble in alcohol solutions (70%–80%) and represent a minor protein fraction in pulses, contributing less than 2%–4% of the total protein content. They act as storage proteins, but their functional significance in pulses is limited compared to albumins and globulins. In cereals, prolamins play a larger role, whereas in pulses, they exist in trace amounts. Prolamins are deficient in essential amino acids, particularly lysine and tryptophan, which limits their overall nutritional value. Their low concentration in pulses further reduces their contribution to the amino acid profile. While prolamins are the dominant proteins in cereals (e.g., gliadin in wheat), they appear only in trace amounts in pulses. Although nutritionally inferior, studies emphasise that prolamins' solubility properties can be harnessed in specialised food formulations requiring alcohol-soluble proteins. Their minor role in pulses suggests limited applications beyond specific food processing needs.

Glutelins

Glutelins are soluble in dilute acid or alkali solutions and constitute approximately 3%–5% of the total protein content in pulses. They provide structural support to the seed and contribute to seed stability. As storage proteins, they are less significant in pulses compared to globulins but still play a role in overall seed stability. Glutelins exhibit moderate protein quality and are less digestible than albumins and globulins. Their amino acid profile includes moderate levels of glutamic acid and arginine, which are essential for metabolic processes and structural functions. While glutelins are dominant in cereals, such as glutenin in wheat, their presence in pulses is limited to trace amounts. Research highlights that although glutelins are not the primary proteins in pulses, they contribute to the overall protein stability and texture of seed-based food products. This makes them relevant in certain food formulations requiring structural integrity. These protein fractions differ in their solubility, functional roles, nutritional significance, and proportions within the total protein content of pulses which is represented in Table 3.7.

STRUCTURE OF PULSE PROTEINS

Figure 3.2 illustrates the representation of chemical structures of different pulse proteins. Figures are depicting amino acids, peptide bonds, and secondary structures found in major storage proteins, viz., globulins, albumins, legumins, glutelin, and

TABLE 3.7
Pulse Protein Their Solubility, Functions, Proportion, and Nutritional Significance

Protein Type	Solubility	Proportion	Function	Nutritional Significance
Albumins	Water	10%–20%	Enzymes, defence, and metabolic roles	Rich in lysine; high nutritional quality
Globulins	Salt solutions	50%–70%	Major storage proteins	Rich in lysine but low in sulphur amino acids
Prolamins	Alcohol (70%–80%)	<2%–4%	Minor storage proteins	Poor nutritional quality; lysine deficient
Glutelins	Dilute acid/alkali	3%–5%	Structural and minor storage roles	Moderate protein quality

Source: Boye et al. (2010).

prolamins. The structural proteins in some common pulses with their proportion are shown in Table 3.8. Globulins have multiple polypeptide chains, often with disulphide bonds linking them, while albumin has three homologous domains that come together to create a heart-shaped protein. Each homologous domain is composed of two subdomains that have a common structural design. Legumin has a hexameric structure consisting of six subunits. Each subunit has an acidic alpha chain, which is further linked to a basic beta chain with the help of disulphide bonds (Amagliani and Schmitt, 2017).

Glutelins are made up of two polypeptide units which have molecular weight of 30–40 and 19–23 kDa. Prolamin has a beta helix structure. Prolamins have great quantities of the amino acids proline and glutamine (Lam et al., 2018; Shevkani et al. 2015; National Center for Biotechnology Information, 2025). The structural variability of these five pulse proteins can be visually distinguished from the figures given below.

PROPERTIES OF PULSE PROTEIN

Pulse proteins exhibit diverse physical (solubility, emulsifying, foaming, and gelling) and chemical (amino acid composition, molecular structure, thermal stability) properties that make them suitable for various food applications. Their functional versatility and nutritional profile enable their use as sustainable alternatives to animal-based proteins in food systems.

Physical Properties of Pulse Proteins

Solubility: Solubility is a fundamental property that affects all other functional properties of proteins, including emulsification, foaming, and gelation. It refers to the ability of proteins to dissolve or disperse uniformly in a solvent, typically water.

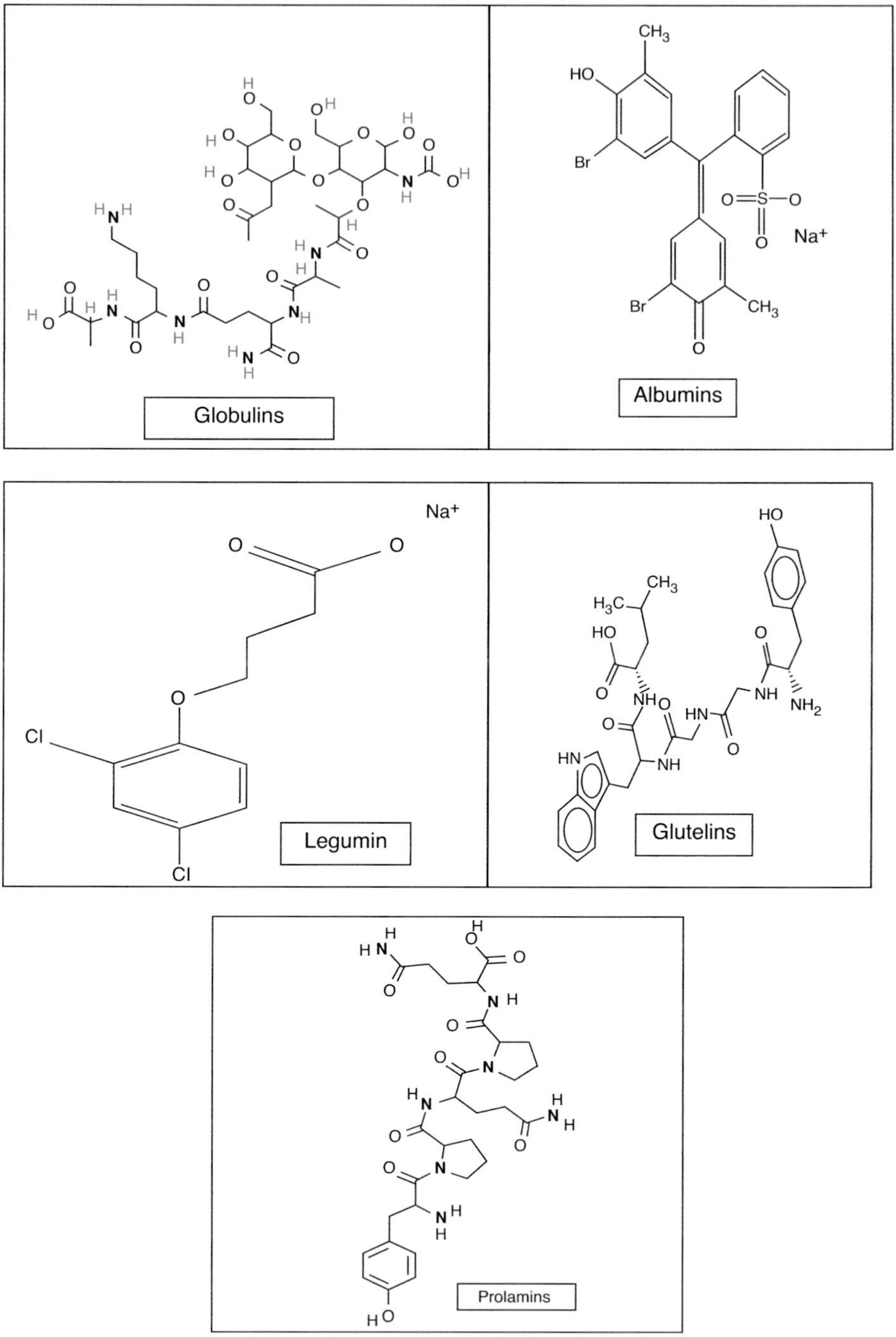

FIGURE 3.2 The structure of the different proteins of pulses. (Lam et al., 2018; Shevkani et al., 2015; National Center for Biotechnology Information, 2025.)

TABLE 3.8
Structural Proteins in Some Common Pulses

Pulse	Albumin (%)	Globulin (%)	Prolamin (%)	Glutelin (%)	Reference
Pea	15–25	50–60	1	12	Gueguen and Barbot (1988); Boye et al. (2010)
Chickpea	8–12	53–60	3–6	19–24	Dhawan et al. (1991)
Lentil	16.8	44.8	3.5	11.2	Gupta and Dhillon (1993); Boye et al. (2010)
Bean	10–30	45–70	1	10	Sathe (1984); Boye et al. (2010)
Soybean	10	85–95	–	–	Montoya et al. (2010)
Kidney beans	12–30	54–79	2–4	20–30	Montoya et al. (2010)
Cow pea	45	51	1	3	Montoya et al. (2010)
Jack bean	30–36	52–60	2–4	7–8	Montoya et al. (2010)
Faba bean	20	65	–	15	Montoya et al. (2010)

Isoelectric Point (pI): Pulse proteins generally have minimal solubility near their isoelectric point (pH 4–5), where the net surface charge is near zero. At this point, electrostatic repulsion is minimised, promoting protein aggregation and precipitation.

pH Dependence: At acidic or alkaline pH values away from the pI, solubility improves due to increased electrostatic repulsion between protein molecules. For example, pea protein isolates show significantly enhanced solubility at $pH > 7$, as the increased negative charge leads to protein unfolding and better dispersion.

Influence of Ionic Strength: Moderate ionic strength can shield charged groups and enhance solubility. However, high salt concentrations may lead to "salting out," reducing protein solubility due to competition for water molecules.

Processing Enhancements: Techniques such as heat treatment, enzymatic hydrolysis, and pH-shift processing are employed to improve solubility. Hydrolysis breaks down proteins into smaller peptides, increasing their hydrophilicity and water dispersion.

Water- and Oil-Holding Capacity (WHC/OHC): These capacities determine the ability of proteins to bind water or oil, impacting product texture, mouthfeel, and stability.

- **Water-Holding Capacity:** This is critical for hydration, viscosity, and moisture retention, especially in products like dough and meat analogues. Pulse proteins possess good WHC due to polar side chains and hydrophilic domains that interact with water molecules.
- **Oil-Holding Capacity:** Proteins can interact with lipids through hydrophobic interactions. A high OHC supports applications in emulsified and fatty food products.
- **Example:** Lentil proteins exhibit a WHC of 3–4 g water/g protein and an OHC of 1.5–2.0 g oil/g protein, making them suitable for use in bakery formulations, patties, and plant-based meat analogues.

Emulsifying Properties: Pulse proteins can stabilise oil-in-water emulsions due to their amphiphilic nature—possessing both hydrophilic and hydrophobic regions.

- **Mechanism:** Proteins adsorb at the oil–water interface, reducing interfacial tension and forming a protective layer around oil droplets, preventing coalescence.
- **Factors Influencing Emulsification:**
  - **Emulsifying Capacity (EC):** The ability to form emulsions.
  - **Emulsion Stability (ES):** The ability to maintain emulsions over time.
  - These properties are influenced by protein type, pH, ionic strength, and protein concentration.
- **Examples:**
  - Pea and chickpea protein isolates exhibit high emulsifying capacity at neutral pH due to optimal solubility and unfolding behaviour.
  - Chickpea proteins have been reported to outperform soy proteins in emulsifying ability, making them promising alternatives for plant-based emulsions.

Foaming Capacity and Stability: Foaming properties are important in products such as whipped toppings, mousses, and baked goods.

- **Foaming Capacity (FC):** The ability of proteins to trap air and form foam.
- **Foaming Stability (FS):** The ability to maintain foam structure over time without collapse.
- **Mechanism:** At the air-liquid interface, proteins denature and form a cohesive film that stabilises air bubbles. Improved flexibility and partial unfolding aid in forming stable foams.
- **pH Effect:** Alkaline pH enhances both FC and FS due to increased protein solubility and molecular flexibility.
- **Example:** Faba bean proteins exhibit strong foaming properties at neutral to alkaline pH, making them valuable for bakery and aerated product formulations.

Gelation: Gelation is the ability of proteins to form a semi-solid or solid matrix through network formation, typically upon heating or pH/ionic modification.

- **Mechanism:** Heating leads to protein unfolding, followed by aggregation through hydrophobic interactions and hydrogen bonding, forming a three-dimensional gel matrix.
- **Functional Implications:** Gelling is crucial for textural integrity in meat substitutes, desserts, and processed foods.
- **Example:** Lentil and pea proteins can form gels at concentrations of 10%–12% when subjected to heating, making them suitable for developing structured foods such as analogue meats or puddings.

Chemical Properties of Pulse Proteins

Amino Acid Composition: Pulse proteins are rich in lysine, an essential amino acid often limited in cereal proteins, but they are deficient in sulphur-containing amino acids such as methionine and cysteine. For example, chickpea and lentil proteins contain approximately 6%–7% lysine but less than 1% methionine, as reported by FAO/WHO (2013). This complementary amino acid profile enhances the nutritional value when combined with cereal proteins.

Molecular Structure: Pulse proteins are primarily classified into two major fractions:

- Albumins (water-soluble proteins) constitute about 10%–20% of the total protein content.
- Globulins (salt-soluble proteins) make up 70%–80% of the total protein and include two main sub-fractions: Vicilin (7S globulin), Legumin (11S globulin), which has a molecular weight of approximately 300–400 kDa.

Pea proteins, for instance, predominantly consist of legumin and vicilin fractions, which contribute to their functional properties.

Surface Hydrophobicity: Surface hydrophobicity (SH) is an indicator of the exposure of hydrophobic amino acid residues due to protein unfolding, influencing protein interactions and functionality, such as emulsification and gelation. Heat treatment and pH adjustments can increase surface hydrophobicity, enhancing the emulsifying and gelation properties of pulse proteins.

Denaturation and Thermal Stability: Denaturation involves the structural unfolding of proteins under heat or other stress conditions, leading to changes in functionality. Pulse proteins exhibit denaturation temperatures ranging from 70°C to 90°C, depending on the protein source.

- Vicilin (7S fraction) is more heat-stable compared to legumin (11S fraction).
- For instance, lentil proteins exhibit a denaturation temperature of approximately 75°C, making them suitable for thermal processing applications.

Functional Groups and Chemical Modifications: Pulse proteins contain reactive functional groups such as amino (–NH_2), carboxyl (–COOH), and thiol (–SH) groups, which can be chemically modified to improve functional properties:

- Acetylation and phosphorylation are common chemical modifications that enhance solubility, emulsification, and overall functionality of pulse proteins.
- These modifications alter the surface charge, hydrophobicity, and molecular structure, improving protein performance in food systems.

EFFECTS OF PROCESSING ON PULSE PROTEINS

Pulse proteins are being employed more and more in the food sector because of their nutritional and functional characteristics. Processing of the pulse proteins plays a very important role in developing their structural and functional characteristics, which are of utmost importance for use in foods. Various processes, like defatting, protein extraction, protein isolation, drying, and enzymatic hydrolysis have varied effects on pulse proteins.

Defatting and Its Impact

Defatting refers to the elimination of lipids from pulse flour, thereby eliminating unwanted flavours such as grassy or beany notes. Research shows that the defatting solvent, for instance, hexane or ethanol, plays a key role in determining the functional properties of the protein. Hexane-defatted chickpea proteins, for example, have shown enhanced emulsifying and foaming ability through exposed hydrophobic amino acids getting to interact with lipophilic solvents (Toews & Wang, 2013). But it decreases solubility since hydrophobic amino acids aggregate (Shevkani et al., 2019). Astonishingly, defatting does not affect the amino acid profile of pulse proteins (Yue et al., 2021). While defatting is performed in chickpea and lupine proteins, defatting is not performed in low-lipid pulses such as lentils and beans.

Methods of Protein Extraction

Protein extraction isolates proteins from the starches, fibres, and other seed materials. The two principal processes, dry extraction and wet extraction, have opposite effects on the functionality of proteins. Dry Extraction: Dry extraction is an energy-efficient physical, non-water, and solvent-free process that maintains the natural protein structure. It is best suited for applications where little structural modification is necessary. Dry-process proteins are lower in purity and contain more starch, restricting them from high-protein applications (Schuck et al., 2015). Air classification, a standard dry extraction method, has been found to be promising for retaining solubility and gelation characteristics despite its increased starch content (Vogelsang-O'Dwyer et al., 2020). Wet Extraction: Wet extraction uses the solubility difference to distinguish proteins from other non-protein compounds. Alkaline extraction is most prevalent because it has the capacity to recover proteins with high purity and enhanced solubility and emulsification power (Singhal et al., 2016). Acid extraction isolates lower solubility and foam stability proteins (Feyzi et al., 2018). Micellar precipitation and salt extraction are two additional wet methods that might assist in the preservation of native protein structures but whose success would be heavily influenced by the pulse source and conditions of processing (Stone et al., 2015).

Protein Isolation Techniques

Protein purification further cleanses the extracted proteins and affects their activity and composition. There are two principal techniques that are widely used: Isoelectric Precipitation (IEP): IEP separates proteins on the basis of solubility at their isoelectric point (pI). Globulin proteins are preferentially precipitated in IEP, and albumins are retained in the supernatant. IEP can be used for the preparation of highly purified isolates but may result in unfavourable functional attributes, i.e., gelation and solubility, by the absence of albumins (Shevkani et al., 2019). Ultrafiltration (UF): UF is a membrane filtration process that holds back globulin and albumin fractions. It has been established that UF-rich isolates possess better solubility, fat- and water-binding properties, and gelation characteristics than IEP isolates (Papalamprou et al., 2009b; Alonso-Miravalles et al., 2019). The holding back of albumins that are rich in cysteine residues contributes to better emulsification and gel elasticity.

Drying Processes

The process of drying is essential for the stabilisation of pulse protein isolates and reduction of transportation costs. However, the drying methods can affect the structure of proteins and their functionality. Freeze-drying, also known as lyophilisation, generally preserves the protein in their natural form due to minimal use of heat. Proteins dried in this way are very effective at solubility, gelation, and emulsification capabilities (Ghribi et al., 2015a). Freeze-drying is time-consuming and expensive and hence less suitable for mass production. Spray-drying is inexpensive and fast but requires high-temperature processing, which results in partial protein denaturation. Although spray-dried proteins exhibit decent solubility and emulsifying properties, functionality is based on drying conditions and pulse type (Cepeda et al., 1998). Vacuum-drying involves lower temperatures to limit thermal denaturation. It can lead to aggregation and decreased solubility (Feyzi et al., 2018). New technologies such as refractance window drying have possible applications in maintaining protein functionality while conserving energy (Raghavi et al., 2018).

Enzymatic Hydrolysis

Enzymatic hydrolysis is an optional process that changes protein structure by cleaving peptide bonds, thus changing functionality. Enzymatic hydrolysis relies on enzyme type, degree of hydrolysis (DH), and filtration after hydrolysis. Low hydrolysis improves emulsification, foaming capacity, and solubility by opening up hydrophilic surfaces and decreasing molecular weight (Ghribi et al., 2015b). Hydrolysis is effective in enhancing solubility around the isoelectric point and thus hydrolysates are utilised in acidic drinks (Wani et al., 2015). Excessive hydrolysis would lead to aggregation and compromised functionality as a result of enhanced hydrophobic interaction (Avramenko et al., 2013). Filtration following hydrolysis experiments normally provides exaggerated functional properties, as insoluble aggregates are filtered out (Xu et al., 2021). Differences in methodology make it challenging to quantify enzymatic hydrolysis effectiveness.

APPLICATIONS OF PULSE PROTEINS

Pulse proteins such as peas, chickpeas, lentils, and beans are gaining prominence in food, feed, and industrial uses. Pulse proteins exhibit functional properties such as water and fat binding, emulsification, foaming, and gelation, which have made them a generic ingredient in various applications. Their fairly high protein level as well as their superior amino acid profile further make them an asset as well as a sustainable option in meeting world nutrition and functionality requirements. The diversified applications of pulse proteins in different industries are discussed below beginning with its food applications.

Pulse proteins are highly valuable in the food industry due to their functional and nutritional properties. These functions have led to their use in a variety of foods, making them richer nutritionally as well as structurally. Pulse proteins are being widely utilised in infant foods as a dense plant-based source of nutrition. Pulse proteins contain growth and development essential amino acids to support the protein requirements of infants and toddlers. Pulse proteins are even lower in allergenicity than some animal proteins, and therefore they are suitable for infants with dairy or soy allergy. They are also readily digestible, which is helpful to infants and toddlers. Inclusion of pulse proteins in infant foods like cereals, purees, and snacks in order to enhance the protein level of these foods assists in offering nourishing, balanced diets, mainly in regions where malnourishment is prevalent.

Plant-based simulation milks are gaining popularity due to the growing incidence of lactose intolerance and milk allergy. Suitable pulse proteins to be utilised in such milks are those with high protein content along with functional properties like solubility and emulsification. Pea protein is a good ingredient to be used in non-dairy milks, for instance, as it delivers a creamy texture without possessing an off-flavour profile. Pulse imitation milks also fit as part of the growing demand for green and vegan foods. Vitamin- and mineral-fortified urd and labanee milks become a healthier option to traditional dairy and appeal to health food consumers. Tofu (bean curd) is a very traditional use of pulse proteins, especially of soybeans. Bean curd making is a process of coagulation of pulse proteins, and this provides a firm, healthy product. This use demonstrates the gel-forming property of pulse proteins. Tofu has been valued for its adaptability in various cuisines, for its ability to absorb flavours, and for its nutrient-rich composition of protein value, and thus tofu has become a familiar character in vegetarianism and veganism worldwide. Process improvements have also witnessed the arrival of non-soy pulses such as peas as a means of access for the substitute bean curds.

Pulse proteins are used extensively as extenders, enhancers, or replacers in meat. Because of their water- and fat-binding characteristics, they contribute to the texture, juiciness, and mouthfeel of meat products like burgers, sausages, and meatballs. For instance, lentil or chickpea protein may be used to reduce the fat level and improve the protein concentration of processed meats. In addition, pulse proteins contribute to the development of plant meat substitutes. With their gelation and emulsification properties that are comparable to the physical form and beauty of conventional meat, they attract the increasing number of vegetarian and flexitarian consumers. Lastly, adding pulses to meat would reduce production cost and complement green food strategies. Pulse proteins are used as functional food ingredients in baked products like bread, cakes, cookies, and muffins. They contribute to protein content, dough functional properties, and crumb texture and softness. It has been demonstrated that the use of pulse protein flour fortifiers has the potential to increase the nutritional value of bakery foods with desirable sensory attributes (Boye et al., 2010). Besides, their water-binding capacity and compatibility with other food ingredients make them capable of improved moisture retention and product stability. Pulse proteins play a pivotal role in the development of baked foods since they enhance the nutritional value and functionality of baked products. Addition of pulse proteins into bread, cookies, and cakes increases their protein levels to fulfil health-oriented consumer demands. Pulse proteins enhance dough strength, improve moisture retention, and increase shelf life. In frostings and glazes, they contribute texture, stability, and appearance to avoid the final product from spoiling when stored and handled. As replacers for nut butters or spreads, the proteins in pulses improve spread quality and texture and nutritional quality.

Pulse proteins are used in extrusion to make foods including pasta, noodles, and puffed snack foods. Addition of pulse proteins to pasta and noodle foods improves them in protein and fibre content and provides a healthier food compared to wheat food. These food products are for the consumer who wants functional foods to complement dietary requirements like high-protein or gluten-free diets. Pulse proteins also find application in the production of puffed snack foods such as crisps and cereal extrudates. With their ability to produce crispy, light textures and high protein content, they are the snack food choice of the health-conscious consumer market. They also have a delivery system to provide flavour and nutrition for satisfaction of different consumer segments. Growing demand for vegetarian foods has resulted in extensive application of pulse proteins for the preparation of meat analogues. Pulse proteins are used for the preparation of such products as vegetarian nuggets, burgers, and sausages because they have good gelation and emulsification capacities. As an example, pea protein isolates are utilised to a great extent in an effort to replicate the mouthfeel and texture of natural meat, providing a healthy and environmentally friendly option for vegetarians and vegans alike (McCurdy and Knipfel, 1990). Pulse proteins are increasingly utilised as substitutes for milk, substitutes for yogurt, and substitutes for cheese. Emulsification properties help to form stable suspensions in order to maintain plant-based beverages, such as pea or chickpea milk, in a liquid state and uniform in consistency and appearance. They are especially appealing to lactose intolerants or those wishing for a vegan choice (Flink and Christiansen, 1973).

For the celiac or gluten-intolerant patient, pulse proteins can serve as a good replacement for wheat gluten-containing ingredients. The foaming and gelling capacity of pulse proteins renders pulse proteins a good replacement for gluten's structural, textural characteristic in baked products like cakes, pasta, and bread. Pulse proteins, thus, are an asset towards the creation of high-quality, gluten-free products (Gupta and Dhillon, 1993). Pulse proteins find extensive application in extruded snack foods, protein bars, and soups. Their water-binding characteristic improves the product's texture as well as moisture retention and makes them a preferred option to health-conscious consumers due to high protein content. Pulse protein isolates are also applied in the processing of ready-to-eat food products, with nutritional balance coupled with convenience being provided (Boye et al., 2010).

Feed Applications: Pulse proteins are not only confined to human nutrition, but they also have a very crucial role in animal nutrition. Their digestibility, protein content, and sustainability make them a great option for feed ingredients.

- Concentrates of pulse proteins are widely applied as a sustainable source of protein for farm animals. Their well-balanced amino acid profiles induce enhanced growth rates, overall health, and productivity of animals. Mixed with cereal grains, proteins in pulses supply an inexpensive and high-protein option in livestock production (Paredes-López et al., 1991).
- In the pet food industry, pulse proteins are utilised for the production of high-protein pet foods for dogs and cats. Of these, pea protein is extensively applied in high-grade pet food due to its easy digestibility and allergen-free nature. It offers pets a balanced diet without the risk of employing common allergens (Chakraborty et al., 1979).
- Pulse proteins have been found to be suitable in aquaculture, as a sustainable fish meal substitute. Lentil and chickpea protein isolates contain an optimal amino acid balance, promoting best growth, health, and feed conversion of farmed finfish species, respectively. The use has benefits regarding sustainable aquaculture since it supports the reduction of the utilisation of conventional fish meal (El Fiel et al., 2002).

Industrial Applications: The applications of pulse proteins extend beyond food and animal feed to other industrial sectors. Their unique properties have rendered them ideal for new uses in materials and environmental applications.

- **Bioplastics:** Pulse proteins find application in the manufacture of biodegradable plastics. Their ability to form films, along with their availability as renewable feedstocks, makes them a natural alternative to petroleum-derived plastics. The protein products find specific applications in packaging and disposable uses, and they play a role in driving the quest for sustainability (Russin et al., 2007).
- **Adhesives and Coatings:** The adhesive properties of the pulse proteins, particularly in those with high globulin content, qualify them for the production of eco-friendly adhesives and coatings. They are

used extensively in the wood, paper, and textile industries for imparting good adhesion and less environmental degradation (Gupta and Dhillon, 1993).

- **Pest Control:** Some proteins from pulses possess bioactive properties that enable them to act as efficient agents in pest control. Chickpea protein fractions, for instance, were found to repel pests from stored grains, and hence offer a natural, sustainable method for pest control in agriculture (Fernandez-Quintela et al., 1997).

Health and Nutritional Applications: Pulse proteins are also important in health-oriented applications, meeting the requirements for several dietary and health applications.

- **Weight Management:** Pulse protein isolates are extensively used in weight management foods because they have a high satiety index. Pea or chickpea protein-fortified shakes and bars suppress appetite, induce weight reduction, and preserve lean muscle mass (Ologhobo et al., 1993).
- **Nutritional Supplements:** Pulse proteins are also in common use as nutritional supplements, particularly for sports persons, older adults, and those with certain dietary requirements. Due to their high lysine content and good amino acid profile, they support muscle maintenance, recovery, and general well-being (Boye et al., 2010).
- **Medical Nutrition:** Pulse protein isolates are employed in hospitals to formulate special nutritional food for malnourished patients or patients suffering from metabolic diseases. They are hypoallergenic and palatable and are ideal for patients with dietary limitations or sensitivities (McCurdy and Knipfel, 1990).
- The summary of different pulse proteins with application is shown in Table 3.9.

CONCLUSION AND FUTURE DIRECTIONS

Pulse proteins are valuable and functional foods with application in food, animal feed, industry, and healthcare. Their nutrient and functional profiles along with allergen-friendliness and environmental sustainability place them at the fore in meeting dietary and environmental requirements of the time. Yet processing technologies have far-reaching effects on structural and functional attributes, determining their application throughout their diverse utilisations. Dry extraction and lyophilisation are favoured by native structure preservation, whereas wet extraction and spray-drying are scalable and yield-enhancing. Functionalisation is enabled by enzymatic hydrolysis but requires to be deeply optimised. Normalising approaches, exploring new processing technologies, and optimising for food application purposes need to be next on the agenda. These advances will make possible the creation of high-quality, sustainable pulse protein ingredients with still broader potential uses and value to the nutrition and sustainability of the world.

TABLE 3.9
Summary of Various Uses of Pulse Proteins

Category	Applications	Examples
Food formulations	Enhancing texture, emulsification, gelation, and foaming properties.	Plant-based meat, dairy alternatives (e.g., almond milk, vegan cheese), gluten-free baked goods.
Protein fortification	Boosting protein content in processed foods and beverages.	Protein bars, shakes, breakfast cereals, soups, and ready-to-eat snacks.
Infant & therapeutic nutrition	Providing digestible, high-quality protein for specialised diets.	Infant formula, medical nutrition products, and food for the elderly or chronically ill.
Functional food products	Delivering bioactive peptides with health benefits such as antioxidant and antihypertensive effects.	Functional beverages, fortified snacks, and health supplements.
Extruded products	Improving texture and protein content in expanded food products.	Protein-enriched extruded snacks and cereal puffs.
Bakery industry	Enhancing dough properties and nutritional value.	Gluten-free bread, cookies, cakes, and pastries.
Plant-based alternatives	Serving as a sustainable protein source in animal-free foods.	Vegan meat substitutes, egg replacements, and non-dairy yogurt.
Feed industry	Providing high-quality protein for animal nutrition.	Livestock feed, aquaculture feed, and pet food.
Health & wellness products	Offering functional properties for weight management and health improvement.	Low-calorie foods, weight management shakes, and meal replacement powders.
Environmentally sustainable food	Supporting sustainable food production with a lower carbon and water footprint compared to animal proteins.	Clean-label and eco-friendly food products aligned with sustainability goals.
Innovative technologies	Used as raw materials in advanced food processing techniques.	Pulse protein isolates and concentrates in new-generation food systems.

REFERENCES

Amagliani, L., & Schmitt, C. (2017). Globular plant protein aggregates for stabilization of food foams and emulsions. *Trends in Food Science & Technology*, *67*, 248–259. 10.1016/j.tifs.2017.07.013.

Anderson, J. W., Baird, P., Davis, R. H., Ferreri, S., Knudtson, M., Koraym, A., Waters, V., & Williams, C. L. (2009). Health benefits of dietary fiber. *Nutrition Reviews, 67*(4), 188–205. https://doi.org/10.1111/j.1753-4887.2009.00189

Andreani, G., Sogari, G., Marti, A., Froldi, F., Dagevos, H., & Martini, D. (2023). Plant-based meat alternatives: Technological, nutritional, environmental, market, and social challenges and opportunities. *Nutrients, 15*(2), 452. https://doi.org/10.3390/nu15020452

Avramenko, N. A., Low, N. H., & Nickerson, M. T. (2013). The effect of limited enzymatic hydrolysis on the physicochemical and emulsifying properties of a lentil protein isolate. *Food Research International, 51*(1), 162–169. https://doi.org/10.1016/j.foodres.2012.12.019

Back, H. I., Yang, J. A., Kim, S. R., Kim, M. G., Jeon, J. Y., Kim, S. Y., Cha, Y. S., & Chae, S. W. (Eds.). (2010). Clinical functionality of Korean fermented foods: Antiobesity and antiatherogenesis effect of Korean traditional soybean fermented foods (Chungkookjang, Doenjang, and Kochujang) supplementation in overweight/obese Korean subjects. In *Proceedings of the International Symposium on Fermented Foods and Their Bright Future* (pp. 55–72). Seoul, South Korea: Ministry of Food, Agriculture, Forestry, and Fisheries.

Bailey, L. B. (2009). *Folate in Health and Disease*. Boca Raton, FL: CRC Press.

Boulter, D., & Croy, R. R. D. (1997). Legume seed storage proteins. *Advances in Botanical Research, 27*, 2–70.

Boye, J. I., Zare, F., & Pletch, A. (2010). Pulse proteins: Processing, characterization, functional properties, and applications in food and feed. *Food Research International, 43*(2), 414–431. https://doi.org/10.1016/j.foodres.2009.09.003

Calles, T., del Castello, R., Baratelli, M., Xipsiti, M., & Navarro, D. K. (2019). *The International Year of Pulses: Final Report*. Rome, Italy: FAO. https://doi.org/10.4060/CA4991EN

Cepeda, E., Cañellas, J., & Rodríguez, S. (1998). Effects of spray-drying conditions on the quality of pea protein concentrates. *International Journal of Food Science & Technology, 33*(6), 529–541. https://doi.org/10.1046/j.1365-2621.1998.00119.x

Chakraborty, P., Jayantha, P., & Perera, M. (1979). Nutritional value and utilization of pulse proteins in pet food formulations. *Journal of Animal Nutrition, 4*(1), 23–30.

Chandler, P. M., Spencer, D., Randall, P. J., & Higgins, T. J. (1984). Influence of sulfur nutrition on developmental patterns of some major pea seed proteins and their mRNAs. *Plant Physiology, 75*(3), 651–657. https://doi.org/10.1104/pp.75.3.651

Chukwuejim, S., Utioh, A., Choi, T. D., & Aluko, R. E. (2024). Lupin seed proteins: A comprehensive review of composition, extraction technologies, food functionality, and health benefits. *Food Reviews International, 40*, 691–714.

Chunhachart, O., Itoh, T., Sukchotiratana, M., Tanimoto, H., & Tahara, Y. (2006). Characterization of γ-glutamyl hydrolase produced by *Bacillus* sp. isolated from Thai thua-nao. *Bioscience, Biotechnology, and Biochemistry, 70*(11), 2779–2782.

Cotacallapa-Sucapuca, M., Vega, E. N., Maieves, H. A., Berrios, J. J., Morales, P., Fernández-Ruiz, V., & Cámara, M. (2021). Extrusion process as an alternative to improve pulses products consumption: A review. *Foods, 10*(5), 1096. https://doi.org/10.3390/foods10051096

Croy, R. R. D., Gatehouse, J. A., & Boulter, D. (1980). The structure of seed storage proteins and their genetic manipulation. *Advances in Biochemical Engineering/Biotechnology, 16*, 37–95.

Dahl, W. J., Foster, L. M., & Tyler, R. T. (2012). Review of the health benefits of pulses. *Cereal Foods World, 57*(3), 105–111. https://doi.org/10.1094/CFW-57-3-0105

Dhawan, K., Singh, N., & Singh, B. (1991). Chemical composition, cooking quality, and protein digestibility of some promising chickpea cultivars. *Journal of Food Science and Technology, 28*(1), 32–34.

El Fiel, H., Ali, O., & El Tinay, A. H. (2002). Protein fractions of cowpea and lentil seeds and their role in functional properties. *Food Chemistry, 77*(3), 261–264. https://doi.org/10.1016/S0308-8146(01)00378-3

El-Adawy, T. A., Rahma, E. H., El-Bedawy, A. A., & El-Beltagy, A. E. (2003). Nutritional potential and functional properties of germinated mung bean, pea and lentil seeds. *Plant Foods for Human Nutrition, 58*(3), 1–13.

Escobedo, A., & Mojica, L. (2021). Pulse-based snacks as functional foods: Processing challenges and biological potential. *Comprehensive Reviews in Food Science and Food Safety, 20*(5), 4678–4702. https://doi.org/10.1111/1541-4337.12809

FAO (Food and Agriculture Organization of the United Nations). (2019). *The State of the World's Biodiversity for Food and Agriculture.* Rome, Italy: FAO Commission on Genetic Resources for Food and Agriculture Assessments.

FAO (Food and Agriculture Organization of the United Nations). (2021). *Pulses: Nutritional Benefits and Food Security.* Rome, Italy: FAO.

Fernandez-Quintela, A., Macarulla, M. T., Del Barrio, A. S., & Martinez, J. A. (1997). Composition and functional properties of protein isolates obtained from chickpea seeds. *Food Chemistry, 62*(3), 325–331. https://doi.org/10.1016/S0308-8146(97)00060-1

Feyzi, S., Milani, E., & Moslehishad, M. (2018). Effect of acid and alkaline pH on functional properties of chickpea protein isolates. *Food Hydrocolloids, 77,* 490–499. https://doi.org/10.1016/j.foodhyd.2017.10.032

Flink, J. M., & Christiansen, N. (1973). Functional and nutritional properties of proteins from pea and chickpea milk. *Journal of Food Science, 38*(5), 816–819.

Frias, J., Peñas, E., & Martinez-Villaluenga, C. (2016). Fermented pulses in nutrition and health promotion. In *Fermented Foods in Health and Disease Prevention* (pp. 385–416) Academic Press. https://doi.org/10.1016/B978-0-12-802309-9.00016-9

Ghribi, A. M., Gafsi, I. M., Blecker, C., Danthine, S., Attia, H., & Besbes, S. (2015a). Effect of drying methods on physico-chemical and functional properties of chickpea protein concentrates. *Journal of Food Engineering, 165,* 179–188. https://doi.org/10.1016/j.jfoodeng.2015.06.012

Ghribi, A. M., Gafsi, I. M., Blecker, C., Danthine, S., Attia, H., & Besbes, S. (2015b). Contribution of enzymatic hydrolysis to structural, functional, and sensory properties of chickpea protein hydrolysates. *Food Chemistry, 187,* 322–330. https://doi.org/10.1016/j.foodchem.2015.04.111

Gueguen, J., & Barbot, J. (1988). Quantitative and qualitative variability of pea (*Pisum sativum* L.) proteins. *Journal of the Science of Food and Agriculture, 42*(3), 209–224. https://doi.org/10.1002/jsfa.2740420305

Gueguen, J., & Cerletti, P. (1994). Proteins of some legume seeds: Soybean, pea, fababean, and lupin. In B. J. F. Hudson (Ed.), *New and Developing Sources of Food Proteins* (pp. 145–193). Durango, CO: Chapman and Hill.

Gupta, Y. P., & Dhillon, T. S. (1993). Chemical composition and protein structure of lentils *(Lens culinaris Medik)*. *Food Chemistry, 47*(2), 177–184. https://doi.org/10.1016/0308-8146(93)90220-B

Gurusamy, S., Vidhya, C. S., Bhosale, Y., & Shanmugam, A. (2022). Pulses for health and their varied ways of processing and consumption in India: A review. *Applied Food Research, 2*, 100171. https://doi.org/10.1016/j.afres.2022.100171

Hallberg, L., & Hulthén, L. (2000). Prediction of dietary iron absorption: An algorithm for calculating absorption and bioavailability of dietary iron. *The American Journal of Clinical Nutrition, 71*(5), 1147–1160. https://doi.org/10.1093/ajcn/71.5.1147

Han, B. Z., Rombouts, F. M., & Nout, M. J. R. (2001). A Chinese fermented soybean food. *International Journal of Food Microbiology,* 65, 1–10.

Hoover, R., & Zhou, Y. (2003). In vitro and in vivo hydrolysis of starches by α-amylase and resistant starch formation in legumes. *Food Research International, 36*(9–10), 921–940. https://doi.org/10.1016/S0963-9969(03)00103-7

Jambunathan, R., & Singh, U. (1990). Present status and prospects for utilization of chickpea. In *Chickpea in the Nineties: Proceedings of the Second International Workshop on Chickpea Improvement* (pp. 41–48). Patancheru, India: ICRISAT.

Jarpa-Parra, M. (2018). Pea protein: A review of fractionation, extraction, and applications. *Food Research International, 109*, 418–428.https://doi.org/10.1016/j.foodres.2018.04.033

Jenkins, D. J. A., Wolever, T. M. S., Taylor, R. H., Barker, H., Fielden, H., Baldwin, J. M., Bowling, A. C., Newman, H. C., Jenkins, A. L., & Goff, D. V. (1981). Glycemic index of foods: A physiological basis for carbohydrate exchange. *The American Journal of Clinical Nutrition, 34*(3), 362–366. https://doi.org/10.1093/ajcn/34.3.362

Joshi, N., Godbole, S. H., & Kanekar, P. (1989). Microbial and biochemical changes during dhokla fermentation with special reference to flavor compounds. *Journal of Food Science and Technology, 26*(2), 113–115.

Jukanti, A. K., Gaur, P. M., Gowda, C. L. L., & Chibbar, R. N. (2012). Nutritional quality and health benefits of chickpea (*Cicer arietinum* L.): A review. *British Journal of Nutrition, 108*(S1), S11–S26. https://doi.org/10.1017/S0007114512000797

Kang, M. G. (2001). *Handbook of Technology for Chinese and Foreign Fermented Foods* (pp. 99–111). Beijing: Chemical Industrial Press.

Kaur, R., & Prasad, K. (2021). Nutritional characteristics and value-added products of chickpea (*Cicer arietinum*) - A review. *Journal of Postharvest Technology, 9*(2), 1–13.

Kimura, A., Fukuda, T., Zhang, M., Motoyama, S., Maruyama, N., & Utsumi, S. (2008). Comparison of physicochemical properties of 7S and 11S globulins from pea, fava bean, cowpea, and French bean with those of soybean—French bean 7S globulin exhibits excellent properties. *Journal of Agricultural and Food Chemistry, 56*, 10273–10279. https://doi.org/10.1021/jf801721b

Kumar, S., Ahuja, A., Tyagi, P., Lal, S., & Pandey, S. (2017). Pulses in India: Retrospect and prospects. *Indian Journal of Agricultural Sciences, 87*(11), 1501–1512.

Lam, A. C. Y., Karaca, A., Tyler, R. T., & Nickerson, M. T. (2018). Pea protein isolates: Structure, extraction, and functionality. *Food Reviews International, 34*, 126–147. https://doi.org/10.1080/87559129.2016.1242135

Lascialfari, M., Magrini, M.-B., & Triboulet, P. (2019). The drivers of product innovations in pulse-based foods: Insights from case studies in France, Italy, and USA. *Journal of Innovation Economics, 28*, 111. https://doi.org/10.3917/jie.028.0111

Lee, C.-H. (2008). Chongkukjang. In K. Kiuchi, T. Nagai, & K. Kimura (Eds.), *Advanced Science on Natto* (pp. 224–226). Tokyo: Kenpakusha (in Japanese).

Longvah, T., Ananthan, R., Bhaskarachary, K., & Venkaiah, K. (2017). *Indian Food Composition Tables*. Hyderabad: National Institute of Nutrition, Indian Council of Medical Research, Department of Health Research, Ministry of Health and Family Welfare, Government of India,.

Maheri-Sis, N., Chamani, M., Sadeghi, A. A., & Mirza-Aghazadeh, A. (2008). Nutritional evaluation of kabuli and desi type chickpeas (*Cicer arietinum* L.) for ruminants using in vitro gas production technique. *African Journal of Biotechnology, 7*(16), 2946–2951.

McCurdy, S. M., & Knipfel, J. E. (1990). Functional properties of pea protein isolates in meat analogues. *Canadian Institute of Food Science and Technology Journal, 23*(5–6), 241–245.

McLean, L. A., Sosulski, F. W., & Youngs, C. G. (1974). Effects of nitrogen and moisture on yield and protein in field peas. *Canadian Journal of Plant Science, 54*, 301–305. https://doi.org/10.4141/cjps74-047

Meng, G. T., & Ma, C. Y. (2001). Fourier-transform infrared spectroscopic study of globulin from *Phaseolus angularis* (red bean). *International Journal of Biological Macromolecules, 29*, 287–294. https://doi.org/10.1016/S0141-8130(01)00178-0

Mertens, C., Dehon, L., Bourgeois, A., Verhaeghe-Cartrysse, C., & Blecker, C. (2012). Agronomical factors influencing the legumin/vicilin ratio in pea (*Pisum sativum* L.) seeds. *Journal of the Science of Food and Agriculture, 92*(12), 1591–1596. https://doi.org/10.1002/jsfa.4738

Montoya, C. A., Lallès, J. P., Beebe, S., & Leterme, P. (2010). Phaseolin diversity as a possible strategy to improve the nutritional value of common beans (*Phaseolus vulgaris*). *Food Research International*, *43*(2), pp.443–449.

Mudryj, A. N., Yu, N., & Aukema, H. M. (2014). Nutritional and health benefits of pulses. *Applied Physiology, Nutrition, and Metabolism, 39*(11), 1197–1204. https://doi.org/10.1139/apnm-2013-0557

Nagai, T., & Tamang, J. P. (2010). Fermented soybeans and non-soybeans legume foods. In J. P. Tamang & K. Kailasapathy (Eds.), *Fermented Foods and Beverages of the World* (pp. 191–224). New York: CRC Press, Taylor & Francis.

Nagai, T., Tran, L.-S., Inatsu, Y., & Itoh, Y. (2000). A new IS4 family insertion sequence, IS4Bsu1, responsible for genetic instability of poly-γ-glutamic acid production in *Bacillus subtilis*. *Journal of Bacteriology, 182*(9), 2387–2392.

National Center for Biotechnology Information. (2025). PubChem Compound Summary for CID 74329879, Globulin G. Retrieved January 6, 2025 from https://pubchem.ncbi.nlm.nih.gov/compound/Globulin-G.

Nout, M. J. R. (2003). Traditional fermented products from Africa, Latin America, and Asia. In T. Boekhout & V. Robert (Eds.), *Yeasts in Food: Beneficial and Detrimental Aspects* (pp. 451–478). https://doi.org/10.1533/9781845698485.451

Nwokolo, E. (1987). Nutritional evaluation of pigeon pea meal. *Plant Foods for Human Nutrition, 37*(4), 283–289.

Okada, N. (2008). Toa-nao. In K. Kiuchi, T. Nagai, & K. Kimura (Eds.), *Advanced Science on Natto* (pp. 212–216). Tokyo: Kenpakusha.

Ologhobo, A. D., Apata, D. F., & Iyayi, E. A. (1993). Pulse protein isolates as functional ingredients in weight management diets. *Nutrition & Metabolism, 4*(2), 87–94.

Osborne, T. B. (1924). *The Vegetable Proteins*. London: Longmans, Green, and Co.

Papalamprou, E. M., Doxastakis, G. I., Biliaderis, C. G., & Kiosseoglou, V. D. (2009). Influence of preparation methods on physicochemical and gelation properties of chickpea protein isolates. *Food Hydrocolloids, 23*(2), 337–343. https://doi.org/10.1016/j.foodhyd.2008.03.009

Paredes-López, O., Ordorica-Falomir, C., & Olivares-Vázquez, M. R. (1991). Chickpea proteins: Chemistry, technology, and nutritional value. *Critical Reviews in Food Science and Nutrition, 30*(1), 1–26. https://doi.org/10.1080/10408399109527534

Parmar, N., Singh, N., Kaur, A., Virdi, A. S., & Shevkani, K. (2017). Protein and microstructure evaluation of harder-to-cook and easy-to-cook grains from different kidney bean accessions. *LWT - Food Science and Technology*, *79*, 487–495. https://doi.org/10.1016/j.lwt.2017.01.027

Prasad, A. S. (2013). Discovery of human zinc deficiency: Its impact on human health and disease. *Advances in Nutrition, 4*(2), 176–190. https://doi.org/10.3945/an.112.003210

Pulse Canada. (2011). Pulse Proteins. Retrieved from https://www.pulsecanada.com

Raghavi, L., Moses, J. A., & Anandharamakrishnan, C. (2018). Refractance window drying of foods: A review. *Journal of Food Engineering, 222,* 267–275. https://doi.org/10.1016/j.jfoodeng.2017.11.019

Reddy, N. R. (2002). Occurrence, distribution, and nutritional implications of phytate in human nutrition. *Food Chemistry, 47*(4), 333–342.

Rehab, F. M. A. (2024). Quality evaluation of snack bars prepared using different proportions of Khudri dates and fried green lentil (*Lens culinaris* Medik.). *Cogent Food & Agriculture*, *10*(1), 2300880. https://doi.org/10.1080/23311932.2023.2300880

Roy, A., Moktan, B., & Sarkar, P. (2007). Microbiological quality of legume-based traditional fermented foods marketed in West Bengal, India. *Food Control*, *18*, 1405–1411. https://doi.org/10.1016/j.foodcont.2006.10.001

Russin, T. A., Boye, J. I., Arcand, Y., & Rajamohamed, S. H. (2007). Pulse proteins as sustainable alternatives for biodegradable plastics. *Trends in Food Science & Technology, 18*(4), 192–202. https://doi.org/10.1016/j.tifs.2006.12.007

Sasaki, M., & Nunomura, N. (2003). Soy (soya) sauce. In B. Caballero, L. C. Trugo, & P. M. Finglas (Eds.), *Encyclopedia of Food Sciences and Nutrition* (2nd ed., pp. 2359–2369). London: Academic Press.

Sathe, S. K., Deshpande, S. S., & Salunkhe, D. K. (1984). Dry beans of Phaseolus: A review. *CRC Critical Reviews in Food Science and Nutrition, 20*(1), 1–46. https://doi.org/10.1080/10408398409527367

Schuck, P., Dolivet, A., Méjean, S., & Jeantet, R. (2015). Recent advances in spray drying and microencapsulation techniques. *Food Research International, 73,* 145–156. https://doi.org/10.1016/j.foodres.2015.02.027

Schwenke, K. D. (2001). Functional properties of food proteins. *European Journal of Lipid Science and Technology, 103*(9), 562–574.

Sharma, N., Handa, S., & Gupta, A. (2013). A comprehensive study of different traditional fermented foods/beverages of Himachal Pradesh to evaluate their nutrition impact on health and rich biodiversity of fermenting microorganisms. *International Journal of Research in Applied, Natural and Social Sciences*, *3*, 19–28.

Shevkani, K., Kaur, A., Kumar, S., & Singh, N. (2015). Cowpea protein isolates: Functional properties and application in gluten-free rice muffins. *LWT - Food Science and Technology*, *63*, 927–933. https://doi.org/10.1016/j.lwt.2015.04.058

Shevkani, K., Singh, N., Chen, Y., & Kaur, A. (2019). Pulses for food and nutritional security. *Trends in Food Science & Technology, 88,* 67–82. https://doi.org/10.1016/j.tifs.2019.02.021

Shewry, P. R. (2007). Improving the protein content and composition of cereal grain. *Journal of Cereal Science, 46*(3), 239–250. https://doi.org/10.1016/j.jcs.2007.06.006

Shobana, S., Krishnaswamy, K., Sudha, V., Malleshi, N. G., Anjana, R. M., Palaniappan, L., & Mohan, V. (2013). Finger millet (*Eleusine coracana* L.): A review of its nutritional properties, processing, and plausible health benefits. *Advances in Food and Nutrition Research*, *69*, 1–39.

Singh, U., & Jambunathan, R. (1982). Distribution of seed protein fractions and amino acids in different anatomical parts of chickpea (*Cicer arietinum* L.) and pigeonpea (*Cajanus cajan* L.). *Plant Foods for Human Nutrition, 31*(4), 347–354.

Singhal, A., Bosqui, A. M., & Chang, S. K. C. (2016). Functional properties of pulse proteins: Applications in food systems. *Journal of Food Science, 81*(9), R2216–R2226. https://doi.org/10.1111/1750-3841.13356

Slavin, J. L. (2005). Dietary fiber and body weight. *Nutrition, 21*(3), 411–418. https://doi.org/10.1016/j.nut.2004.08.018

Sosulski, F. W., & Gadan, H. S. (1988). Composition and functional properties of pea flour protein fractions. *Journal of Food Science, 53*(5), 1218–1221.

Stone, A. K., Karalash, A., Tyler, R. T., Warkentin, T. D., & Nickerson, M. T. (2015). Functional properties of protein isolates from different pea cultivars. *Food Science and Technology, 63*(2), 259–265. https://doi.org/10.1016/j.lwt.2015.03.057

Sugawara, E. (2010). Fermented soybean pastes miso and shoyu with reference to aroma. In J. P. Tamang & K. Kailasapathy (Eds.), *Fermented Foods and Beverages of the World* (pp. 225–245). New York: CRC Press, Taylor & Francis Group.

Swanson, B. G. (1990). Pea and lentil protein extraction and functionality. *Journal of the American Oil Chemists' Society, 67*(5), 276–280.

Tamang, J. P. (2010). *Himalayan Fermented Foods: Microbiology, Nutrition, and Ethnic Values*. New York: CRC Press, Taylor & Francis Group.

Tanaka, T. (2008). Pepok. In K. Kiuchi, T. Nagai, & K. Kimura (Eds.), *Advanced Science on Natto* (pp. 218–221). Tokyo: Kenpakusha (in Japanese).

Tang, C. H., & Sun, X. (2010). Physicochemical and structural properties of 8S and/or 11S globulins from mungbean (*Vigna radiata* L. Wilczek) with various polypeptide constituents. *Journal of Agricultural and Food Chemistry*, *58*, 6395–6402. https://doi.org/10.1021/jf904254f

Tang, C. H., Xiao, M. L., Chen, Z., Yang, X. Q., & Yin, S. W. (2009). Properties of cast films of vicilin-rich protein isolates from Phaseolus legumes: Influence of heat curing. *LWT - Food Science and Technology*, *42*, 1659–1666. https://doi.org/10.1016/j.lwt.2009.05.020

Thakur, N., & Bhalla, T. C. (2004). Characterization of some traditional fermented foods and beverages of Himachal Pradesh. *Indian Journal of Traditional Knowledge*, 3(3), 325–335.

Tiwari, B. K., & Singh, N. (2012). *Pulse Chemistry and Technology*. London: Royal Society of Chemistry.

Toews, R., & Wang, N. (2013). Physicochemical and functional properties of protein concentrates from pulses. *Food Research International, 52*(2), 445–451. https://doi.org/10.1016/j.foodres.2012.11.002

USDA (United States Department of Agriculture). (2020). *National Nutrient Database for Standard Reference.* Beltsville: Nutrient Data Laboratory, Beltsville Human Nutrition Research Center, ARS, USDA.

USDA (US Dept of Agriculture). (2022). FoodData Central. Retrieved from https://fdc.nal.usda.gov/

Vioque, J., Alaiz, M., & Giron-Calle, J. (2012). Nutritional and functional properties of *Vicia faba* protein isolates and related fractions. *Food Chemistry*, 132, 67–72. https://doi.org/10.1016/j.foodchem.2011.10.033

Vogelsang-O'Dwyer, M., Petersen, I. L., & Holm, J. (2020). Dry fractionation of pulses: A review of the impact on protein functionality. *Journal of Food Science, 85*(11), 1–13. https://doi.org/10.1111/1750-3841.15133

Wani, I. A., Sogi, D. S., & Gill, B. S. (2015). Physicochemical and functional properties of hydrolyzed pulse protein isolates. *Food Hydrocolloids, 43,* 68–75. https://doi.org/10.1016/j.foodhyd.2014.05.001

Whelton, P. K., He, J., Cutler, J. A., Brancati, F. L., Appel, L. J., Follmann, D., & Klag, M. J. (1997). Effects of oral potassium on blood pressure. *Journal of the American Medical Association, 277*(20), 1624–1632. https://doi.org/10.1001/jama.1997.03540440040033

Winarno, F. G. (1986). Traditional technologies of Indonesia with special attention to fermented foods. In *Traditional Foods: Some Products and Technologies* (pp. 136–147). Mysore: CFTRI.

Xu, B., & Chang, S. K. C. (2008). Effect of soaking, boiling, and steaming on total phenolic content and antioxidant activities of cool season food legumes. *Food Chemistry, 110*(1), 1–13. https://doi.org/10.1016/j.foodchem.2008.01.045

Xu, Y., Chen, X., & Qin, Y. (2021). Challenges in hydrolysis: Functionality of pulse protein hydrolysates. *Food Chemistry, 345,* 128812. https://doi.org/10.1016/j.foodchem.2020.128812

Yue, X., Liu, M., Zhang, J., & Zheng, Y. (2021). Influence of defatting on the physicochemical properties of pulse protein isolates. *Journal of Food Processing and Preservation, 45*(1), e15062. https://doi.org/10.1111/jfpp.15062

4 Oilseed Proteins

Anamika Sharma and Vikas Nanda

INTRODUCTION

Protein ingredients play a vital role in the food industry, offering both essential nutrition and functional properties that enhance food formulations. Their amphipathic nature enables them to act as a highly versatile structural component, to serve as dynamic surface-active agents, facilitating interactions between proteins, carbohydrates, lipids, air, and water at different interfaces within the food matrices (Akharume et al., 2021). From a nutritional standpoint, extensive research has emphasised the positive impact of proteins on overall health and well-being. These benefits include supporting weight management, enhancing satiety, lowering the glycaemic index, promoting heart health, preserving muscle mass, and improving athletic performance (Livesey et al., 2008; Paul et al., 2009). As the global population grows and the demand for sustainable food sources increases, there is a pressing need to explore alternative protein options that are both nutritionally adequate and environmentally friendly. While animal-based proteins have long been the primary source of dietary protein, their production poses significant environmental challenges, including high greenhouse gas emissions, water consumption, loss of biodiversity and other human health hazards such as cardiovascular diseases (López et al., 2018; Alemayehu et al., 2015). This has led to a shift towards plant-based proteins, which offer a more sustainable and health-conscious alternative. However, plant proteins have also been regarded as nutritionally inadequate because of the few essential amino acid deficiency. For example, most cereal proteins are typically deficient in lysine, whereas pulse proteins are deficient in methionine, cystine, and tryptophan (Singh, 2017; Sun-Waterhouse et al., 2014). Despite these limitations, recent medical studies have shown that the differences in bioavailability and digestibility between plant and animal proteins are clinically insignificant, especially when a variety of plant-based diets are consumed (Joshi et al., 2018). A complementary combination of plant-based diets, such as cereals and pulses, can provide all essential amino acids necessary for body maintenance. Among plant-based proteins, oilseeds are a vital category of crops grown primarily for their high oil content, typically ranging between 20% and 50%, depending on the specific type of oilseed. These oilseeds serve as key resources for various industries, including food, cosmetics, pharmaceuticals, and biofuels.

Globally, more than 200 species of oil plants are cultivated, spanning diverse climatic regions. Major producers include the United States, Brazil, India, China, and Indonesia. In the 2004/2005 period, global oilseed production reached an impressive 380 million metric tons, while protein meal production amounted to 207 million metric tons. Among the major contributors to protein meal production, soybean, rapeseed, cottonseed, sunflower seed, and peanuts were the most prominent, accounting

DOI: 10.1201/9781003613909-4

for 69%, 12.4%, 6.9%, 5.3%, and 2.8% of the global output, respectively (Ash & Dohlman, 2006). A wide variety of other oilseeds are also being utilised such as coconut, hempseed, flaxseed, pumpkin seed, sesame seed etc. Oilseeds and their derived oils are gaining increasing popularity as ingredients in food products, leading to a systematic growth in the global cultivation of oil-bearing plants. Oilseeds typically contain at least 15% fat and are also a valuable source of protein. During cold-pressed oil production, oil extraction yields a by-product known as oilcake, which is rich in protein and fibre. Oils derived from these seeds serve as a vital source of energy, essential fatty acids, and fat-soluble vitamins, playing an essential role in human nutrition and various industrial applications (Hidalgo & Zamora, 2006). Oilcakes or meals, the by-products of oil extraction, have been extensively studied as raw materials for producing protein isolates and hydrolysates (Kaushik et al., 2016; Mamone et al., 2019; Raikos et al., 2017). Consequently, oilseeds, high in fibre, sterols, proteins, unsaturated fatty acids, and bioactive peptides, represent a potent source of functional ingredients. In the context of meat products, there is a growing trend to replace portions of the meat-fat fraction with vegetable oil, plant-based proteins, or dietary fibre. This substitution not only enriches the product with polyunsaturated fatty acids or reduces its energy value but also aligns with efforts to increase the use of plant-based protein alternatives in meat products (Zdunczyk & Jankowski, 2013).

With ongoing advancements in food processing technologies and a growing emphasis on sustainability, oilseed proteins hold great promise as a key component of future food systems, addressing both nutritional needs and environmental challenges. These proteins provide a rich source of essential amino acids and also contribute to the circular economy by utilising by-products from oil extraction processes (Zhang et al., 2023). Oilseed proteins offer balanced amino acid profiles, bioactive peptides with antioxidant and cholesterol-lowering properties, and desirable functional attributes such as emulsification, foaming, and gelling. Their abundance, cost-effectiveness, and potential to reduce food waste make them an attractive alternative in various food applications, including meat analogues, dairy substitutes, and bakery products. Advances in extraction, processing, and formulation technologies continue to enhance the utilisation of oilseed proteins, making them an essential component in the future of food and biotechnology. Understanding the fundamental properties and potential applications of oilseed proteins provides valuable insights for food scientists, and industry professionals seeking to harness their full potential (Kotecka-Majchrzak et al., 2020).

COMPOSITION OF DIFFERENT OILSEEDS

Soybean (*Glycine max*)

Soybean is among the most widely cultivated oilseeds, primarily used for its high-quality protein and oil content. Soybean, predominantly contains protein in a range of 38.97%–41.45% with glycinin and β-conglycinin as the major storage protein fractions, followed by carbohydrates (32.80%), and an oil content of 15%–25% (Li et al., 2024). The lipid content is rich in unsaturated fatty acids, notably linoleic acid, oleic acid, and linolenic acid whereas carbohydrates consist of soluble sugars like sucrose (2.5%–8.2%), raffinose (0.1%–1.0%), and stachyose (1.4%–4.1%), along with insoluble polysaccharides contributing to dietary fibre (Sharma et al., 2014). Soybeans also provide essential minerals such as calcium, magnesium, phosphorus, potassium, and iron, and vitamins including vitamin K1, folate, thiamine (B1), and riboflavin (B2) (Table 4.1). Additionally, they contain bioactive compounds like isoflavones (genistein, daidzein, glycitein), saponins, and phytosterols, which are associated with various health benefits (Zhang et al., 2017). The unique composition of soybeans makes them a valuable component in human nutrition, providing a balanced source of essential nutrients and health-promoting compounds.

Chia Seeds (*Salvia hispanica*)

Chia seeds are recognised for their high dietary fibre content, comprising approximately 30–34 g/100 g. The majority of this fibre is insoluble dietary fibre (IDF), making up around 85%–93%, while soluble dietary fibre (SDF) represents approximately 7%–15% of the total fibre content (da Silva Marineli et al., 2015; Reyes-Caudillo et al., 2008). Compared to dried fruits, cereals, and nuts, chia seeds provide a superior dietary fibre content. Their lipid composition is abundant in polyunsaturated fatty acids, with α-linolenic acid (ALA) accounting for approximately 60% of the total fatty acids. Smaller quantities of linoleic, oleic, and palmitic acids are also present. Notably, chia seeds contain higher amounts of omega-3 fatty acids than flaxseeds, along with a beneficial omega-6 to omega-3 ratio of approximately 0.3:0.35 (Ciftci et al., 2012; Nitrayová et al., 2014; Ayerza, 1995; 2011). Chia seeds provide essential minerals, including phosphorus (860–919 mg/100 g), calcium (456–631 mg/100 g), potassium (407–726 mg/100 g), and magnesium (335–449 mg/100 g). They also contain vitamins, predominantly vitamin B1 (0.6 mg/100 g), vitamin B2 (0.2 mg/100 g), and niacin (8.8 mg/100 g) (USDA, 2018; Jin et al., 2012) (Table 4.1).

Additionally, chia seeds are rich in bioactive phytochemicals, particularly polyphenols such as gallic, caffeic, chlorogenic, cinnamic, and ferulic acids, along with flavonoids like quercetin, kaempferol, rutin, apigenin, and p-coumaric acid (Oliveira-Alves et al., 2017; Rahman et al., 2017). Isoflavones, including daidzein, glycitein, genistein, and genistin, are present in smaller quantities (Table 4.1). According to Ciftci et al. (2012) chia seeds contain phytosterols such as campesterol (472 mg/kg lipids), stigmasterol (1,248 mg/kg lipids), β-sitosterol (2057 mg/kg lipids), and Δ5-avenasterol. They also provide tocopherols, including α-tocopherol (8 mg/kg lipids), γ-tocopherol (422 mg/kg lipids), and δ-tocopherol (15 mg/kg lipids) (Kulczyński et al., 2019).

Coconut (*Cocos nucifera*)

Whole coconut (*Cocos nucifera* L.) is a highly versatile fruit with a rich chemical composition that contributes to its nutritional and functional properties. The major components of whole coconut include water, carbohydrates, proteins, lipids, dietary fibre, vitamins, and minerals. Coconut water, found in immature coconuts, contains approximately 94% water, electrolytes such as potassium, sodium, calcium, and magnesium, along with small amounts of proteins, sugars, and vitamins (Yong et al., 2009).

TABLE 4.1
Composition of the Different Oilseeds: Nutritional and Bioactive Components

Oilseeds	Oil Content (%)	Protein (%)	Carbohydrate (%)	Fibre (%)	Major Fatty Acids	Major Bioactive Compounds	Vitamins	Mineral	References
Soybean	15–25	38.97–41.45	32.80	24.96–32.63	Linoleic, oleic, palmitic, linolenic, stearic	Phenolic acids, isoflavones, anthocyanins	Vitamin B, E, K	Sodium, Potassium, Phosphorous, and iron	Li et al. (2024); Sharma et al. (2014); Zhang et al. (2017)
Chia	30–35	16–25	30–40	25–35	Oleic, α-Linolenic, linoleic acids	Phenolic acids (quercetin, kaempferol, p-coumaric acid), isoflavones (daidzein, glycitein, genistein)	Vitamin B1, B2, Niacin	Calcium, magnesium, phosphorus	Dash and Ghosh (2017)
Coconut	60–70	6–8	20–25	13–16	Lauric Acid	Catechins, flavonoids, phenolic acids	Vitamin C, thiamine, and niacin	Potassium, sodium calcium, and magnesium	FRIDA (2019); Appaiah et al. (2014)
Hemp	25–35	20–25	25–30	20–25	Linoleic acid, α-Linolenic acid	Polyphenols, carotenoids	vitamin E	Magnesium, phosphorous, and potassium	Siano et al. (2019)
Flax	35–45	18–25	20–25	20–30	α-Linolenic acid	Ferulic acid, caffeic acid, quercetin, and kaempferol	Vitamin A. C, and E	Potassium, zinc, copper, manganese	Soni et al. (2016)

TABLE 4.1 (*Continued*)
Composition of the Different Oilseeds: Nutritional and Bioactive Components

Oilseeds	Oil Content (%)	Protein (%)	Carbohydrate (%)	Fibre (%)	Major Fatty Acids	Major Bioactive Compounds	Vitamins	Mineral	References
Nigella (black seed)	30–35	20–22	35–40	15–20	Linoleic acid, oleic acid		Thymoquinone	Iron, zinc	Mamun and Absar (2018); Sultan et al. (2009)
Pumpkin	40–50	20–30	10–15	15–20	Linoleic acid, oleic acid	Zeaxanthin, crypto-xanthin	Niacin, pantothenic acid, pyridoxine, riboflavin, thiamine	Magnesium, zinc, phosphorous, selenium	Farida (2019)
Rapeseed	35–50	20–25	25–30	10–15	Oleic acid, linoleic acid	Carotenoids, phenolic compounds	High proportions of all vitamins except pantothenic acid	Calcium, phosphorous	Orsavova et al. (2015); Yoshie-Stark et al. (2008)
Sesame	40–60	18–25	10–15	10–15	Oleic acid, linoleic acid	Ferulic, cinnamic, sesamin	Vitamin A, thiamin, riboflavin, niacin, folic acid	Calcium, copper magnesium, zinc	Dravie et al. (2020)
Sunflower	40–50	20–25	20–25	10–15	Linoleic acid, oleic acid	Ferulic, caffeic, chlorogenic	Vitamin E, folate	Phosphorous, potassium, selenium, magnesium	Tanwar and Goyal (2021)

The coconut kernel, comprising the white edible flesh, is rich in lipids, primarily medium-chain fatty acids like lauric acid, which constitutes about 47% of the total fat content (DebMandal & Mandal, 2011). It also provides proteins, ranging between 3% and 4% of the dry weight, with globulins and albumins as the predominant protein fractions (Seow & Gwee, 1997). The kernel is also a significant source of dietary fibre, ranging from 13% to 16%, contributing to its functional and health-promoting properties (DebMandal & Mandal, 2011). The testa, or brown outer layer of the kernel, exhibits the highest protein concentration, along with polyphenolic compounds, which contribute to antioxidant activity (Vieira et al., 2024). Coconut milk, extracted from the grated kernel, contains water, fat, protein, and small amounts of vitamins such as vitamin C, thiamine, and niacin (Seow & Gwee, 1997). The whole coconut also provides essential minerals like calcium, phosphorus, potassium, and magnesium. The diverse composition of coconut makes it a highly nutritious and functional food with significant applications in the food, cosmetic, and pharmaceutical industries (Table 4.1).

Flaxseed (*Linum usitatissimum*)

Flaxseed, also known as linseed, is a rabi crop grown in more than 50 countries (Canada, India, China, and the USA) with an annual production of 2.92 million tons. Flaxseed is a nutrient-rich oilseed with approximately 41% fat, 20% protein, 28% dietary fibre, and 4% ash content (Morris, 2007; Gopalan et al., 2007). It is particularly valued for its bioactive components, including α-linolenic acid (ALA), lignans, fibre, and phenolic compounds (Oomah & Mazza, 2000). Among oilseeds, flaxseed is distinguished by its exceptionally high ALA content (18:3n-3), an omega-3 fatty acid known for its anti-inflammatory, anti-thrombotic, and anti-arrhythmic properties (Simopoulos, 1999). Flaxseed oil constitutes 35%–45% of the seed's weight, with 45%–52% of the oil composed of ALA, making it one of the richest plant-based sources of omega-3 fatty acids (Bhatty, 1995). The nutritional composition of flaxseed can fluctuate based on genetic variations, growing conditions, and processing techniques (Morris, 2007). Additionally, flaxseed supplies essential vitamins such as vitamins A, C, and E, along with minerals like phosphorus, magnesium, potassium, sodium, iron, copper, manganese, and zinc, contributing to its health-promoting properties (Table 4.1).

Hempseed (*Cannabis sativa*)

Hempseed is recognised for its high nutritional value, containing 35.5% crude oil, 24.8% crude protein, 27.6% crude fibre, and 5.6% ash (Callaway, 2004). Its oil is rich in unsaturated fatty acids, primarily linoleic acid (52%–62%) and α-linolenic acid (12%–23%) (Leizer et al., 2000), with an optimal omega-6 to omega-3 ratio of 3:1 (Erasmus, 1999). The oil contains lower levels of saturated fatty acids such as palmitic acid and stearic acid (Ally & Horrobin, 1980). Hempseed is an excellent source of vitamin E, particularly γ-tocopherol (710–870 ppm), while α-tocopherol is found in smaller amounts (7–80 ppm) (Deferne & Pate, 1996). Although α-tocopherol has better biological activity, γ-tocopherol shows stronger antioxidant properties at lower peroxynitrite concentrations (Leizer et al., 2000). Mineral content in hempseed is significant, with phosphorus and potassium being the predominant minerals with a total ash content of around 5.6%, reflecting its rich mineral composition (Callaway, 2004). The nutrient profile of hempseed makes it a valuable functional food ingredient with potential health benefits (Table 4.1).

Rapeseed (*Brassica napus*)

Canola, derived from rapeseed (*Brassica napus*), is one of the most widely cultivated oilseed crops globally, originating in India over 3,000 years ago (Shahidi, 1990). The seeds contain approximately 3%–4.5% moisture, 40%–45% oil, 19%–28% crude protein, 5%–6% ash, 8%–10% soluble sugars, and 14%–15% crude fibre (Carré et al., 2016). Canola is distinct from traditional rapeseed due to its low erucic acid content and reduced glucosinolate levels, making it more suitable for human consumption (Assadi et al., 2011). Canola oil ranks as the third most consumed vegetable oil worldwide, following palm and soybean oil. It is particularly valued for its high oleic acid (C18:1) content and a balanced omega-6 to omega-3 ratio (2:1), offering various health benefits (Matthaus et al., 2016). Genetic modifications have further enhanced α-linolenic acid content and oxidative stability. The remaining canola meal after oil extraction contains up to 50% protein on a dry basis, along with significant amounts of calcium, phosphorus, potassium, iron, zinc, and selenium (Siddiqui & Wood, 1977). However, the high phytate content of canola meal limits the bioavailability of phosphorus, especially in monogastric animals. Canola meal also contains indigestible carbohydrates like raffinose and stachyose, which cause flatulence and restrict its direct use as food (Wickramasuriya et al. 2015). Nevertheless, proper processing can improve its nutritional quality, making canola a valuable source of proteins, minerals, and bioactive compounds (Table 4.1).

Pumpkin Seeds (*Cucurbita pepo*)

Pumpkin seeds (*Cucurbita pepo*) are highly nutritious, offering a rich source of proteins, oils, minerals, and bioactive compounds. The protein content ranges from 24.50% to 32.00%, with higher concentrations observed in defatted seed flour (55.40%) (Lazos, 1986). Essential amino acids such as arginine, glutamic acid, and aspartic acid are the most abundant, while methionine and tryptophan are the limiting amino acids (Kim et al., 2012; El-Adawy & Taha, 2001). Protein isolates contain higher essential amino acid concentrations compared to concentrates and meals (Mansour et al., 1993). Pumpkin seed oil is rich in unsaturated fatty acids (73.03%), mainly linoleic acid (52.69%) and oleic acid (18.14%) (Alfawaz, 2004). The seeds also contain significant amounts of minerals like potassium, magnesium, and phosphorus, making them a valuable dietary source (Aliu et al., 2016). However, the presence of phytates can reduce the bioavailability of minerals such as zinc. Pumpkin seeds are also abundant in bioactive compounds, including phytosterols, tocopherols, and phenolic compounds, which exhibit antioxidant, antimicrobial, and anti-inflammatory properties (Ovca et al., 2011). The seeds contain γ-tocopherol as the predominant vitamin E isomer, with smaller amounts of α-tocopherol (Murkovic et al., 1996). Moreover, pumpkin seeds provide phytonutrients like cucurbitins, lignans, and cardiac glycosides, contributing to their health-promoting effects (Rahman et al., 2013). Their balanced composition of proteins, lipids, and bioactive compounds makes them a promising ingredient for functional foods and nutraceutical applications (Table 4.1).

Sesame Seeds (*Sesamum indicum*)

Sesame seeds (*Sesamum indicum*) are highly nutritious oilseeds valued for their rich composition of proteins, lipids, vitamins, minerals, and bioactive compounds. They contain approximately 21.9% protein and 45%–57% lipids, due to which it is known as "Queen of Oil" (Hu et al., 2007). The protein fraction in sesame is diverse, primarily consisting of globulin, albumin, prolamin, and glutelin, with globulin as the dominant fraction (Morris et al., 2021). Sesame protein is considered complete, with an essential amino acid profile closely resembling human requirements, particularly rich in arginine, leucine, and phenylalanine. The protein digestibility of sesame protein isolate reaches 89.57%, making it suitable for enriching foods, especially in regions where protein deficiency is prevalent (Fasuan et al., 2018). Lipids in sesame seeds are mainly unsaturated fatty acids like oleic acid (18:1) and linoleic acid (18:2), which account for around 80% of the total fat content. These fatty acids play crucial roles in cholesterol metabolism, cardiovascular health, and immune regulation. However, excessive sesame consumption may cause weight gain, gastrointestinal discomfort, or hypotension due to its high unsaturated fat content and omega-3 fatty acids (Ozdemir et al., 2018). Sesame seeds are a good source of vitamins, especially vitamin E (γ-tocopherol), with concentrations reaching 50.4 mg/100 g in black sesame seeds. Additionally, sesame seed also contain vitamins A, B-complex (thiamine, riboflavin, niacin) (Hegde, 2012). Carbohydrates are primarily found in the sesame hull, consisting of hemicellulose, cellulose, and pectic polysaccharides, along with sugars like raffinose and stachyose (Liu et al., 2021). Sesame seeds are also rich in minerals such as calcium, iron, magnesium, phosphorus, and zinc. However, their bioavailability is reduced due to the presence of phytic acid and oxalic acid, which act as antinutrients by binding to minerals and proteins, thereby limiting absorption (Elleuch et al., 2007). Cooking, sprouting, and fermentation processes can significantly reduce these antinutritional factors and enhance mineral availability. The nutritional composition of sesame seeds makes them a valuable functional food with multiple health benefits. However, proper processing methods are recommended to maximise their bioavailability and mitigate potential adverse effects (Table 4.1).

Sunflower (*Helianthus annuus*)

Sunflower seeds are widely cultivated for high oil and protein content and is being recognised as one of the best sources of vegetable proteins, with their nutritional and functional properties explored extensively (Venktesh & Prakash, 1993; Ravindran & Blair, 1992). According to USDA (2008), 100 g of sunflower seeds contain approximately 20.78 g of protein, 51.46 g of total lipids, 3.02 g of ash, 20 g of carbohydrates, and 8.6 g of fibre, providing a total energy value of 2,445 kJ. The chemical composition of whole sunflower seeds includes 5.50% moisture, 18.72% protein, 37.47% crude fat, 28.30% crude fibre, 3.49% ash, and 6.11% carbohydrates. Sunflower cake, a by-product of oil extraction, varies in composition depending on the extent of dehulling. Whole sunflower cake contains 5.80% moisture, 23.60% protein, 11.01% crude fat, 30.18% crude fibre, 5.66% ash, and 23.75% carbohydrates. Partially dehulled sunflower cake shows 5.60% moisture, 25.28% protein, 21.38% crude fibre, 6.89% ash, and 19.34% carbohydrates. In contrast, fully dehulled sunflower cake consists of 7.59% moisture, 29.00% protein, 30.25% crude fat, 8.60% crude fibre, 7.50% ash, and 11.17% carbohydrates (Srilatha & Krishnakumari, 2003). Rodrıguez et al. (1998) reported the chemical composition of hulled full-fat sunflower seeds as 33.2 g/kg moisture, 26.0 g/kg ash, 212.3 g/kg crude protein, 457.1 g/kg ether extract, and 134.0 g/kg crude fibre on a fresh weight basis. Sunflower seeds are also rich in minerals that offer health benefits, including the potential to reduce the risk of certain cancers due to their selenium content (WCRF, 1997). According to USDA (2008), 100 g of sunflower seeds contain 78 mg calcium, 5.25 mg iron, 325 mg magnesium, 660 mg phosphorus, 645 mg potassium, 9 mg sodium, 5 mg zinc, 1.80 mg copper, 1.95 mg manganese, and 53.0 mcg selenium (Table 4.1).

PROTEIN CONTENT AND AMINO ACID PROFILE OF DIFFERENT OILSEEDS

Oilseeds are a great source of high-quality proteins. They are rich in both essential and non-essential amino acids, as shown in Tables 4.2 and 4.3, respectively.

Soybean

Soybean protein is regarded as one of the highest quality plant-based proteins, comparable to proteins from meat, milk, and eggs (Kudełka et al., 2021). Among plant-derived proteins, soybean is recognised for its well-balanced amino acid profile and acknowledged for having a significant biological value. The protein content of soybean ranges between 35% and 40% of dry weight, making it one of the richest sources of plant proteins (Medic et al., 2014). The amino acid composition of soybean protein closely resembles that of animal proteins, particularly in terms of essential amino acids, including phenylalanine, methionine, threonine, valine, isoleucine, leucine, tryptophan, and lysine. According to the data reported, an estimated amount of 8 g leucine, 6.5 g lysine, 5 g valine, 5 g isoleucine, and 4 g phenylalanine is present in 100 g of soybean. However, the concentration of sulphur-containing amino acid is quite low (methionine and cysteine), which makes soybean protein inferior in comparison to other animal proteins that exhibit balanced amino acid composition. Besides poor sulphur-containing amino acid content, soybean protein is nutritionally valued by vegetarians because of higher digestibility and its ability to fulfil human protein needs. About 70% of the total proteins in soybeans are globulins, which include glycinin (11S) and β-conglycinin (7S) as the major fractions. These proteins play an important role in contributing to the functional properties of soybean proteins, such as foaming, emulsification, and gelation (Modgil et al., 2021). In comparison to

TABLE 4.2
Essential Amino Acid Profile of Various Oilseeds (g/100 g Oilseed)

Oilseeds/Amino Acid	Ile	Leu	Lys	Met	Phe	Thr	Trp	Val	Reference
Chia	0.80	1.37	0.98	0.80	1.10	0.76	0.79	0.95	Ziemichód et al. (2019)
Coconut	3.9	7.2	3.5	2.2	5.2	3.4	1.38	5.8	Danish Food Composition Databank (2019)
Hemp	1.29	2.16	1.28	0.93	1.45	1.27	0.37	1.78	Tanwar and Goyal (2021)
Flax	4.5	5.5	2.8	1.9	5.3	3.4	2.0	5.5	U.S Department of Agriculture (2019)
Nigella (black seed)	0.64	1.11	0.63	0.23	0.64	0.63	0.12	0.82	Al-Gaby (1998)
Pumpkin	1.29	2.49	1.36	0.73	1.81	1.04	0.61	1.71	Danish Food Composition Databank (2019)
Rapeseed	3.47	6.19	5.92	1.94	4.06	4.27	1.33	4.97	Tanwar and Goyal (2021)
Sesame	0.9	1.7	0.7	0.7	1.1	0.9	0.2	1.1	Danish Food Composition Databank (2019)
Sunflower	0.92	1.40	0.86	0.53	1.05	0.81	0.35	1.11	Danish Food Composition Databank (2019)
Soybean	4.3	7.2	5.5	1.2	4.6	3.3	1.1	4.4	Kudełka et al.(2021)

Ile, Isoleucine; Leu, Leucine; Lys, Lysine; Met, Methionine; Phe, Phenylalanine; Thr, Threonine; Trp, Tryptophan; Val, Valine.

the normal reference protein, soybeans have an Essential Amino Acid Index (EAAI) of approximately 86%–90% and Protein Digestibility Corrected Amino Acid Score (PDCAAS) of 0.92, showcasing its excellent nutritional value (Burssens et al., 2011). The usage of soybean proteins can be advantageous for plant-based goods, food formulations, and nutritional supplements because of its high protein content, balanced amino acid profile, and functional properties (Chen et al., 2012).

Chia

Chia seeds account for 18%–24% of the total mass as proteins, which makes chia seed a potential plant-based protein source. Chia seeds possess higher protein concentration in contrast to barley, oats, corn, rice, and wheat, which contain approximately 9.2%, 15.3%, 14%, 8.5%, and 14 protein content, respectively. Moreover, chia protein is also

TABLE 4.3
Non-Essential Amino Acid Profile of Various Oilseeds (g/100 g Oilseed)

Oilseeds/ Amino Acid	Ala	Arg	Asp	Cys	Glu	Tyr	Gly	Pro	Ser	His	Reference
Chia	1.04	2.14	1.69	0.53	3.50	0.57	0.94	0.78	1.05	0.53	Ziemichod et al.(2019)
Coconut	4.3	14.3	9.0	1.62	19.2	3.0	4.5	3.7	4.8	2.0	Danish Food Composition Databank (2019)
Hemp	1.28	3.10	2.78	0.41	4.57	0.82	1.14	1.15	1.27	0.71	Tanwar and Goyal (2021)
Flax	4.4	10.8	10.2	1.1	18.5	2.6	4.8	3.8	4.7	2.2	Peng et al. (2022)
Pumpkin	1.5	5.2	3.2	0.39	6.3	1.19	1.8	1.3	1.7	0.8	Danish Food Composition Databank (2019)
Rapeseed	4.36	6.62	7.25	2.29	18.14	2.50	4.92	5.97	4.00	3.39	Tanwar and Goyal (2021)
Sesame	1.1	3.1	2.1	0.43	4.6	0.78	1.2	0.9	1.2	0.6	Ahmed et al. (2020)
Sunflower	1.11	2.40	2.44	0.45	5.57	0.57	1.46	1.18	1.07	0.63	Danish Food Composition Databank (2019)
Soybean	3.8	6.7	10.2	11	16.4	3.3	3.7	4.5	4.6	2.3	Kudełka et al. (2021)

Ala, Alanine; Arg, Arginine; Asp, Aspartic acid; Cys, Cysteine; Glu, Glutamic acid; Tyr, Tyrosine; Gly, Glycine; Pro, Proline; Ser, Serine; His, Histidine.

highly digestible, with a digestion rate of about 78.9%, similar to that of casein (88.6%) and beans (77.5%) and higher than proteins of maize (66.6%), rice (59.4%), and wheat (52.7%) (Sandoval-Oliveros & Paredes-López, 2013; Grobelnik-Mlakar et al., 2009; Betancur-Ancona et al., 2004). The digestibility of chia protein serves as a key indicator of its nutritional quality, potentially linked to its chemical structure, which influences its susceptibility to proteolytic enzyme activity (López et al., 2018). Chia seeds offer a complete amino acid profile, containing all essential amino acids necessary for human nutrition, including isoleucine, leucine, lysine, methionine, phenylalanine, threonine, tryptophan, histidine, and valine (Sandoval-Oliveros & Paredes-López, 2013). Among these, glutamine is the most abundant, while histidine is present in the lowest amount. The chemical score of chia protein meets 100% of the recommended intake for sulphur-containing amino acids, while coverage for other essential amino acids ranges between 52% and 76%. The proportion of essential amino acids in chia seed flour varies from 66% to 126%, with lysine identified as the limiting amino acid due to its lowest coverage of nutritional requirements (Table 4.3) (FAO/WHO/UNU, 2008; Sandoval-Oliveros & Paredes-López, 2013). Their gluten-free nature makes them a suitable option for individuals with celiac disease.

Coconut

Coconut (*Cocos nucifera* L.) is widely known for its oil-rich composition, yet it also provides a significant amount of protein across various parts. The protein content varies, with coconut water containing approximately 0.12%–0.72%, coconut milk ranging from 2% to 4%, copra comprising 8.80%–10.2%, and testa (brown outer thin layer) exhibiting the highest protein concentration, ranging between 9.3% and 32.22% (Mat et al., 2022). Coconut proteins exhibit moderately balanced amino acid compositions with glutamic acid, arginine, and aspartic acid as the dominant amino acids. The predominant protein fractions in coconut endosperm are globulin and albumin accounting for 40% and 21% of the total proteins, respectively (Balachandran et al., 1985). The globulin fraction of coconut protein is rich in charged amino acids, including aspartic acid, glutamic acid, arginine, and lysine, while the albumin fraction contains a high proportion of amino acids with polar side chains (Garcia et al., 2005). These distinct protein fractions influence both the functional and nutritional quality of the protein. Amount or proportion of protein present in coconut meal is further affected by changes in maturation stage, climatic conditions, and the raw materials (Patil & Benjakul, 2017).

Flaxseed

Flaxseed is a great source of protein ranging from 20% to 30% of the total seed mass, the majority of which is present in cotyledons (70%), and the remaining is present in seed coat and endosperm (30%) (Hall et al., 2006; Dev et al., 1986). The essential storage proteins in flaxseed are globulins, which account for roughly 80% of the entire protein, comprising linin and conlinin as the main divisions (Madhusudhan, 2009; Hall et al., 2006). Arginine, aspartic, and glutamic acid are among the dominant amino acids available in flaxseed with lysine as the limiting amino acid (Singh et al., 2011). The significant amounts of cysteine and methionine contributes to flaxseed's antioxidant properties, potentially lowering cancer risk (Oomah & Mazza, 1993). According to the Food and Agriculture Organization (FAO), the protein score of flaxseed meal is 82, surpassing that of soybean meal (Sarwar et al., 1973). The Lysine-to-Arginine ratio of flaxseed protein is 0.37, considerably lower than that of soybean protein (0.88), indicating that flaxseed protein has lower lipidemic and atherogenic potential (Oomah & Mazza, 2000). This balanced amino acid profile makes flaxseed a promising protein source. Flaxseed protein comprises of globulin and albumin as major fractions (linin and conlinin), differentiated by their degree of solubility in salt solution and water. Globulins, which represent 80% of the protein, have a high molecular weight (252–298 kDa), contain 3% α-helical and

17% β-sheet structures, and are particularly rich in sulphur-containing amino acids, branched-chain amino acids, and amide-rich amino acids such as Glu/Gln, Asp/Asn, and Arg (Wu et al., 2019).

Hempseed

Hempseed, derived from industrial hemp, is a high-protein non-legume seed with 20%–25% protein content (Aiello et al., 2016). It provides all essential amino acids in balanced proportions, making it comparable to high-quality proteins like soy and casein (Tang et al., 2006; Deferne & Pate, 1996). The amino acid composition of hempseed meets the FAO/WHO nutritional requirements for 2 to 5 years old children. However, lysine is the first limiting amino acid, followed by leucine and tryptophan (House et al., 2010). Hempseed protein is particularly rich in arginine (12%), significantly higher than in other food proteins like soy, wheat, and egg white (Callaway, 2004). Arginine is a precursor for nitric oxide, which supports blood pressure regulation and cardiovascular health (Wu & Meininger, 2002). The Arg/Lys ratio of hemp protein (3.0–5.5) is much higher than soy (1.41) and casein (0.46), making it a valuable ingredient for cardiovascular health-promoting foods. Hemp protein also contains high levels of sulphur-containing amino acids like cysteine and methionine (3.5%–5.9%), essential for antioxidant functions (Russo & Reggiani, 2013; Erasmus 1999). The protein fractions include globulin (edestin) accounting for 60%–80% and albumin, with globulins being richer in sulphur-containing amino acids, hydrophobic amino acids, and aromatic amino acids. The Arg/Lys ratio of globulins (4.37) is higher than that of albumins (1.74), enhancing their potential for cardiovascular health applications (Malomo & Aluko, 2015a). Hempseed protein hydrolysates, obtained through enzymatic treatment, are known to exhibit antioxidant and hypotensive properties, making them ideal for functional foods and nutraceuticals (Aiello et al., 2016).

Rapeseed

Canola seeds are a valuable oilseed crop containing around 17%–26% protein and 40% oil, while de-oiled canola meal, a by-product of oil extraction, holds up to 50% protein on a dry basis (Uppstrom, 1995). The predominant proteins in canola are cruciferin (60%) and napin (20%), which serve as storage proteins, along with oleosin, a structural protein associated with oil bodies (Ghodsvali et al., 2005; Wu & Muir, 2008). Canola protein is well known for balanced amino acid profile, which includes significant levels of sulphur-containing amino acids like cysteine and methionine along with high amounts of arginine, glutamine, and leucine (Aider & Barbana, 2011). Canola protein is suitable for human nutrition because of these amino acids, which make it comparable to different vegetable proteins. Its greater protein efficiency ratio (2.64) than soybean protein (2019) suggests its better protein utilisation. Furthermore, the FAO/WHO guidelines for essential amino acids—particularly those that contain sulphur—are met by canola protein (Ohlson & Anjou, 1979). However, the inclusion of glucosinolates, phytates, and phenolic compounds—which lead to toxicity, poor digestibility, and unpleasant flavours—restricts its use. Despite these limitations, adequate extraction and purification methods improve their nutritional and functional attributes, positioning rapeseed protein as a valuable ingredient in food formulations.

Pumpkin

Pumpkin seeds have a protein concentration of 24.50 to 32.00 g/100 g in raw seed, and up to 55.40 g/100 g in their defatted seed flour (Kim et al., 2012; Elinge et al., 2012; Aliu et al., 2016; Nyam et al., 2009; Lazos, 1986). According to Sharma and Lakhawat (2017), roasted pumpkin seed flour has a protein level of 29.65%, and oil extraction raises it to 65% in pumpkin seed cake. Cucurbita pepo has the greatest content of arginine (63.99 mg/kg) among the nine essential amino acids found in pumpkin seed proteins (El-Adawy & Taha, 2001; Patel, 2013; Kim et al., 2012). The essential amino acid content of protein isolates is higher (40.92–42.59 g/16 g N) than that of concentrates and meals (36.13 g/16 g N) due to the loss of non-essential amino acids during processing (Mansour et al., 1993). The primary protein fraction is 12S globulins (cucurbitins), comprising 59% of total protein, with each molecule weighing 325 kDa and consisting of six identical subunits (54 kDa) made of two polypeptide chains (33 and 22 kDa) linked by disulphide bonds (Blagrove et al., 1981; Fruhwirth & Hermetter, 2007). Minor fractions include 2S albumins and 18S globulins, which form dimers of 12S components. Pumpkin seed proteins are particularly rich in arginine and glutamic acid, contributing to blood pressure regulation and antioxidant activity, respectively (Chen et al., 2022). However, they are deficient in lysine, making them complementary to sulphur-containing amino acids found in milk and eggs (Chen et al., 2022). The true digestibility coefficient (TD) of pumpkin seed protein is 83.1%, comparable to soybean meal, although its lower protein efficiency ratio (PER: 1.01) can improve with lysine supplementation (Zdunczyk et al., 1999).

Sesame

Sesame seeds (*Sesamum indicum*) are not only renowned for their high oil content but also serve as an excellent source of protein, accounting for 18%–40% of their composition (El-Khier et al., 2008). The primary protein fractions in sesame seeds include albumin (8.6%), globulin (67.3%), prolamin (1.4%), and glutelin (6.9%) (Achouri et al., 2012). Among these, α-globulin and β-globulin are the dominant storage proteins, making up 80%–90% of the total seed protein content (Tai et al., 2001). However, the high proportion of 13S globulin (95%), known for its hydrophobic nature, limits the functional properties of sesame proteins in liquid and beverage formulations (Sharma et al., 2016). Sesame seed protein exhibits a well-balanced combination of essential and non-essential amino acids, particularly rich in sulphur-containing amino acids, valine, and tryptophan, though it is deficient in lysine (Achouri & Boye, 2013). Mehta (2000) reported the amino acid composition (g/100 g) of dehulled sesame seeds, highlighting the presence of arginine (12–13 g), cysteine (2 g), histidine (2.4–2.8 g), isoleucine (3.3–4.9 g), leucine

(6.5–8.9 g), lysine (2.5–3.5 g), methionine (2.0–4 g), phenylalanine (4.2–4.5 g), threonine (3.4–3.6 g), tryptophan (1.9–2.4 g), and valine (4.2–4.5 g). In another study, Maneemegalai and Prasad (2011) analysed commercially available sesame meal and found valine, isoleucine, arginine, aspartic acid, and glutamine in high concentrations. The solubility of sesame protein is influenced by pH, with higher solubility observed in alkaline conditions than in acidic environments, reaching its peak solubility at pH 10. This characteristic makes sesame protein more suitable for certain food applications requiring solubility under alkaline conditions.

Sunflower

Sunflower seeds are widely cultivated as an oilseed crop and are recognised for their nutritional significance, particularly their protein content ranging from 20% to 40% (Sammour et al., 1995). Sunflower seed proteins are composed of approximately 38.33% albumins, 39.04% globulins, 5.54% prolamins, and 17.09% glutelins, making them a valuable protein source (Sammour et al., 1995). The amino acid profile of sunflower proteins is crucial for evaluating their nutritional quality. Sunflower proteins are rich in sulphur-containing amino acids as compared to soybean proteins with around 2.1% sulphur-containing amino acid in sunflower protein and only 1.3% in soybean protein, although they are deficient in lysine, which can be compensated by blending with pulse proteins. Sunflower proteins are less preferred for human consumption primarily due to two factors. First, protein denaturation occurs during the oil extraction process, leading to diminished functionality. Second, the phenolic compounds, especially chlorogenic acid (CGA), bind to proteins and result in discolouration of the final products (Wildermuth et al., 2016). Storage proteins account for approximately 85% of the total sunflower protein, mainly consisting of globulins. These proteins are synthesised during seed development and serve as nitrogen reservoirs during germination (Tai et al., 2001). The Osborne classification system categorises sunflower proteins into albumins, globulins, prolamins, and glutelins based on their solubility properties (Osborne, 1994). Albumins are water-soluble, globulins dissolve in dilute salt solutions, prolamins are soluble in alcohol, and glutelins dissolve in alkaline solutions. According to Venktesh and Prakash (1993), genetic variants, environmental conditions, and extraction techniques all affect the relative quantities of these fractions. Sedimentation coefficients are another way that sunflower proteins are divided into several fractions. The two main protein fractions are sunflower albumins (SFAs), also known as 2S albumins, and 11S globulin (helianthinin) (Sammour et al., 1995). The predominant storage protein is helianthinin, while the lesser storage proteins are the smaller 2S albumins. The pH, ionic strength, and type of buffer all affect how sunflower proteins sediment, producing different amounts of protein fractions (Rodrıguez et al., 1998). The functional uses of sunflower proteins are restricted because of their vulnerability to denaturation and phenolic interactions, notwithstanding their nutritional potential. The use of sunflower proteins in food compositions may be improved by more study on improving protein extraction techniques and lowering CGA-protein interactions.

COMPOSITION OF DIFFERENT OILSEED-BASED FOOD PRODUCTS

Chia Seed

Chia seeds (*Salvia hispanica* L.) are a nutrient-dense oilseed widely used in food products due to their high content of omega-3 fatty acids, proteins, dietary fibre, and bioactive compounds (Ayerza & Coates, 2005). Their diverse applications in food formulations make them a valuable ingredient in functional and health-promoting foods. Whole chia seeds are a storehouse of 30%–35% fat, mainly α-linolenic acid, along with 16%–24% protein and 30%–40% dietary fibre (Ixtaina et al., 2011). Chia seed oil on the other hand primarily composed of unsaturated fatty acids, with omega-3 (55%–65%) and omega-6 (18%–25%), along with natural antioxidants like polyphenols and tocopherols (Capitani et al., 2012). Defatted chia meal is protein-rich (30%–40%) and an excellent source of dietary fibre (50%–60%) and essential minerals (Olivos-Lugo et al., 2010). The nutritional profile of chia-infused baked goods, such as bread, cookies, and crackers, is improved by their higher fibre and protein content, demonstrating the adaptability of chia in a variety of culinary applications (Coelho & Salas-Mellado, 2015). Furthermore, drinks with chia included, including smoothies and dairy substitutes, offer dietary fibre, omega-3 fatty acids, and a modest amount of protein, all of which support general health advantages (Grancieri et al., 2019) (Table 4.4).

Coconut

According to Guarte et al. (1996), the coconut kernel is a very nutrient-dense part of the fruit that is high in proteins, lipids, carbs, amino acids, vitamins, and minerals. The main component is carbohydrate; with dry coconut (copra) containing twice as much of this substance as the wet kernel, making it a substantial source of energy. The proteins in coconut kernels, which include essential, non-essential, and conditionally essential amino acids that are vital for human nutrition, range from 8.1% to 10.2% when dried and 6.2% to 7.5% when fresh (Appaiah et al., 2014). Among these, leucine, lysine, and valine are essential, while alanine and glutamic acid fall under non-essential amino acids. Arginine and cysteine are considered conditionally essential, playing a role in metabolic processes. Lipids are the most abundant macronutrient in coconut kernel, with copra containing 59.0%–63.6% fat, whereas the wet kernel has a lower fat content of 34.7%–38.8%. Copra white kernel has the highest fat concentration. Lauric acid is the dominant fatty acid in coconut oil, along with varying levels of myristic, palmitic, oleic, and linoleic acids (Bhatnagar et al., 2009). The coconut kernel also provides essential vitamins and minerals. Among water-soluble vitamins, ascorbic acid (Vitamin C) is most abundant (3.3 mg/100 g), along with niacin, pantothenic acid, thiamine, pyridoxine, folate, and riboflavin. On the hand, it contains α-tocopherol (Vitamin E) and phylloquinone (Vitamin K) as oil-soluble vitamins. Additionally, coconut kernel is a rich source of potassium (107.8–124.1 mg/100 g), followed by sodium (15.5–29.8 mg), calcium (14.0–18.1 mg), iron (1.5–7.9 mg), and zinc (1.6–3.0 mg), making it a valuable dietary component (Table 4.5). Overall, coconut kernel offers a balanced profile of macronutrients and micronutrients, supporting its role as a nutritious food ingredient (Appaiah et al., 2014).

TABLE 4.4
Composition of Different Oilseeds-Based Food Products from Chia Seeds

Food Product	Moisture (%)	Protein (%)	Fat (%)	Carbohydrates (%)	Dietary Fibre (%)	Other Components	References
Chia seeds (whole)	5–7	16–24	30–35	40–45	30–40	Ash: 4%–5%	Ayerza and Coates (2005); Ixtaina et al. (2011)
Chia oil	–	–	98–99	–	–	Omega-3(ALA): 55%–65%; Omega-6(LA): 18%–25%; Tocopherols: 15–30mg/100g; Polyphenols: 500–700mg/kg	Capitani et al. (2012); Munoz et al.(2013)
Defatted meal	5–8	30–40	5–8	35–45	50–60	Minerals: 4%–6% (Ca, Mg, Fe, Zn)	Ixtaina et al. (2011); Olivos-Lugo et al.(2010)
Bakery products	8–12	8–12	5–15	50–60	10–20	Ash: 2%–4%	Coelho and Salas-Mellado (2015); Steffolani et al.(2014)
Chia-added beverages	85–90	2–5	3–6	5–10	2–6	Omega-3: 500–800mg/serving	Grancieri et al. (2019); Capitani et al. (2012)

Rapeseed

Rapeseed is a most commonly grown oilseed that is used for the production of edible oil, protein meals, and other food products (Shahidi, 1990). It is especially valued for its balanced protein profile, high oleic and linoleic acid content, and bioactive substances including phytosterols and tocopherols (Przybylski, 2005). In addition to 18%–24% protein and glucosinolates, which have both antinutritional and nutritional advantages, whole rapeseeds contain 40%–45% oil, which is high in monounsaturated and polyunsaturated fatty acids (Tan et al., 2011). Rapeseed oil, often known as canola oil, is a low-saturated fat oil that has a high percentage of oleic acid (55%–65%) and linoleic acid (20%–30%). It possesses heart-healthy promoting benefits since it includes tocopherols, phytosterols, and omega-3 fatty acids (8%–12%). A by-product of oil extraction, rapeseed meal (defatted flour) has 35%–40% protein and 30%–40% fibre, which makes it suitable for use in animal feed. According to Wanasundara (2011), rapeseed protein isolate is a high-protein product that contains 85%–90% protein and has a good amino acid profile which is suitable for the

TABLE 4.5
Composition of Different Oilseeds-Based Food Products from Coconut

Sample	Moisture (%)	Fat (%)	Protein (%)	Carbohydrates (%)	Crude Fibre (%)	Ash (%)	Reference
Copra whole	4.3	59.8	10.2	24.3	7	1.4	Appaiah et al. (2014)
Copra white kernel	3.8	63.6	8.1	22.4	6.6	2.1	Solangi and Iqbal (2011)
Copra testa	4	59	9.3	26.3	11.6	1.4	Appaiah et al. (2014)
Wet coconut whole	42.2	37	7.5	12.3	14.3	1	Appaiah et al. (2014)
Wet coconut white kernel	43.5	38.8	6.2	10.6	11.7	0.9	Appaiah et al. (2014)
Wet coconut testa	32.9	34.7	7.1	24.6	17.2	0.7	Appaiah et al. (2014)

development of plant-based protein beverages and meat substitutes. Apart from these applications, rapeseed incorporated bread, biscuits, and crackers tend to provide better protein quality, increased fibre content, and a more balanced omega fatty acid profile. Similar to this, rapeseed-based drinks, including protein shakes and dairy substitutes, offer a plant-based protein source in addition to antioxidants and polyphenols, which makes them a great option for customers who are concerned about their health (Aider & Barbana, 2011) (Table 4.6).

Flaxseed

Flaxseed (*Linum usitatissimum* L.) is a nutrient-dense oilseed valued for its high content of α-linolenic acid, dietary fibre, lignans, and protein (Kajla et al., 2023). It is widely incorporated into various food products due to its benefits for cardiovascular health, digestion, and functional food applications. Below is an overview of different flaxseed-based food products and their compositions (Table 4.7). Whole flaxseeds are rich in fat (35%–45%), with α-linolenic acid comprising 50%–60% of the total fat. They also contain 18%–25% protein and 25%–35% dietary fibre, along with lignans that exhibit antioxidant and oestrogenic properties (Oomah, 2001). Flaxseed oil consists almost entirely of fat (98%–99%), with a high proportion of α-linolenic acid (50%–60%) and essential polyunsaturated fatty acids, making it a key ingredient in dietary supplements and functional foods (Coskuner & Karababa, 2007). Flaxseed Meal (defatted flour) is a by-product of oil extraction, offering 30%–40% protein, 40%–50% fibre, and essential minerals like calcium and magnesium (Ciftci et al., 2012). Flaxseed-enriched bakery products enhance the omega-3 and fibre content of baked goods such as bread, cookies, and crackers, while also improving antioxidant properties (Ganorkar & Jain, 2013). Flaxseed-based functional beverages provide dietary fibre, α-linolenic acid, and polyphenols, supporting heart health and digestive well-being (Kajla et al., 2023).

TABLE 4.6
Composition of Different Oilseeds-Based Food Products from Rapeseeds

Food Product	Moisture (%)	Protein (%)	Fat (%)	Carbohydrates (%)	Dietary Fibre (%)	Other Components	References
Whole rapeseeds	5–8	18–24	40–45	20–30	10–15	Ash: 3%–5%; Rich in tocopherols (~20–40mg/100g), glucosinolates	Khattab et al. (2010); Wanasundara (2011)
Rapeseed oil	–	–	98–99	–	–	Oleic acid: 55%–65%; Linoleic acid: 20%–30%; Omega-3: 8%–12%; Phytosterols	Shahidi (2013); Przybylski (2005)
Rapeseed defatted flour	5–8	35–40	3–7	30–40	30–40	High in glucosinolates, polyphenols, and minerals (Ca, Mg, Zn)	Khattab et al. (2010); Tan et al. (2011)
Rapeseed protein isolate	5–7	85–90	2–4	2–5	–	High lysine content, excellent digestibility	Wanasundara (2011); Tan et al. (2011)
Rapeseed-enriched bakery products	8–12	8–12	5–15	50–60	10–20	Improved fibre, antioxidants, and balanced omega fatty acid profile	Niu et al. (2014); Aider and Barbana (2011)
Rapeseed-based beverages	85–90	5–8	2–6	5–10	2–5	Rich in plant-based protein, antioxidants, and polyphenols	Aider and Barbana (2011); Shahidi (2013)

Sunflower

The proximate composition of whole sunflower seeds, sunflower cooking oil, and deoiled meal underscores their nutritional value and industrial significance (Le Clef & Kemper, 2015). Sunflower cooking oil has been traditionally favoured for its oxidative stability, attributed to its low α-linolenic acid content (<1%). Additionally, its high levels of γ- and δ-tocopherols, compared to α- and β-tocopherols, enhance its

TABLE 4.7
Composition of Different Oilseeds-Based Food Products from Flax Seeds

Food Product	Moisture (%)	Protein (%)	Fat (%)	Carbohydrates (%)	Dietary Fibre (%)	Other Components	References
Flax seeds (whole)	5–8	18–25	35–45	25–30	25–35	Ash: 2%–4%; Rich in lignans and antioxidants	Oomah, (2001); Kajla et al. (2023)
Flaxseed oil	–	–	98–99	–	–	ALA: 50%–60%; Linoleic acid: 15%–25%; Tocopherols: 20–40mg/100 g	Coskuner and Karababa (2007); Morris (2007)
Flaxseed Meal (Defatted Flour)	5–9	30–40	5–8	35–45	40–50	High in lignans, calcium, magnesium, and iron	Morris (2007); Ciftci et al. (2012)
Flaxseed-Enriched Bakery Products	8–12	8–14	5–12	50–60	10–25	Improved omega-3content, fibre, and antioxidant properties	Ganorkar and Jain (2013); Wolever et al.(2011)
Flaxseed-Based Functional Beverages	85–90	2–5	3–7	5–12	2–6	Contains ALA, dietary fibre, and polyphenols	(Ciftci et al., 2012; Kajla et al., 2023)

stability during high-temperature processing (Tanwar & Goyal, 2021). Whole sunflower seeds contain 4.73%–6.7% moisture, which increases to 11%–12% in deoiled meal due to processing conditions. The protein content varies significantly, ranging from 14%–20.78% in whole seeds, while deoiled meal, a concentrated protein source, contains 38%–40%. Sunflower seeds are naturally rich in fat, with a lipid content of 40%–51.46%, which is drastically reduced to 0.5% in deoiled meal following oil extraction (Petraru et al., 2021). Carbohydrates make up approximately 20% of whole seeds, though specific values for deoiled meal remain unreported. Crude fibre, essential for digestion, ranges from 8.6%–20% in whole seeds and increases to 14%–17% in deoiled meal. Ash content, reflecting the mineral composition, is 3.02%–3.5% in whole seeds and rises to 6% in deoiled meal (De Oliveira et al., 2021). Due to its high protein and fibre content, deoiled sunflower meal serves as a valuable by-product, particularly in animal feed formulations (Table 4.8).

Soybean

Soybean is widely utilised for human welfare due to its diverse range of products. Its primary product is soybean oil, which is extracted using solvent extraction. This oil is predominantly composed of triglycerides, along with phosphatides, free fatty acids,

TABLE 4.8
Composition of Different Oilseeds-Based Food Products from Sunflower

Food Product	Moisture (%)	Protein (%)	Fat (%)	Carbohydrates (%)	Crude Fibres (%)	Ash (%)	Reference
Whole seeds/ kernels	4.73	20.78	51.46	20.00	8.6	3.02	U.S Department of Agriculture (2019)
Whole seeds/ kernels	6.5–6.7	14–18	40–45	ND	13–20	3.5	Le Clef and Kemper (2015)
Deoiled meal (fully dehulled)	11–12	38–40	0.5	ND	14–17	6	Le Clef and Kemper (2015)

and essential minerals such as iron and copper (Perkins, 1995). The remaining soybean meal, a by-product of oil extraction, serves as a valuable ingredient in various applications, including food, animal feed, and industrial products, as illustrated in Table 4.9.

CLASSIFICATION OF OILSEED PROTEINS

Based on the biological functions, seed proteins can be divided into three categories: storage proteins, structural proteins, and physiologically active proteins. When compared to storage proteins, biologically active proteins, such as lectins, enzymes, and enzyme inhibitors, are found in a smaller quantity but contain a more regulated amino acid, which makes them important for nutrition. Storage proteins' main function during seed germination is to serve as a reservoir of nutrients, providing essential proteins for the developing plant. Seed proteins can be classified into four main types based on their solubility characteristics as: albumins (soluble in water), globulins (soluble in salt), prolamins (soluble in alcohol), and glutelins (soluble in diluted acids, alkalis, or detergents like SDS) (Osborne, 1994). Furthermore, seed proteins can also be categorised based on sedimentation factors as proposed by Mandal and Mandal (2000), providing further insights into their structural and functional properties.

Classification Based on Solubility

A commonly used classification of categorising seed proteins based on how soluble they are in different solvents is the Osborne classification. Thomas Burr Osborne developed the solubility-based categorisation system in 1924, which separates proteins into four major fractions: albumin, globulin, prolamin, and glutelin:

Albumin: Albumins are the water-soluble proteins present in plants that are involved in performing various biological functions such as defence mechanism and enzymatic activity but are found in a smaller quantity than storage

TABLE 4.9
Composition of Different Oilseeds-Based Food Products from Soybean (100g of Soybean) (Perkins, 1995)

Soy Food Products	Water (g)	Protein (g)	Fat (g)	Carbohydrate (g)	Energy (Kcal)	Crude Fibre (G)	Riboflavin (mg)	Thiamin (mg)	Niacin (mg)	Calcium (mg)	Zinc (mg)	Iron (mg)
Soybean	8.5	36.5	19.9	30.2	416	5.0	0.87	0.87	1.62	277	4.89	15.70
Soy protein isolate	5.0	80.7	3.4	7.4	338	0.3	0.10	0.18	1.44	178	4.03	14.50
Soy protein concentrate	5.8	58.1	0.5	31.2	332	3.8	0.14	0.32	0.72	363	4.40	10.78
Soy sauce	66.0	10.5	0.1	5.6	60	0.0	0.15	0.06	3.95	20	0.43	2.38
Defatted soy flour	7.3	47.0	1.2	38.4	329	4.3	0.25	0.70	2.61	241	2.46	9.24
Full-fat soy flour	5.2	34.5	20.6	35.2	436	4.7	1.16	0.58	4.32	6.37	3.92	6.37
Tempeh	55.0	19.0	7.7	17.0	199	3.0	0.11	0.13	4.63	93	1.81	2.26
Natto	55.0	17.7	11.0	14.4	212	1.6	0.19	0.16	0.00	217	3.03	8.60
Miso	41.5	11.8	6.1	28.0	206	2.5	0.25	0.10	0.86	66	3.32	2.74

protein (Arrutia et al., 2020). However, albumins account for up to 50% of the overall protein content in other oilseeds, such as rapeseed, whereas they only make approximately 10% to 20% in sunflower and soybean (Chéreau et al., 2016). Albumin proteins available in different oilseed are: lectins and enzyme inhibitors in soybeans, defensin protein in sunflower, and napin protein in rapeseed.

Globulin: Globulins are proteins that dissolve in diluted salt solutions and precipitate when diluted or conducted by dialysis. They are the main storage proteins found in sunflower seeds and soybeans, as well as other oilseeds. The main types of globulins in soybeans are 7S (β-conglycinin) and 11S (glycinin), which contribute to an important proportion of the overall protein content (Chéreau et al., 2016). The nutritional and functional properties of food products made from oilseeds are dependent on those proteins, which act as nitrogen reserves for seed germination (Nishinari et al., 2014). As an example, sunflower involves helianthinin, rapeseed has cruciferin, and soybeans have glycinin (11S) and β-conglycinin (7S).

Prolamin: Prolamins are alcohol soluble proteins, found largely in cereals (wheat, barley, and maize). Prolamins do not make up a significant proportion of the protein content in oilseeds due to their minor quantities. According to Shewry et al. (1995), prolamins make up a tiny protein fraction in oilseeds such as sunflower and rapeseed, accounting for just 5% of their total protein composition.

Glutelin: Glutelins function as storage proteins in oilseeds and are soluble in both acidic and alkaline ($pH > 11$) conditions. They are less prevalent in oilseeds but more abundant in grains. For example, sunflower and rapeseed contain 15% and 10% glutelins, respectively, while peas and faba beans contain around 15%–20%. These proteins contribute to the structural integrity of seeds during development (Shewry et al., 1995; Chéreau et al., 2016).

Classification Based on Sedimentation Coefficient

2S Albumins: 2S albumins are widely present in dicot seeds and have been thoroughly investigated in the Cruciferae family, particularly in oilseed rape, where they are known as napins. These proteins consist of two polypeptide chains, with molecular weights of around 9,000 and 4,000 Da, connected through interchain disulphide bonds (Ericson et al., 1986). Initially produced as a single precursor protein, they undergo proteolytic cleavage, which removes a linker peptide along with small segments from both the N-terminal and C-terminal regions. This structural form is considered the typical configuration of 2S albumins (Shewry et al., 1995). Different species exhibit variant forms of 2S albumins. In sunflower seeds, these proteins persist as a single, uncleaved polypeptide chain and in peas, they lack interchain disulphide bonds (Kortt & Caldwell, 1990). In spite of the differences in subunit structure, all 2S albumins possess a secure globular structure with conserved cysteine residues, which is essential for their structural stability (Hara-Hishimura et al., 1993; Shewry et al., 1995).

7S Globulins: 7S proteins, generally referred to as vicilin protein, represent an important category of storage proteins in legumes and oilseeds. These proteins serves as nutrition reserve during seed germination and in the initial phases of plant growth, which makes them essential to the plant life cycle. Their solubility in diluted saline solutions puts them under the category of salt-soluble globulins. Helianthinin in sunflower seeds, vicilin in peas and faba beans, and β-conglycinin in soybeans are some of the well-studied vicilins found in many plant species. 7S proteins have structural differences among species, yet they also have similarities. With a molecular weight of 300–350 KDa, they typically form trimeric complexes of proteins consisting of three subunits free of interchain disulphide linkages, which sets them apart structurally from other globulins like 11S legumins (Gatehouse et al., 1984; Casey et al., 1986). The high protein concentration and essential amino acid content make them nutritionally relevant, although they lack sulphur-containing amino acids like cysteine and methionine. This limits their nutritional value use, particularly when included into human diets (Chéreau et al., 2016). However, they are extremely beneficial in food formulations due to their abundance and capacity to form stable protein structures, which enhance the texture, emulsification, and water-binding qualities of protein-based products.

11S Globulins: In oilseeds and legumes, the 11S globulins—also referred to as legumins—are one of the primary storage proteins. Being an essential source of amino acids, these proteins also contribute in seed germination and the early stages of plant growth. Similar to 7S vicilin, they are classified as salt-soluble globulins, although they differ in structure and nutrition. Oilseeds including soybeans (Glycinin), rapeseed (Cruciferin), and sunflower seeds are the main source of 11S globulins. They produce hexameric protein complexes, each of which is made up of two trimers of six subunits that connect non-covalently (Chéreau et al., 2016). Compared to 7S vicilin, the protein structure is more compact and stable because each of these subunit pairs are made up of an acidic subunit of MW ~ 40,000 and a basic subunit of MW ~ 20,000 linked by interchain disulphide interactions (Gatehouse et al., 1984). Disulphide bonds are another factor that contributes to their gel-forming and heat stability. When it comes to nutrition, 11S globulins provide a better ratio of important amino acids than 7S vicilin, especially because they include more sulphur-containing amino acids like cysteine and methionine. As a result, they are more acceptable for human consumption and become increasingly important in plant-based protein diets. But when compared to animal proteins, their lysine content is still low (Duranti et al., 1988). Because of their superior gelation, emulsification, and water-holding properties, 11S globulins are useful in food applications such as plant-based substitutes for milk, protein concentrates, and substitutes for meat. Their exceptional textural qualities and capacity for generating heat-induced gels further strengthen their use in the development of food products. Furthermore, the functional and nutritional quality of protein isolates is strongly impacted by the ratio of 7S to 11S proteins in oilseeds

(Thanh & Shibasaki, 1979). To formulations for food, it is important to balance both protein types because a larger concentration of 11S globulins usually produces increased gelation qualities but can also hinder emulsification ability. 11S globulins are a major focus of study on plant-based protein substitutes and their uses in the food and pharmaceuticals sectors because of their significance in nutrition and function.

STRUCTURE OF OILSEED PROTEINS

The molecular structure of oilseed protein is described by the arrangement of amino acids, which will influence the functional properties, nutritional value, and behaviour in different food systems. Depending on their type and composition, oilseed proteins have different structural diversity. The four hierarchical levels of primary, secondary, tertiary, and quaternary structures are commonly used to categorise these proteins.

Primary Structure

The basic primary structure of protein is a distinct linear arrangement of amino acids joined together by peptide bonds to form a polypeptide chain. In a sequential process of reacting carboxyl group (-COOH) of one amino acid with the amino group ($-NH_2$) of another amino acid, one molecule of water is released. The formation of a bond between carbonyl carbon (C=O) of one amino acid and the nitrogen (N) of other amino acid is known as peptide bond as shown in Equation 4.1.

$$H_2N-CH(R_1)-COOH+H_2N-CH(R_2)-COOH \rightarrow H_2N-CH(R_1)-CO-NH-CH(R_2)-COOH+H \quad (4.1)$$

This sequence is essential since it determines ultimate three-dimensional shape and function of protein.

The covalent bond forms between the sulphur atoms of two different cysteine residues, which stabilises the proteins structure by linking different parts of the polypeptide chain or even connecting separate polypeptide chains. However, the quantity and arrangement of amino acid residues in each protein are unique. For example, sesame and sunflower, have a higher methionine content, making them valuable in protein nutrition. However, certain oilseeds, such as flax and rapeseed, may exhibit deficiencies in lysine or sulphur-containing amino acids, which can impact their overall protein quality (Lehninger, 2004).

Secondary Structure

Protein secondary structure implies to the local, and repetitive folding patterns of the polypeptide chain, which are stabilised by hydrogen bonds. α-Helix, β sheets, and β turns are the most commonly identified types of secondary structures.

With a right-handed coiled shape, the α-Helix is one of the common secondary structure in proteins. The inner side of the helix is formed by the tightly curled polypeptide backbone, while the side chains extend outward from the helical axis. The α-Helix tends to be stable because of the hydrogen bonds formed by the oxygen in the CO group of the fourth residue and the hydrogen in the NH group. Each helix has about 3.6 amino acid residues per turn, with a helical pitch (distance along the axis per turn) of 0.54 nm. This regular hydrogen bonding pattern helps to make the helix more stable overall (Dunitz, 2001).

β-Sheets structure on the other hand also represents secondary structure in proteins, where two or more extended polypeptide chains align side by side. Hydrogen bonds form between the –NH group of one peptide chain and the C=O group of an adjacent chain. In β-sheets, all peptide chains are interconnected through hydrogen bonds, which are nearly perpendicular to the chain's long axis. This structure displays repeating units along the length of the chain (Richardson, 1981).

β-Sheets can be classified into two types:

- **Parallel β-Sheets:** Polypeptide chains run in the same direction, with their N-terminal aligned.
- **Antiparallel β-Sheets:** Neighbouring chains run in opposite directions, with alternating N-terminal and C-terminal.

Polypeptide regions that do not adopt α-helix or β-sheet conformations are known as random coils. Together with α-helices and β-sheets, these asymmetrical structures are commonly seen in globular proteins. The β-turn, a crucial component of random coils, aids in the formation of compact protein structures and encourages chain folding (Chothia, 1984).

β-Turn, also known as hairpin structure, consists of four consecutive amino acid residues. This structure allows the polypeptide backbone to fold back nearly by 180°. The turn is stabilised when a hydrogen bond is formed between the amide hydrogen (N–H) of the fourth residue and the carbonyl oxygen (C=O) of the first residue. The dihedral angles of the second and third amino acid residues are the main factors that govern the conformation of the β-turn (Pauling & Corey, 1951).

Tertiary Structure

A number of interactions, such as ionic, disulphide, and hydrophobic bonds, maintain the three-dimensional structure of the polypeptide chain in oilseed storage proteins. Compact globular conformations are adopted by proteins like albumins and globulins, which enhance their functionality. These proteins are useful in culinary applications because of their hydrophobic core and surface-exposed hydrophilic residues, which improve solubility and emulsifying qualities. Furthermore, disulphide linkages are essential for preserving structural stability, especially in proteins present in sunflower and sesame seeds (Sim et al., 2021).

Quaternary Structure

Proteins found in oilseeds usually form functional complexes made up of many polypeptide subunits. The majority of these proteins are categorised as either albumins or globulins (11S and 7S) because they form oligomeric structures. Globulins, the main storage proteins found in oilseeds including rapeseed, pumpkin, and hemp, come in two main forms: 7S proteins (vicilin-like), which have a trimeric structure, and 11S proteins (legumin-like), which have a hexameric structure. On the other hand, albumins are water-soluble, smaller proteins that contain essential amino acids. According to Lampart-Szczapa (2001), the oligomeric structure of these proteins is essential for improving their stability and functionality. It also contributes to characteristics like gel formation and enhanced structural integrity in food applications.

Structure of Rapeseed Proteins

Cruciferin: Cruciferin, a key storage protein found in canola (*Brassica napus*) seeds, is a member of cupin superfamily, which makes up to 60% of the total protein content in matures seeds. (Hoglund et al., 1992). Similar to other 11S or 12S globulins, cruciferin exhibits a hierarchical structural organisation, encompassing primary, secondary, tertiary, and quaternary structures. Glycinin, the predominant 11S storage protein in soybean (Glycine max), shares significant structural similarity with cruciferin, with its three-dimensional structure extensively studied. Cruciferin (~300 kDa) adopts a hexameric quaternary structure composed of six subunits or protomers (Schwenke et al., 1983; Tandang-Silvas et al., 2010). Due to the involvement of multiple genes in its expression, the subunits that form the tertiary structure may exhibit slight variations. In *B. napus*, the primary sequence of cruciferin ranges from 465 to 509 amino acid residues depending on the gene expressed, resulting in the identification of five distinct subunits: CRU1, CRU2, CRU3, CRUA, and CRU5 (Wanasundara, 2011; Sjödahl et al., 1991; Withana-Gamage, 2013). Each protomer consists of two polypeptide chains: an acidic α-chain and a basic β-chain. Structural modelling of rapeseed pro-cruciferin, based on X-ray diffraction data, revealed 25–27 β-sheets, seven α-helices, and three to four 310-helices, resembling the A3B4 protomer structure of soybean glycinin (Tandang-Silvas et al., 2010). Three cruciferin subunits assemble into trimers, which further stack to form hexamers. Recent work by Nietzel et al. (2013) suggested that in its native state within protein storage vacuoles (PSVs), cruciferin may form an octameric barrel-shaped complex (~420 kDa). 11S cruciferin proteins accumulate within the same PSV in cruciferous plants however, the interactions between these proteins during storage or complex formation remain unclear (Gruis et al., 2002; Otegui et al., 2006).

Napin: Napin is a storage protein found in *Brassica napus* (canola/rapeseed) and belongs to the 2S albumin protein family. It plays a crucial role in providing amino acids during seed germination. Similar to cruciferin, the expression of napin in *Brassica napus* is controlled by multiple genes, with 10–16 different napin-encoding genes identified (Shewry et al., 1995). Various isoforms have been reported, including Napin-1, Napin-2, Napin-3, Napin-1A, Napin-B, and Nap1, with molecular masses ranging from 12.5 to 14.5 kDa.

The mature napin protein consists of two polypeptide chains: a small chain (~4 kDa) and a large chain (~9 kDa), connected by two interchain disulphide bonds. Additionally, the large chain contains two intra-chain disulphide bonds (Rico et al., 1996).

Structure of Soybean Proteins

Soybean contains four main protein fractions referred to as 2S, 7S, 11S, and 15S fractions, each of which possesses unique physicochemical, nutritional, and functional characteristics. About 40% of the total protein content of soybeans is made up of the 7S fraction (β-conglycinin), which has a molecular weight of 150–200 kDa. This is smaller than the 11S fraction (glycinin), which makes up about 25% of soybean proteins and has a molecular weight of 350 kDa (Staswick et al., 1981; Thanh & Shibasaki, 1979). Three acidic chains (~35 kDa) and three basic chains (~20 kDa) make up the six polypeptides that build up the quaternary structure of the 11S protein (glycinin) (Tay et al., 2006). On the other hand, the quaternary structure of the 7S protein (β-conglycinin) is made up of three polypeptides: α′ (60 kDa), α (67 kDa), and β (71 kDa), together forming a quaternary structure (Thanh & Shibasaki, 1979).

Structure of Peanut Proteins

Arachin and conarachin are the two main storage proteins found in peanut seeds that accounts for around 55% and 44.56%, respectively, of the total protein. However, an analysis of 40 peanut varieties revealed a considerable variation in protein composition, with arachin ranging between 46.4% and 62.3%, while conarachin content varied from 37.3% to 53.5%. Conarachin proteins consist of two subtypes, type I and type II polypeptides, with average contents of 25.99% and 18.6%, respectively. According to SDS-PAGE analysis under reducing conditions, arachin comprises four primary polypeptides of 24, 36, 38, and 41 kDa, while conarachin is composed of 16, 17, 18, and 61 kDa polypeptides (Wang et al., 2014a).

Structure of Sunflower Proteins

Sunflower seeds primarily contain two major proteins: salt-soluble globulins (helianthinin) and water-soluble albumins, which make up approximately 50%–70% and 20%–40% of the total protein content, respectively, depending on the variety and growing conditions (Gonzalez-Perez & Vereijken, 2007; Guéguen et al., 1996). Helianthinin belongs to the 11S protein family, exhibiting a typical hexameric quaternary structure with a molecular mass of 300–350 kDa. It consists of a disulphide-linked acidic β polypeptide (32–44 kDa) and a basic polypeptide (21–27 kDa) (Schwenke et al., 1979; Dalgalarrondo et al., 1984). Meanwhile, albumins are classified as 2S storage proteins, comprising a single polypeptide chain (10–18 kDa) with one intramolecular disulphide bond, and are rich in methionine and cysteine (Slabi et al., 2020; Youle & Huang, 1981).

PROPERTIES OF OILSEED PROTEINS

Soybean

Surface Hydrophobicity: The 7S (β-conglycinin) fraction exhibits higher surface hydrophobicity compared to the 11S (glycinin) fraction. This characteristic positively affects the binding of polyphenolic compounds and improves emulsion properties (Diarrassouba et al., 2013; Jia et al., 2022; Fukushima, 2001). Additionally, surface hydrophobicity has been shown to correlate positively with the emulsification properties of soybean protein isolate (SPI) (Yan et al., 2021).

Thermal Properties: Different soybean protein fractions have significantly different thermal characteristics. The denaturation temperature (95.6°C) and enthalpy (3.91 J/g) of the 11S (glycinin) fraction are greater, suggesting a more tightly packed and organised quaternary structure with stronger protein–protein interactions. The large concentration of disulphide bonds is responsible for this stability. As an indication of a less stable structure, the 7S (β-conglycinin) fraction shows a lower denaturation temperature (74°C) and enthalpy (0.67 J/g) (Zhu et al., 2022).

Nutritional Quality: Nutritional differences between the two fractions are distinguishable such that 11S (glycinin) fraction contains 3–4 times higher methionine and cysteine as compared to the 7S (β-conglycinin) fraction, which is low in these essential amino acids. The higher content of methionine and cysteine contribute to the rich nutritional profile of glycinin (Chmielewska et al., 2021).

Gelation Properties: Gelling ability of soybean proteins is affected by the ratio of 11S to 7S fractions which signifies that a higher 11S:7S ratio (1.6–2.5) will improve gelation properties (Malaki Nik et al., 2011; Yagasaki et al., 2000). However, the lack of 11S subunit in glycinin results in harder gels with increased water-holding capacity.

Foaming Properties: The composition of soybean proteins subunits determines their foaming attributes. When the 7S (β-conglycinin) fraction lacks both α and α′ subunits, it becomes less soluble and has a reduced ability to foam. In contrast to the wild type, the α subunit presumably plays a crucial role in foam production because its removal alone increases solubility and foaming qualities. Surface charge intensity and hydrophobicity have a positive correlation with foaming capability (Li et al., 2022).

Solubility: Soybean protein solubility is pH-dependent, reaching its minimum at pH 4–5 (the isoelectric point) and increasing above and below this range. Higher solubility is observed at pH >7 compared to pH <4. Solubility values are approximately 50% at pH 3, 80% at pH 7, and **92%** at pH 9. Despite similar surface charge values at pH 3 and pH 7, the higher solubility at neutral pH suggests that negatively charged residues promote stronger interactions with the hydrophilic environment (Ge et al., 2021).

Water- and Oil-Holding Capacity: Water-holding capacity of SPI is higher at pH 3 and pH 7 than at pH 5, corresponding to solubility patterns. However, oil-holding capacity is highest at pH 3, consistent with solubility trends at acidic pH (Ge et al., 2021).

Emulsifying Properties: Emulsifying properties of SPI are also pH-dependent. The emulsifying activity index remains similar (9–10 m^2/g) across pH 3, 7, and 9. However, emulsion stability is highest at pH 9, likely due to higher surface charge and stronger repulsion between oil droplets (Ge et al., 2021).

Gel Formation: The gelation capacity of SPI improves with increasing pH. Higher protein concentrations are required for gelation at pH 3 (12%) compared to pH 7 (10%) and pH 9 (8%). The enhanced gelling ability at higher pH values is attributed to increased solubility, better hydration, and greater structural flexibility of the polypeptides (Ge et al., 2021).

Rapeseed Protein

Thermal Properties: Cruciferin displays a denaturation temperature of 90.7°C and enthalpy of 12.5 J/g, while napin exhibits higher values (109.9°C and 15.9 J/g, respectively). This indicates that napin has a more conserved, folded structure than cruciferin. The addition of mercaptoethanol reduces both temperature and enthalpy of denaturation for both proteins, confirming the dissociation of oligomeric protein structures by converting disulphide bonds into sulfhydryl groups (Wu & Muir, 2008). Additionally, the presence of NaCl increases the thermal denaturation temperature of canola protein isolates, which delays structure development and weakens protein network formation (Léger & Arntfield, 1993).

Emulsification Properties: Cruciferin forms superior emulsions with a mean oil droplet size of 1.4 μm, compared to 26.5 μm for napin. The emulsion stability of cruciferin (98%) is significantly higher than that of napin (77%), reflecting its more flexible structure, which facilitates better unfolding and encapsulation of oil droplets (Wu & Muir, 2008). Canola protein isolates produced by electro-activated protein isolation (EAPI) generate smaller oil droplet sizes at acidic and alkaline pH values than those produced by pI precipitation, indicating less structural damage during electro-activated extraction (Gerzhova et al., 2015).

Solubility: Canola protein solubility follows a U-shaped pattern, with minimum values at pH 4–5 and increased solubility at acidic and alkaline pH. Salt-extracted isolates show higher solubility (≥94%) at pH 3 compared to pI-precipitated isolates, confirming that pI precipitation induces protein aggregation and reduces solubility (Li et al., 2023). The presence of lipids in cold-pressed canola meals interferes with protein dissolution during alkaline extraction, while salt extraction produces higher solubility by dissolving canola globulins without emulsifying lipids (Li et al., 2023).

Water and Fat Absorption Capacity: Water absorption capacity is similar between EAPI (1.25 g/g) and conventional protein isolate (CPI) (1.30 g/g), while fat absorption capacity is 0.97 g/g for EAPI and 1.11 g/g for CPI. These results suggest that both extraction methods expose similar amounts of hydrophilic and hydrophobic groups (Gerzhova et al., 2015).

Foaming Properties: Foaming capacity increases with pH, with EAPI showing higher values than CPI at pH 4 (28.92 vs. 19.62%), pH 7 (60.46 vs.

57.83%), and pH 9 (73.64 vs. 66.91%). Higher negative charge density in EAPI proteins enhances structural flexibility, leading to better unfolding at the air-water interface and improved foam formation (Gerzhova et al., 2015). However, protein isolates from hexane-defatted meals fail to form foams at pH 5, likely due to excessive protein denaturation caused by the defatting process (Li et al., 2023).

Gelation Properties: Only albumin forms gels at pH 4, attributed to its lower isoelectric point (pH 3) compared to globulin (pH 4.7) and CPI (pH 4.1). At pH 7 and 9, globulin forms firmer gels due to optimal charge ratios, while excessive charge density in albumin and CPI inhibits network formation (Gerzhova et al., 2015). According to Tan et al. (2014) and Cheung et al. (2014), the addition of NaCl reduces protein–protein interactions and screens surface charges thereby weakening the gel strength of proteins.

Peanut Protein

Thermal Properties: Thermal transitions of two major peanut proteins (arachin and conarachin) fractions measured through differential scanning calorimetry, revealed significant thermal changes in arachin (105°C) and conarachin (86°C) fractions of peanut protein isolates prepared using pI precipitation technique (Chen et al., 2019a). Because arachin has a better thermal stability than conarachin, its denaturation enthalpies were 6.3 and 5.2 J/g, respectively which is consistent with the higher polypeptide size of arachin, which probably encourages stronger protein–protein interactions within its quaternary structure.

Gelation Properties: The protein content of peanut protein isolate affects its gelation characteristics. Gel quality and arachin content are negatively correlated, although conarachin and gel quality are positively correlated (Chen et al., 2019b). Compared to the larger arachin polypeptides (24–41 kDa), the smaller conarachin polypeptides (16–18 kDa) facilitate better gel network formation. Similarly, a conarachin-enriched extract increased amount of free sulfhydryl groups which resulted in stronger gels (Basse et al., 2020). Despite having comparable disulphide bond concentrations, conarachin, which was extracted from peanuts that were enriched with selenium, also showed a higher concentration of sulfhydryl groups than arachin (Chen et al., 2022).

Solubility Properties: Concerning the solubility of arachin and conarachin, lowest solubility was observed at pH 5 with a U-shaped solubility pattern. However, solubility increased as pH was shifted from the isoelectric point, with conarachin exhibiting higher solubility at alkaline pH range of 9–11 than arachin (Hu et al., 2021; Zhang et al., 2020).

Emulsification Properties: Emulsification properties of conarachin were found to be better than arachin which is ascribed to its increased solubility, surface hydrophobicity, and sulfhydryl group content, all of which enhance the development of rigid interfacial membranes and more robust interactions with oil droplets. Furthermore, it was reported that arachin produced larger oil droplet (1.77 μm initially and 2.62 μm after storage) than conarachin at pH 7 (1.31 μm initially and 1.68 μm after 20 days of storage) (Hu et al., 2021).

Foaming Properties: Peanut protein conarachin fraction demonstrated enhanced both foaming capacity and stability at pH 7.4 which is contributed by the reduced polypeptide size and increased structural flexibility, which allow for better unfolding at the air-water interface, are probably the causes of this improved performance (Hu et al., 2021).

EFFECT OF PROCESSING ON OILSEED PROTEINS

The quality of the oilseed and the method used for oil extraction have a significant impact on the extraction yield and techno-functional characteristics of proteins since various processing strategies alter the functionality (Wanasundara et al., 2024). In reference to the techniques employed for oil extraction, it was observed that cold-pressed rapeseed press cake showed the highest protein recovery yield (45%) compared to hot-pressed rapeseed meal (26%) and solvent-extracted rapeseed meal (5.1%) due to lower heat exposure (Östbring et al., 2019). Hot-pressed rapeseed meal, subjected to temperatures above protein denaturation onset (77.9°C for rapeseed proteins, 82.2°C for cruciferin, and 96.8°C for napin), had reduced solubility in alkaline conditions. The proteins in the hot-pressed samples were partly denatured by heat during pressing and could not be solubilised by the alkali pH during extraction to the same extent as proteins in cold-pressed rapeseed press cake (Wu & Muir, 2008). Extensive heat treatment (107°C for 1 hour) in solvent extraction further lowered protein recovery. Similar Fetzer et al. (2018) observed higher protein solubility in cold-pressed RPC (52.3%) than in hot-pressed rapeseed meal (36.7%). Furthermore, mechanical screw pressing yields higher-quality oils and meals compared to conventional solvent extraction and high-temperature methods, which can cause oil darkening and the degradation of heat-sensitive minor components (Maciel et al., 2020). Emulsifying properties cold-pressed rapeseed cake were found to be better than heat-induced rapeseed press cake (Östbring et al., 2019). Moreover, sesame seed meal produced from cold-pressed sesame seed exhibited better proximate composition with total protein content of 35% as compared to sesame seed (22%) (Sá et al., 2022). Similarly, hempseed deoiled with diethyl ether in Soxhlet apparatus gave rise to deoiled hemp seed flour with better proximate composition of 55% protein, 28% carbohydrate (Karabulut & Yemiş, 2022).

Amino acid composition of expeller pressed canola seed reveals an increased total amino acid content of 280.9 g/kg than cold-pressed canola seed (229.5 g/kg) (Grageola et al., 2013). Besides the effect of oil extraction techniques, processes such as germination, fermentation, soaking, roasting, autoclaving alter the functional properties of oilseed proteins. Most native proteins do not inherently possess the functional properties required for food applications. However, their nutritional value and functionality, particularly solubility, can be enhanced through various modification techniques (Kilara et al., 1986; Panyam & Kilara, 1996). These modifications offer additional benefits, such as slowing down deterioration, eliminating toxic or inhibitory compounds, and incorporating nutrients and additives via covalent bonding. Such alterations lead to structural and conformational changes at multiple levels, enabling the optimisation of key properties like size, surface characteristics, charge distribution, hydrophobicity-to-hydrophilicity ratio, and molecular flexibility (Feeney et al., 1985). Protein modifications can be achieved by either altering their composition (removing or introducing hetero constituents) or modifying their molecular size.

Proteins that have been extracted can undergo chemical, thermal, or enzymatic treatments to enhance their functional properties. Among these, chemical modifications are frequently used to investigate the relationship between structure, stability, and functionality (El-Adawy et al., 2000; Gruener & Ismond, 1997). Many studies on food proteins focus on modifying the ε-amino groups of lysine residues, a process that influences charge distribution and overall net charge of protein molecules. Adjustments in conformation, intra- and intermolecular interactions can further modify protein hydrophobicity (Feeney et al., 1985).

Certain physical treatments, such as high-pressure processing, unfolding, and exposing hydrophobic sites, improve protein functionality (Molina et al., 2002). The addition of sugar enhances whipping properties in oilseed proteins (Khalil et al., 1985). Polysaccharides, when combined with proteins, can form biopolymers, which help stabilise foams, emulsions, and dispersions against thermal or mechanical stress (Benichou et al., 2002). Ultrasound processing of hempseed protein isolate showed higher emulsification activity as well as stability which is attributed by to the release of inner hydrophobic groups of protein molecules (Karabulut & Yemiş, 2022). Similar results were observed for soy protein isolate, and sunflower protein isolate (Jambrak et al., 2009; Malik & Saini, 2017).

Chemical modifications can be achieved using alkylation, oxidation, acylation, esterification, or amide formation. Succinylation increases viscosity in unmodified protein isolates, enhances the solubility, emulsification, foaming properties (Beuchat, 1977; Shahidi & Wanasundara, 1998), whereas, acetylation and phosphorylation enhance pseudoplastic flow properties (Krause, 2002). Acid hydrolysis at high temperatures, or in the presence of catalysts, selectively removes amide groups, improving protein functionality while minimising peptide bond hydrolysis (Shih, 1987). Deamidation (removal of amide groups) enhances the food applications of oilseed proteins, while in non-food industries, disulphide bond rearrangement using β-mercaptoethanol has been explored as a cost-effective method for improving protein physical properties (Làsztity et al., 1998).

Another strategy involves the use of enzymes, where peptide bonds are cleaved to generate peptides with specific size, charge, and surface properties to increase protein solubility. Trypsin-treated proteins exhibit greater solubility and water retention capacity compared to their untreated counterparts (Jones & Tung, 1983). Protease hydrolysis on the other hand enhances foaming, stability, and gelation capacity in flour (Hrckova et al., 2002; Taha et al., 2002). The enzymatic hydrolysis of soybean proteins using pepsin, papain, and trypsin has been well-documented for reducing antinutritional factors.

As a result, the application of trypsin to canola protein isolate, soy, and sunflower protein concentrates enhances water-holding capacity and solubility, with soy protein concentrate exhibiting the highest water hydration capacity. The impact of trypsin is particularly notable in soy and sunflower proteins due to their carbohydrate content. Additionally, the presence of fibre residues can hinder further water-holding capacity improvements (Jones & Tung, 1983).

APPLICATION OF OILSEED PROTEINS

Food Application

Meat Analogues: Proteins are vital nutrients needed for the human body's growth and development, playing a crucial role in boosting lean body mass, bone density, and general health. The rise in global population highlights the need for alternative protein sources than animal proteins in the form meat products (Sharma et al., 2025). These meat products pose problem to animal welfare as well as human health. Therefore, to deal with animal protein related health risks, numerous plant-based meat analogues have been produced to fulfil the protein need with reduced health concerns.

Among the oilseed proteins, soy products such as soy protein concentrates, and isolates are commonly utilised in meat analogues because of their distinctive functional attributes, such as gelation, water-holding capacity, and emulsification properties (Lan et al., 2020). It was reported by Hughes et al. (2011) that soy protein isolates exhibit all essential amino acid needed by the human body with PDCAAS value of 1.0, similar to that of animal proteins. Soy protein-based meat alternatives are produced using defatted soy flour, soy protein concentrate, and soy protein isolate, with protein contents of 50%–90% (Asgar et al., 2010). The soy protein isolates to wheat starch ratio (9:1) and soy protein isolates, wheat starch, and wheat gluten (60:40:5) affect fibrous structure and juiciness (Lin et al., 2002; MacDonald et al., 2009; Liu & Hsieh, 2008). Soy protein concentrate-based formulations extrude more easily, while spy protein concentrate to wheat gluten (7:3) enhances texture (Chiang et al., 2019). In addition to soy protein product diversity, other oilseed protein is also being utilised as meat alternative. Rapeseed 2S albumin napin possesses the ability to replace casein protein to improve texture in sausage batter and sauces when added at a concentration of 2% (Schwartz et al., 2015; Yoshie-Stark et al., 2008). Chia seeds have been introduced in ham-like products which enhances the functional attributes of ham (Ding et al., 2018). Moreover, peanut based meat alternatives have been produced with better sensory and compositional properties than soy. Despite the numerous advantages of oilseed protein-based meat alternatives, they often have undesirable flavour and sensory attributes. These analogues typically exhibit poorer chewiness and reduced succulence compared to real meat. Additionally, soy-based ingredients contain inherent off-flavours, such as grassy and beany notes, which negatively impact consumer acceptance.

Bakery and Confectionary Products: Baked goods are widely consumed staple foods across the globe. They encompass a variety of flour-based products prepared in an oven, including bread, biscuits, cakes, and pastries.

Incorporation of oilseed protein in bakery goods increase the nutritional and functional value of the end product. In accordance to improved functionality, rapeseed proteins, napin (Supertein) and cruciferin (Puratein) constitute 2% of the bakery products (bread, rolls, cakes, muffins, etc.) and up to 50% of protein bars (GRAS, 2010). Research has demonstrated that rapeseed protein concentrates can completely replace eggs in cake formulations without compromising quality (Von Der Haar et al., 2014). Additionally, studies indicate that up to 18% of wheat gluten can be substituted with rapeseed protein concentrates with minimal impact on dough properties (Wanasundara et al., 2024). Moreover, soy protein has been employed as an ingredient in bread making which enhanced the crumb formation of bread (Masure et al., 2019). Masoumeh and Nouri (2020) summarised that sesame proteins can add value to the gluten-free bread. The incorporation of sunflower protein isolates to gluten-free bread at different concentrations produced superior bread with better acceptability (Zorzi et al., 2020).

Beverages: Oilseed proteins are being extensively used for beverage applications in addition to their use as meat substitutes. GRAS (2010) states that rapeseed albumin and globulin proteins can be added to a variety of fruit and vegetable juices including orange juice and tomato juice, up to 10%. Furthermore, protein drinks and quick milkshakes can be added with 9% or more rapeseed protein hydrolysates. Oilseed proteins not only work as meat alternative but also as promising dairy alternative. Particularly soy milk, which is well known as a plant-based alternative to milk that comes from animals because of its high protein content. These plant-based milk offers reduced risk of various human ailments such as cardiovascular diseases, human ageing, and cancer on their regular consumption (Xu et al., 2010). Moreover, groundnut milk is found to exhibit higher nutritional composition (high protein content) and better sensory attributes as compared to cow milk. Groundnut yogurt prepared from groundnut milk offers yogurt with improved mouthfeel and thicker consistency. The increase in production of groundnut based dairy alternatives aligns with the growing demand for nutrient-rich food products (Gamli & Atasoy, 2018).

Non-Food Application

Oilseed proteins have long been valued for their nutritional qualities, have drawn a lot of attention for their various non-food uses because of their functional versatility, biodegradability, and sustainability. As the need for sustainable and renewable materials increases, oilseed proteins—especially those derived from sources like soybean, canola, and sunflower—are being investigated in industrial sectors due to their film-forming, cross-linking, and emulsification capacity. Canola oilseed protein conjugates with poly (glycidylmethacrylate) can be utilised in adhesives to improve bonding strength and water resistance (Wang et al., 2014b). Soy protein has also been utilised as a renewable resource in the production of adhesives, serving as an ecologically friendly alternative to synthetic adhesives derived from petrochemicals (Kumar et al., 2002). Calcium-induced cruciferin particles may be utilised as potential carriers for bioactive substances and nutraceuticals in carrier and delivery systems due to their resistance to gastrointestinal enzymes (Akbari & Wu, 2016; Bandara et al., 2018). In addition to cruciferin, soy protein is also among the promising source of delivery systems for nutraceuticals (Wan et al., 2015). Use of protein in delivery systems alter the functionality of protein (solubility, surface hydrophobicity, surface activity, thermal stability etc.). A study demonstrated that the interfacial tension reduced at soy/phosphate buffer interface with increased rutin concentration (Cui et al., 2014). Protein-based films derived from rapeseed protein, especially when modified with gelatin, chitosan, or succinylation, improve mechanical, thermal, and water barrier properties, making them viable for food packaging (Jang et al., 2011; Chang & Nickerson, 2014; Shi & Dumont, 2014; Zhang et al., 2019; He et al., 2019). Soy protein-based films incorporated with carvacrol demonstrated antimicrobial activity against *Listeria grayi* and *E. coli* K12 upon direct contact. Additionally, the carvacrol vapour released from these films effectively inhibited the growth of *L. grayi* (Tao et al., 2022; Gomes & Sobral, 2021). Similarly, exposing a sesame protein isolate-based film-forming solution to ultraviolet radiation led to cross-linking, resulting in enhanced mechanical properties of the film (Fathi et al., 2018). In biodegradable plastics, canola protein-based materials demonstrate superior thermal stability and can be used for packaging, compostable containers, and even automotive components (Zhang et al., 2018; Manamperi et al., 2010; Manamperi & Pryor, 2012; Bandara et al., 2018).

REFERENCES

Achouri, A., & Boye, J. I. (2013). Thermal processing, salt and high pressure treatment effects on molecular structure and antigenicity of sesame protein isolate. *Food Research International*, *53*(1), 240–251.

Achouri, A., Nail, V., & Boye, J. I. (2012). Sesame protein isolate: Fractionation, secondary structure and functional properties. *Food Research International*, *46*(1), 360–369.

Ahmed, I. A. M., AlJuhaimi, F., Özcan, M. M., Ghafoor, K., Şimşek, Ş., Babiker, E. E., … & Salih, H. A. (2020). Evaluation of chemical properties, amino acid contents and fatty acid compositions of sesame seed provided from different locations. *Journal of Oleo Science*, *69*(8), 795–800.

Aider, M., & Barbana, C. (2011). Canola proteins: Composition, extraction, functional properties, bioactivity, applications as a food ingredient and allergenicity–A practical and critical review. *Trends in Food Science & Technology*, *22*(1), 21–39.

Aiello, G., Fasoli, E., Boschin, G., Lammi, C., Zanoni, C., Citterio, A., & Arnoldi, A. (2016). Proteomic characterization of hempseed (*Cannabis sativa* L.). *Journal of Proteomics*, *147*, 187–196.

Akbari, A., & Wu, J. (2016). Cruciferin nanoparticles: Preparation, characterization and their potential application in delivery of bioactive compounds. *Food Hydrocolloids*, *54*, 107–118. https://doi.org/10.1016/j.foodhyd.2015.09.017.

Akharume, F. U., Aluko, R. E., & Adedeji, A. A. (2021). Modification of plant proteins for improved functionality: A review. *Comprehensive Reviews in Food Science and Food Safety*, *20*(1), 198–224.

Alemayehu, F. R., Bendevis, M. A., & Jacobsen, S. E. (2015). The potential for utilizing the seed crop amaranth (*Amaranthus* spp.) in East Africa as an alternative crop to support food security and climate change mitigation. *Journal of Agronomy and Crop Science*, *201*(5), 321–329.

Alfawaz, M. A. (2004). Chemical composition and oil characteristics of pumpkin (*Cucurbita maxima*) seed kernels. *Food Science and Agriculture*, *2*(1), 5–18.

Al-Gaby, A. M. (1998). Amino acid composition and biological effects of supplementing broad bean and corn proteins with *Nigella sativa* (black cumin) cake protein. *Nahrung*, *42*(5), 290–294.

Aliu, S., Rusinovci, I., Salihu, S., Fetahu, S., Kika, A., & Zeka, D. (2016). Nutritive values and seed chemical variability of pumpkin (*Cucurbita pepo* L.) populations. *Journal of Food, Agriculture & Environment*, *14*(2), 28–33.

Ally, A. I., & Horrobin, D. F. (1980). Thromboxane A2 in blood vessel walls and its physiological significance: Relevance to thrombosis and hypertension. *Prostaglandins and Medicine*, *4*, 431–438

Appaiah, P., Sunil, L., Prasanth Kumar, P. K., & Gopala Krishna, A. G. (2014). Composition of coconut testa, coconut kernel and its oil. *Journal of the American Oil Chemists' Society*, *91*(6), 917–924.

Arrutia, F., Binner, E., Williams, P., & Waldron, K. W. (2020). Oilseeds beyond oil: Press cakes and meals supplying global protein requirements. *Trends in Food Science & Technology*, *100*, 88–102.

Asgar, M. A., Fazilah, A., Huda, N., Bhat, R., & Karim, A. A. (2010). Nonmeat protein alternatives as meat extenders and meat analogs. *Comprehensive Reviews in Food Science and Food Safety*, *9*(5), 513–529.

Ash, M., & Dohlman, E. (2006). *Oil Crops Situation and Outlook Yearbook. Electronic Outlook Report from the Economic Research Service*. Washington, DC: United States Department of Agriculture.

Assadi, E., Janmohammadi, H., Taghizadeh, A., & Alijani, S. (2011). Nutrient composition of different varieties of full-fat canola seed and nitrogen-corrected true metabolizable energy of full-fat canola seed with or without enzyme addition and thermal processing. *Journal of Applied Poultry Research*, *20*(1), 95–101.

Ayerza, R. (1995). Oil Content and fatty acid composition of Chia (*Salvia hispanica* L.) from five northwestern locations in Argentina. *Journal of the American Oil Chemists' Society*, *72*, 1079–1081.

Ayerza, R. (2011). The seed's oil content and fatty acid composition of chia (*Salvia hispanica* L.) Var. Iztac 1, grown under six tropical ecosystems conditions. *Interciencia*, *36*(8), 620–624.

Ayerza, R., & Coates, W. (2005). *Chia: Rediscovering a Forgotten Crop of the Aztecs*. Tucsin, AZ: University of Arizona Press.

Balachandran, C., Arumughan, C., & Mathew, A. G. (1985). Distribution of major chemical constituents and fatty acids in different regions of coconut endosperm. *Journal of the American Oil Chemists' Society*, *62*(11), 1583–1586.

Bandara, N., A. Akbari, Y. Esparza, and J. Wu. 2018. Canola protein: A promising protein source for delivery, adhesive, and material applications. *Journal of the American Oil Chemists' Society*, *95*(8), 1075–1090. https://doi.org/10.1002/aocs.12039

Basse, B., Bosc, V., Saiter, J. M., Chan-Huot, M., Dupas, J. P., Maillard, M. N., & Menut, P. (2020). Combined effects of ionic strength and enzymatic pre-treatment in thermal gelation of peanut proteins extracts. *Food Research International*, *137*, 109362. https://doi.org/10.1016/j.foodres.2020.109362

Benichou, A., Aserin, A., & Garti, N. (2002). Protein-polysaccharide interactions for stabilization of food emulsions. *Journal of Dispersion Science and Technology*, *23*(1–3), 93–123.

Betancur-Ancona, D., Gallegos-Tintoré, S., & Chel-Guerrero, L. (2004). Wet-fractionation of phaseolus lunatus seeds: Partial characterization of starch and protein. *Journal of the Science of Food and Agriculture*, *84*(10), 1193–1201. https://doi.org/10.1002/jsfa.1804

Beuchat, L. R. (1977). Functional and electrophoretic characteristics of succinylated peanut flour protein. *Journal of Agricultural and Food Chemistry*, *25*(2), 258–261.

Bhatnagar, A. S., Prasanth Kumar, P. K., Hemavathy, J., & Gopala Krishna, A. G. (2009). Fatty acid composition, oxidative stability, and radical scavenging activity of vegetable oil blends with coconut oil. *Journal of the American Oil Chemists' Society*, *86*, 991–999. https://doi.org/10.1007/s11746-009-1435-y

Bhatty, R. S. (1995). Nutritional composition of whole flaxseed and flaxseed meal. In Cunnane, S. C., & Thompson, L. H. (Eds.), *Flaxseed in Human Nutrition* (pp. 22–45). Champaign, IL: AOCS Press.

Blagrove, R. J., Lilley, G. G., & Kortt, A. A. (1981). Partial tryptic digestion of cucurbitin from pumpkin seed. *Functional Plant Biology*, *8*(6), 507–513.

Burssens, S., Pertry, I., Ngudi, D. D., Kuo, Y. H., Van Montagu, M., & Lambein, F. (2011) Soya, human nutrition and health. In Yin, Y., Fatufe, A. A., & Blachier, F. (Eds.), *Soya Bean Meal and Its Extensive Use in Livestock Feeding and Nutrition* (pp 157–180). London: INTECH Open Access Publisher.

Callaway, J. (2004). Hempseed as a nutritional resource: An overview. *Euphytica*, 140, 65–72.

Capitani, M. I., Spotorno, V., Nolasco, S. M., & Tomás, M. C. (2012). Physicochemical and functional characterization of by-products from chia (*Salvia hispanica* L.) seeds of Argentina. *LWT - Food Science and Technology, 45*(1), 94–102.

Carré, P., Citeau, M., Robin, G., & Estorges, M. (2016). Hull content and chemical composition of whole seeds, hulls and germs in cultivars of rapeseed (*Brassica napus*). *OCL*, *23*(3), A302.

Casey, R., Domoney, C., & Ellis, N. (1986). Legume storage proteins and their genes. In Miflin, B. J. (Ed.), *Oxford Surveys of Plant Molecular and Cell Biology*, Vol. 3. (pp. 1–95). Oxford: Oxford University Press.

Chang, C., & Nickerson, M. T. (2014). Effect of plasticizer-type and genipin on the mechanical, optical, and water vapor barrier properties of canola protein isolate-based edible films. *European Food Research and Technology*, *238*(1), 35–46. https://doi.org/10.1007/s00217-013-2075-x

Chen, K. I., Erh, M. H., Su, N. W., Liu, W. H., Chou, C. C., & Cheng, K. C. (2012). Soyfoods and soybean products: From traditional use to modern applications. *Applied Microbiology and Biotechnology*, *96*, 9–22.

Chen, L., Ettelaie, R., & Akhtar, M. (2019a). Improved enzymatic accessibility of peanut protein isolate pre-treated using thermosonication. *Food Hydrocolloids*, *93*, 308–316. https://doi.org/10.1016/j.foodhyd.2019.02.050

Chen, Q., Chen, X., Li, S., Ning, F., Xiong, H., & Zhao, Q. (2022). Preparation, characterization, and in vitro antioxidant activities of natural selenium-enriched peanut protein fractions. *Food Bioscience*, *49*, 101923. https://doi.org/10.1016/j.fbio.2022.101923

Chen, Q., Lyu, Y., Bi, J., Wu, X., Jin, X., Qiao, Y., … & Lyu, C. (2019b). Quality assessment and variety classification of seed-used pumpkin by-products: Potential values to deep processing. *Food Science & Nutrition*, *7*(12), 4095–4104.

Chéreau, D., Videcoq, P., Ruffieux, C., Pichon, L., Motte, J. C., Belaid, S., … & Lopez, M. (2016). Combination of existing and alternative technologies to promote oilseeds and pulses proteins in food applications. *OCL*, *23*(4), D406.

Cheung, L., Wanasundara, J., & Nickerson, M. T. (2014). The effect of pH and NaCl levels on the physicochemical and emulsifying properties of a cruciferin protein isolate. *Food Biophysics*, *9*(2), 105–113. https://doi.org/10.1007/s11483-013-9323-2

Chiang, J. H., Loveday, S. M., Hardacre, A. K., & Parker, M. E. (2019). Effects of soy protein to wheat gluten ratio on the physicochemical properties of extruded meat analogues. *Food Structure*, *19*, 100102.

Chmielewska, A., Kozłowska, M., Rachwał, D., Wnukowski, P., Amarowicz, R., Nebesny, E., & Rosicka-Kaczmarek, J. (2021). Canola/rapeseed protein–nutritional value, functionality and food application: A review. *Critical Reviews in Food Science and Nutrition*, *61*(22), 3836–3856.

Chothia, C. (1984). Principles that determine the structure of proteins. *Annual Review of Biochemistry*, *53*(1), 537–572.

Ciftci, O. N., Przybylski, R., & Rudzińska, M. (2012). Lipid components of flax, perilla, and chia seeds. *European Journal of Lipid Science and Technology*, *114*, 794–800.

Coelho, M. S., & Salas-Mellado, M. M. (2015). Chemical characterization of chia (*Salvia hispanica* L.) for the improvement of breads. *Journal of Food Science and Technology, 52*(10), 6894–6899.

Coskuner, Y., & Karababa, E. (2007). Some physical properties of flaxseed (*Linum usitatissimum* L.). *Journal of Food Engineering, 78*(3), 1067–1073.

Cui, Z., Kong, X., Chen, Y., Zhang, C., & Hua, Y. (2014). Effects of rutin incorporation on the physical and oxidative stability of soy protein-stabilized emulsions. *Food Hydrocolloids*, *41*, 1–9.

da Silva Marineli, R., Lenquiste, S. A., Moraes, E. A., & Marostica Jr, M. R. (2015). Antioxidant potential of dietary chia seed and oil (*Salvia hispanica* L.) in diet-induced obese rats. *Food Research International*, *76*, 666–674.

Dalgalarrondo, M., Raymond, J., & Azanza, J. L. (1984). Sunflower seed proteins: Characterization and subunit composition of the globulin fraction. *Journal of Experimental Botany*, *35*(11), 1618–1628.

Danish Food Composition Databank. (2019). *Food Data: Technical University of Denmark Version 4*. National Food Institute. Available at: https://frida.fooddata.dk/?lang=en/ (accessed 23 September 2025).

Dash, P., & Ghosh, G. (2017). Proteolytic and antioxidant activity of protein fractions of seeds of *Cucurbita moschata*. *Food Bioscience*, *18*, 1–8. https://doi.org/10.1016/j.fbio.2016.12.004

de Oliveira Filho, J. G., & Egea, M. B. (2021). Sunflower seed byproduct and its fractions for food application: An attempt to improve the sustainability of the oil process. *Journal of Food Science*, *86*(5), 1497–1510.

DebMandal, M., & Mandal, S. (2011). Coconut (*Cocos nucifera* L.: Arecaceae): In health promotion and disease prevention. *Asian Pacific Journal of Tropical Medicine*, *4*(3), 241–247.

Deferne, J. L., & Pate, D. W. (1996) Hemp seed oil: A source of valuable essential fatty acids. *Journal of the International Hemp Association*, *3*, 4–7.

Dev, D. K., Quensel, E., & Hansen, R. (1986). Nitrogen extractability and buffer capacity of defatted linseed (*Linum usitatissimum* L.) flour. *Journal of the Science of Food and Agriculture*, *37*(2), 199–205.

Diarrassouba, F., Remondetto, G., Liang, L., Garrait, G., Beyssac, E., & Subirade, M. (2013). Effects of gastrointestinal pH conditions on the stability of the β-lactoglobulin/vitamin D_3 complex and on the solubility of vitamin D_3. *Food Research International*, *52*, 515–521. https://doi.org/10.1016/j.foodres.2013.02.026

Ding, Y., Lin, H. W., Lin, Y. L., Yang, D. J., Yu, Y. S., Chen, J. W., … & Chen, Y. C. (2018). Nutritional composition in the chia seed and its processing properties on restructured ham-like products. *Journal of Food and Drug Analysis*, *26*(1), 124–134.

Dravie, E. E., Kortei, N. K., Essuman, E. K., Tettey, C. O., Boakye, A. A., & Hunkpe, G. (2020). Antioxidant, phytochemical and physicochemical properties of sesame seed (*Sesamum indicum* L). *Scientific American*, *8*, e00349. https://doi.org/10.1016/j.sciaf.2020.e00349

Dunitz, J. D. (2001). Pauling's left-handed α-helix. *Angewandte Chemie International Edition*, *40*(22), 4167–4173.

Duranti, M., Guerrierl, N., Takajashl, T., & Cerletti, P. (1988). The legumin-like storage proteins of Lupinus albus seeds. *Phytochemistry*, *27*, 15–23.

El-Adawy, T. A., & Taha, K. M. (2001). Characteristics and composition of watermelon, pumpkin, and paprika seed oils and flours. *Journal of Agricultural and Food Chemistry*, *49*(3), 1253–1259.

El-Adawy, T. A., Rahma, E. H., El-Bedawy, A. A., & Sobihah, T. Y. (2000). Effect of soaking process on nutritional quality and protein solubility of some legume seeds. *Food/Nahrung*, *44*(5), 339–343.

Elinge, C. M., Muhammad, A., Atiku, F. A., Itodo, A. U., Peni, I. J., Sanni, O. M., & Mbongo, A. N. (2012). Proximate, mineral and anti-nutrient composition of pumpkin (*Cucurbita pepo* L) seeds extract. *International Journal of Plant Research*, *2*(5), 146–150.

El-Khier, K. S., Ishag, K. E. A., & Yagoub, A. E. A. (2008). Chemical composition and oil characteristics of sesame seed varieties grown in Sudan. *Research Journal of Agriculture and Biological Sciences*, *14*, 1–6.

Elleuch, M., Besbes, S., Roiseux, O., Blecker, C., & Attia, H. (2007). Quality characteristics of sesame seeds and by-products. *Food Chemistry*, *103*, 641–650.

Erasmus, U. (1999). *Fats that Heal, Fats that Kill (1993)*. Berkeley, CA: Alive Books.

Ericson, M. L., Rödin, J., Lenman, M., Glimelius, K., Josefsson, L. G., & Rask, L. (1986). Structure of the rapeseed 1.7 S storage protein, napin, and its precursor. *Journal of Biological Chemistry*, *261*(31), 14576–14581.

FAO/WHO/UNU. (2008). *Protein and Amino Acids Requirements in Human Nutrition*. Geneva, Switzerland: WHO Library Cataloguing. Available at: https://apps.who.int/iris/bitstream/handle/10665/43411/WHO_TRS_935_eng.pdf?ua=1

Fasuan, T. O., Gbadamosi, S. O., & Omobuwajo, T. O. (2018). Characterization of protein isolate from *Sesamum indicum* seed: In vitro protein digestibility, amino acid profile, and some functional properties. *Food Science & Nutrition*, *6*, 1715–1723.

Fathi, N., Almasi, H., & Pirouzifard, M. K. (2018). Effect of ultraviolet radiation on morphological and physicochemical properties of sesame protein isolate based edible films. *Food Hydrocolloids*, *85*, 136–143.

Feeney, R. E., Whitaker, J. R., Wong, W. D., Osuga, D. T., & Gershwin, M. E. (1985). Chemical reactions of proteins. In *Chemical Changes in Food during Processing* (pp. 255–287). New York: Van Nostrand Reinhold Company.

Fetzer, A., Herfellner, T., Stäbler, A., Menner, M., & Eisner, P. (2018). Influence of process conditions during aqueous protein extraction upon yield from pre-pressed and cold-pressed rapeseed press cake. *Industrial Crops and Products*, *112*, 236–246.

Fruhwirth, G. O., & Hermetter, A. (2007). Seeds and oil of the Styrian oil pumpkin: Components and biological activities. *European Journal of Lipid Science and Technology*, *109*(11), 1128–1140.

Fukushima, D. (2001). Recent progress in research and technology on soybeans. *Food Science and Technology Research*, *7*, 8–16. https://doi.org/10.3136/fstr.7.8

Gamli, Ö. F., & Atasoy, A. F. (2018). Physico-chemical and sensorial properties of groundnut milk and it's yoghurt. *Journal of Food Measurement and Characterization*, *12*(3), 1997–2004. https://doi.org/10.1007/s11694-018-9814-4

Ganorkar, P. M., & Jain, R. K. (2013). Flaxseed—A nutritional punch. *International Food Research Journal, 20*(2), 519–525.

Garcia, R. N., Arocena, R. V., Laurena, A. C., & Tecson-Mendoza, E. M. (2005). 11S and 7S globulins of coconut (*Cocos nucifera* L.): Purification and characterization. *Journal of Agricultural and Food Chemistry*, *53*(5), 1734–1739.

Gatehouse, J. A., Croy, R. R., Boulter, D., & Shewry, P. R. (1984). The synthesis and structure of pea storage proteins. *Critical Reviews in Plant Sciences*, *1*(4), 287–314.

Ge, J., Sun, C. X., Mata, A., Corke, H., Gan, R. Y., & Fang, Y. (2021). Physicochemical and pH-dependent functional properties of proteins isolated from eight traditional Chinese beans. *Food Hydrocolloids*, *112*, 106288. https://doi.org/10.1016/j.foodhyd.2020.106288

Gerzhova, A., Mondor, M., Benali, M., & Aider, M. (2015). Study of the functional properties of canola protein concentrates and isolates extracted by electro-activated solutions as non-invasive extraction method. *Food Bioscience*, *12*, 128–138. https://doi.org/10.1016/j.fbio.2015.10.002

Ghodsvali, A., Khodaparast, M. H., Vosoughi, M., & Diosady, L. L. (2005). Preparation of canola protein materials using membrane technology and evaluation of meals functional properties. *Food Research International*, *38*(2), 223–231.
Gomes, A., & Sobral, P. J. D. A. (2021). Plant protein-based delivery systems: An emerging approach for increasing the efficacy of lipophilic bioactive compounds. *Molecules*, *27*(1), 60.
González-Pérez, S., & Vereijken, J. M. (2007). Sunflower proteins: Overview of their physicochemical, structural and functional properties. *Journal of the Science of Food and Agriculture*, *87*(12), 2173–2191.
Gopalan, C., Ramasastri, B. V., & Subramanian, S. C. (2007). *Nutritive Value of Indian Foods*. Hyderabad: National Institute of Nutrition.
Grageola, F., Landero, J. L., Beltranena, E., Cervantes, M., Araiza, A., & Zijlstra, R. T. (2013). Energy and amino acid digestibility of expeller-pressed canola meal and cold-pressed canola cake in ileal-cannulated finishing pigs. *Animal Feed Science and Technology*, *186*(3–4), 169–176.
Grancieri, M., Martino, H. S. D., & Gonzalez de Mejia, E. (2019). Chia seed (*Salvia hispanica* L.) as a source of proteins and bioactive peptides. *Journal of Food Science, 84*(8), 1751–1764.
GRAS. (2010). GRAS notice No. 327. GRAS notification for cruciferin-rich and napin-rich protein isolates derived from canola/rapeseed (Puratein® and SuperteinTM). Available at:https://hfpappexternal.fda.gov/scripts/fdcc/index.cfm?set=grasnotices&id=327 (accessed 29 October 2019).
Grobelnik-Mlakar, S., Turinek, M., Jakop, M., Bavec, M., & Bavec, F. (2009). Nutrition value and use of grain amaranth: Potential future application in bread making. *Agricultura*, *6*, 43–53.
Gruener, L., & Ismond, M. A. H. (1997). Effects of acetylation and succinylation on the functional properties of the canola 12S globulin. *Food Chemistry*, *60*(4), 513–520.
Gruis, D., Selinger, D. A., Curran, J. M., & Jung, R. (2002). Redundant proteolytic mechanisms process seed storage proteins in the absence of seed-type members of the vacuolar processing enzyme family of cysteine proteases. *The Plant Cell*, *14*, 2863–2882.
Guarte, R. C., Mühlbauer, W., & Kellert, M. (1996). Drying characteristics of copra and quality of copra and coconut oil. *Postharvest Biology and Technology*, *9*(3), 361–372.
Guéguen, J., Popineau, Y., Anisimova, I. N., Fido, R. J., Shewry, P. R., & Tatham, A. S. (1996). Functionality of the 2S albumin seed storage proteins from sunflower (*Helianthus annuus* L.). *Journal of Agricultural and Food Chemistry*, *44*(5), 1184–1189.
Hall III, C., Tulbek, M. C., & Xu, Y. (2006). Flaxseed. *Advances in Food and Nutrition Research*, *51*, 1–97.
Hara-Hishimura, I., Takeuchi, Y., Inoue, K., & Nishimura, M. (1993). Vesicle transport and processing of the precursor to 2S albumin in pumpkin. *The Plant Journal*, *4*(5), 793–800.
He, R., Dai, C., Li, Y., Wang, Z., Li, Q., Zhang, C., … & Yuan, J. (2019). Effects of succinylation on the physicochemical properties and structural characteristics of edible rapeseed protein isolate films. *Journal of the American Oil Chemists' Society*, *96*(10), 1103–1013. https://doi.org/10.1002/aocs.12264
Hegde, D. M. (2012). Sesame. In *Handbook of Herbs and Spices* (2nd ed., vol. 2, pp. 449–486). Amsterdam, the Netherlands: Elsevier.
Hidalgo, F. J., & Zamora, R. (2006). Peptides and proteins in edible oils: Stability, allergenicity, and new processing trends. *Trends in Food Science & Technology*, *17*, 56–63. https://doi.org/10.1016/j.tifs.2005.10.006
Hoglund, A. S., Rodin, J., Larsson, E., & Rask, L. (1992). Distribution of napin and cruciferin in developing rape seed embryos. *Plant Physiology*, *98*(2), 509–515.

House, J. D., Neufeld, J., & Leson, G. (2010). Evaluating the quality of protein from hemp seed (*Cannabis sativa* L.) products through the use of the protein digestibility-corrected amino acid score method. *Journal of Agricultural and Food Chemistry, 58*, 11801–11807.

Hrckova, M., Rusnakova, M., & Zemanovic, J. (2002). Enzymatic hydrolysis of defatted soy flour by three different proteases and their effect on the functional properties of resulting protein hydrolysates. *Czech Journal of Food Sciences, 20*(1), 7–14.

Hu, X., Amakye, W. K., He, P., Wang, M., & Ren, J. (2021). Effects of microfluidization and transglutaminase cross-linking on the conformations and functional properties of arachin and conarachin in peanut. *LWT - Food Science and Technology, 146*, 111438. https://doi.org/10.1016/j.lwt.2021.111438

Hu, Y. M., Ye, W. C., Yin, Z. Q., & Zhao, S. X. (2007). Chemical constituents from flos *Sesamum indicum* L. *Yao xue bao= Acta pharmaceutica Sinica, 42*, 286–291.

Hughes, G. J., Ryan, D. J., Mukherjea, R., & Schasteen, C. S. (2011). Protein digestibility-corrected amino acid scores (PDCAAS) for soy protein isolates and concentrate: Criteria for evaluation. *Journal of Agricultural and Food Chemistry, 59*(23), 12707–12712.

Ixtaina, V. Y., Nolasco, S. M., & Tomás, M. C. (2011). Physical and chemical properties of chia (*Salvia hispanica* L.) seeds from Argentina. *Journal of Agricultural and Food Chemistry, 59*(10), 2718–2726.

Jambrak, A. R., Lelas, V., Mason, T. J., Krešić, G., & Badanjak, M. (2009). Physical properties of ultrasound treated soy proteins. *Journal of Food Engineering, 93*(4), 386–393.

Jang, S. A., Lim, G. O., & Song, K. B. (2011). Preparation and mechanical properties of edible rapeseed protein films. *Journal of Food Science, 76*(2), C218–C223. https://doi.org/10.1111/j.1750-3841.2010.02026.x.

Jia, Y., Yan, X., Huang, Y., Zhu, H., Qi, B., & Li, Y. (2022). Different interactions driving the binding of soy proteins (7S/11S) and flavonoids (quercetin/rutin): Alterations in the conformational and functional properties of soy proteins. *Food Chemistry, 396*, 133685. https://doi.org/10.1016/j.foodchem.2022.133685

Jin, F., Nieman, D. C., Sha, W., Xie, G., Qiu, Y., & Jia, W. (2012). Supplementation of milled chia seeds increases plasma ALA and EPA in postmenopausal women. *Plant Foods for Human Nutrition, 67*, 105–110.

Jones, L. J., & Tung, M. A. (1983). Functional properties of modified oilseed protein concentrates and isolates. *Canadian Institute of Food Science and Technology Journal, 16*(1), 57–62.

Joshi, S., Shah, S., & Kalantar-Zadeh, K. (2018). Adequacy of plant-based proteins in chronic kidney disease. *Journal of Renal Nutrition, 29*(2), 112–117. https://doi.org/10.1053/j.jrn.2018.06.006

Kajla, P., Goyal, N., Bangar, S. P., Chaudhary, V., & Lorenzo, J. M. (2023). Flaxseed proteins (Linum usitassimum): Thermal, Functional and spectroscopic characterization. Food Analytical Methods, 16(2), 459–467.

Karabulut, G., & Yemiş, O. (2022). Modification of hemp seed protein isolate (*Cannabis sativa* L.) by high-intensity ultrasound treatment. Part 1: Functional properties. *Food Chemistry, 375*, 131843.

Kaushik, P., Dowling, K., McKnight, S., Barrow, C. J., Wang, B., & Adhikari, B. (2016). Preparation, characterization and functional properties of flax seed protein isolate. *Food Chemistry, 197*, 212–220. https://doi.org/10.1016/j.foodchem.2015.09.106

Khalil, M., Ragab, M., & Hassanien, F. R. (1985). Some functional properties of oilseed proteins. *Food/Nahrung, 29*(3), 275–282.

Khattab, R., Arntfield, S. D., & Nyachoti, C. M. (2010). Nutritional quality of legume seeds as affected by some physical treatments. *Food Chemistry, 121*(2), 310–316.

Kilara, A., Sharkasi, T. Y., & Morr, C. V. (1986). Effects of temperature on food proteins and its implications on functional properties. *Critical Reviews in Food Science & Nutrition, 23*(4), 323–395.

Kim, M. Y., Kim, E. J., Kim, Y. N., Choi, C., & Lee, B. H. (2012). Comparison of the chemical compositions and nutritive values of various pumpkin (*Cucurbitaceae*) species and parts. *Nutrition Research and Practice*, *6*(1), 21–27.

Kortt, A. A., & Caldwell, J. B. (1990). Low molecular weight albumins from sunflower seed: Identification of a methionine-rich albumin. *Phytochemistry*, *29*(9), 2805–2810.

Kotecka-Majchrzak, K., Sumara, A., Fornal, E., & Montowska, M. (2020). Oilseed proteins–Properties and application as a food ingredient. *Trends in Food Science & Technology*, *106*, 160–170.

Krause, J. P. (2002). Comparison of the effect of acylation and phosphorylation on surface pressure, surface potential and foaming properties of protein isolates from rapeseed (*Brassica napus*). *Industrial Crops and Products*, *15*(3), 221–228.

Kudełka, W., Kowalska, M., & Popis, M. (2021). Quality of soybean products in terms of essential amino acids composition. *Molecules*, *26*(16), 5071.

Kulczyński, B., Kobus-Cisowska, J., Taczanowski, M., Kmiecik, D., & Gramza-Michałowska, A. (2019). The chemical composition and nutritional value of chia seeds—Current state of knowledge. *Nutrients*, *11*(6), 1242.

Kumar, R., Choudhary, V., Mishra, S., Varma, I. K., & Mattiason, B. (2002). Adhesives and plastics based on soy protein products. *Industrial Crops and Products*, *16*(3), 155–172.

Lampart-Szczapa, E. L. E. O. N. O. R. A. (2001). Legume and oilseed proteins. *Chemical and Functional Properties of Food Proteins*, *2*, 407–436.

Lan, T., Dong, Y., Zheng, M., Jiang, L., Zhang, Y., & Sui, X. (2020). Complexation between soy peptides and epigallocatechin-3-gallate (EGCG): Formation mechanism and morphological characterization. *LWT - Food Science and Technology*, *134*, 109990.

Làsztity, R., Tömösközi, S., Szilägyi, E., & Gaugecz, J. (1998). Modification of functional properties by rearrangement of disulfide bonds and change of ratio of protein fractions. *Food/Nahrung*, *42*(03–04), 210–212.

Lazos, E. S. (1986). Nutritional, fatty acid, and oil characteristics of pumpkin and melon seeds. *Journal of Food Science*, *51*(5), 1382–1383.

Le Clef, E., & Kemper, T. (2015). Sunflower seed preparation and oil extraction. In Martínez-Force, E., Dunford, N. T., & Salas, J. J. (Eds.), *Sunflower: Chemistry, Production, Processing, and Utilization* (pp. 187–226). Urbana, IL: AOCS Press.

Léger, L. W., & Arntfield, S. D. (1993). Thermal gelation of the 12S canola globulin. *Journal of the American Oil Chemists' Society*, *70*, 853–861. https://doi.org/10.1007/BF02545343

Lehninger, A. L. (2004). *Lehninger Principles of Biochemistry: David L. Nelson, Michael M. Cox*. New York: Recording for the Blind & Dyslexic.

Leizer, C., Ribnicky, D., Poulev, A., Dushenkov, S., & Raskin, I. (2000). The composition of hemp seed oil and its potential as an important source of nutrition. *Journal of Nutraceuticals, Functional & Medical Foods*, *2*, 35–53.

Li, C., Shi, D., Stone, A. K., Wanasundara, J. P., Tanaka, T., & Nickerson, M. T. (2023). Effect of canola meal fermentation and protein extraction method on the functional properties of resulting protein products. *Journal of the American Oil Chemists' Society*, *100*, 437–448. https://doi.org/10.1002/aocs.12701

Li, J., Yang, X., Swallah, M. S., Amin, K., Fu, H., Ji, L., … & Liu, S. (2022). Structure and rheology of foams stabilized by different soybean varieties deficient in β-conglycinin subunits trimers. *Food Hydrocolloids*, *131*, 107749. https://doi.org/10.1016/j.foodhyd.2022.107749

Li, S., Chen, J., Hao, X., Ji, X., Zhu, Y., Chen, X., & Yao, Y. (2024). A systematic review of black soybean (*Glycine max* (L.) Merr.): Nutritional composition, bioactive compounds, health benefits, and processing to application. *Food Frontiers*, *5*(3), 1188–1211.

Lin, S., Huff, H. E., & Hsieh, F. (2002). Extrusion process parameters, sensory characteristics, and structural properties of a high moisture soy protein meat analog. *Journal of Food Science*, *67*(3), 1066–1072.

Liu, H. M., He, M. K., Yao, Y. G., Qin, Z., Cai, X. S., & Wang, X. D. (2021). Pectic polysaccharides extracted from sesame seed hull: Physicochemical and functional properties. *International Journal of Biological Macromolecules*, *192*, 1075–1083.

Liu, K., & Hsieh, F. H. (2008). Protein–protein interactions during high-moisture extrusion for fibrous meat analogues and comparison of protein solubility methods using different solvent systems. *Journal of Agricultural and Food Chemistry*, *56*(8), 2681–2687.

Livesey, G., Taylor, R., Hulshof, T., & Howlett, J. (2008). Glycemic response and health—A systematic review and meta-analysis: Relations between dietary glycemic properties and health outcomes. *The American Journal of Clinical Nutrition*, *87*(1), 258S–268S.

López, D. N., Galante, M., Robson, M., Boeris, V., & Spelzini, D. (2018). Amaranth, quinoa and chia protein isolates: Physicochemical and structural properties. *International Journal of Biological Macromolecules*, *109*, 152–159.

MacDonald, R. S., Pryzbyszewski, J., & Hsieh, F. H. (2009). Soy protein isolate extruded with high moisture retains high nutritional quality. *Journal of Agricultural and Food Chemistry*, *57*(9), 3550–3555.

Maciel, L. G., Ribeiro, F. L., Teixeira, G. L., Molognoni, L., Dos Santos, J. N., Nunes, I. L., & Block, J. M. (2020). The potential of the pecan nut cake as an ingredient for the food industry. *Food Research International*, *127*, 108718.

Madhusudhan, B. (2009). Potential benefits of flaxseed in health and disease-A perspective. *Agriculturae Conspectus Scientificus*, *74*(2), 67–72.

Malaki Nik, A., Alexander, M., Poysa, V., Woodrow, L., & Corredig, M. (2011). Effect of soy protein subunit composition on the rheological properties of soymilk during acidification. *Food Biophysics*, *6*(1), 26–36.

Malik, M. A., & Saini, C. S. (2017). Polyphenol removal from sunflower seed and kernel: Effect on functional and rheological properties of protein isolates. *Food Hydrocolloids*, *63*, 705–715.

Malomo, S. A., & Aluko, R. E. (2015a). A comparative study of the structural and functional properties of isolated hemp seed (*Cannabis sativa* L.) albumin and globulin fractions. *Food Hydrocolloids*, *43*, 743–752.

Mamone, G., Picariello, G., Ramondo, A., Nicolai, M. A., & Ferranti, P. (2019). Production, digestibility and allergenicity of hemp (*Cannabis sativa* L.) protein isolates. *Food Research International*, *115*, 562–571. https://doi.org/10.1016/j.foodres.2018.09.017

Mamun, M. A., & Absar, N. (2018). Major nutritional compositions of black cumin seeds – Cultivated in Bangladesh and the physicochemical characteristics of its oil. *International Food Research Journal*, *25*(6), 2634–2639.

Manamperi, W. A., Chang, S. K., Ulven, C. A., & Pryor, S. W. (2010). Plastics from an improved canola protein isolate: Preparation and properties. *Journal of the American Oil Chemists' Society*, *87*(8), 909–915.

Manamperi, W. A., & Pryor, S. W. (2012). Properties of canola protein-based plastics and protein isolates modified using SDS and SDBS. *Journal of the American Oil Chemists' Society*, *89*(3), 541–549.

Mandal, S., & Mandal, R. K. (2000). Seed storage proteins and approaches for improvement of their nutritional quality by genetic engineering. *Current Science*, *79*, 576–589.

Maneemegalai, S., & Reena Prasad, R. P. (2011). Evaluation of amino acid composition and protein solubility profile of commercially available sesame and groundnut seed meal. *Asian Journal of Food and Agro-Industry (AJOFAI)*, *4*, 161–166.

Mansour, E. H., Dworschák, E., Lugasi, A., Barna, É., & Gergely, A. (1993). Nutritive value of pumpkin (*Cucurbita pepo Kakai* 35) seed products. *Journal of the Science of Food and Agriculture*, *61*(1), 73–78.

Masure, H. G., Wouters, A. G., Fierens, E., & Delcour, J. A. (2019). Impact of egg white and soy proteins on structure formation and crumb firming in gluten-free breads. *Food Hydrocolloids*, *95*, 406–417.

Mat, K., Abdul Kari, Z., Rusli, N. D., Che Harun, H., Wei, L. S., Rahman, M. M., ... & Goh, K. W. (2022). Coconut palm: Food, feed, and nutraceutical properties. *Animals*, *12*(16), 2107.

Matthaus, B., Özcan, M. M., & Al Juhaimi, F. (2016). Some rape/canola seed oils: Fatty acid composition and tocopherols. *Zeitschrift für Naturforschung C*, *71*(3–4), 73–77.

Medic, J., Atkinson, C., & Hurburgh, C. R. (2014). Current knowledge in soybean composition. *Journal of the American Oil Chemists' Society*, *91*, 363–384.

Mehta, B. V. (2000). *Sea Millennium Handbook 2000*, 7th edition. Mumbai: The Solvent Extractors Association of India.

Modgil, R., Tanwar, B., Goyal, A., & Kumar, V. (2021). Soybean (glycine max). In *Oilseeds: Health Attributes and Food Applications* (pp. 1–46). Singapore: Springer Singapore.

Molina, E., Defaye, A. B., & Ledward, D. A. (2002). Soy protein pressure-induced gels. *Food Hydrocolloids*, *16*(6), 625–632.

Morris, D. H. (2007). *Flax—A Health and Nutrition Primer*, 4th edition. Winnipeg, MB: Flax Council of Canada. Available at: https://www.flaxcouncil.ca

Morris, J. B., Wang, M. L., & Tonnis, B. D. (2021). Variability for oil, protein, lignan, tocopherol, and fatty acid concentrations in eight sesame (*Sesamum indicum* L.) genotypes. *Industrial Crops and Products*, *164*, 113355.

Munoz, L. A., Cobos, A., Diaz, O., & Aguilera, J. M. (2013). Chia seed (*Salvia hispanica*): An ancient grain and a new functional food. *Food Reviews International, 29*(4), 394–408.

Murkovic, M., Hillebrand, A., Winkler, J., & Pfannhauser, W. (1996). Development of pumpkin seed for production of edible oil: Distribution of tocopherols in breeding lines. *Food Chemistry*, *57*(1), 58.

Nietzel, T., Dudkina, N. V., Haase, C., Denolf, P., Semchonok, D. A., Boekema, E. J., ... & Sunderhaus, S. (2013). The native structure and composition of the cruciferin complex in *Brassica napus*. *Journal of Biological Chemistry*, *288*, 2238–2245.

Nishinari, K., Fang, Y., Guo, S., & Phillips, G. O. (2014). Soy proteins: A review on composition, aggregation and emulsification. *Food Hydrocolloids*, *39*, 301–318.

Nitrayová, S., Brestenský, M., Heger, J., Patráš, P., Rafay, J., & Sirotkin, A. (2014) Amino acids and fatty acids profile of chia (*Salvia hispanica* L.) and flax (*Linum usitatissimum* L.) seed. *Slovak Journal of Food Sciences*, *8*, 72–76.

Niu, Y., Liu, L., Wan, Y., & Chen, X. (2014). Functional properties and applications of rapeseed protein in food. *Food and Bioprocess Technology, 7*(1), 57–70.

Nyam, K. L., Tan, C. P., Lai, O. M., Long, K., & Man, Y. C. (2009). Physicochemical properties and bioactive compounds of selected seed oils. *LWT - Food Science and Technology*, *42*(8), 1396–1403.

Ohlson, R., & Anjou, K. (1979). Rapeseed protein products. *Journal of the American Oil Chemists' Society*, *56*(3Part3), 431–437.

Oliveira-Alves, S. C., Vendramini-Costa, D. B., Cazarin, C. B. B., Júnior, M. R. M., Ferreira, J. P. B., Silva, A. B., ... & Bronze, M. R. (2017). Characterization of phenolic compounds in chia (*Salvia hispanica* L.) seeds, fiber flour and oil. *Food Chem.* 2017, *232*, 295–305.

Olivos-Lugo, B. L., Valdivia-López, M. Á., & Tecante, A. (2010). Thermal and physicochemical properties of chia seed (*Salvia hispanica* L.) *Journal of Food Science, 75*(1), C43–C48.

Oomah, B. D. (2001). Flaxseed as a functional food source. *Journal of the Science of Food and Agriculture, 81*(9), 889–894.

Oomah, B. D., & Mazza, G. (1993). Flaxseed proteins—A review. *Food Chemistry*, *48*(2), 109–114.

Oomah, B. D., & Mazza, G. (2000). Bioactive components of flaxseed: Occurrence. In *Phytochemicals Phytopharmaceuticals* (p. 106). Champaign, IL: The American Oil Chemists' Society (AOCS) Press.

Orsavova, J., Misurcova, L., Ambrozova, J. V., Vicha, R., & Mlcek, J. (2015). Fatty acids composition of vegetable oils and its contribution to dietary energy intake and dependence of cardiovascular mortality on dietary intake of fatty acids. *International Journal of Molecular Sciences*, *16*, 12871–12890. https://doi.org/10.3390/ ijms160612871

Osborne, J. W. (1994). Some similarities and differences among phenomenological and other methods of psychological qualitative research. *Canadian Psychology/Psychologie Canadienne*, *35*(2), 167.

Östbring, K., Malmqvist, E., Nilsson, K., Rosenlind, I., & Rayner, M. (2019). The effects of oil extraction methods on recovery yield and emulsifying properties of proteins from rapeseed meal and press cake. *Foods*, *9*(1), 19.

Otegui, M. S., Herder, R., Schulze, J., Jung, R., & Staehelin, L. A. (2006). The proteolytic processing of seed storage proteins in Arabidopsis embryo cells starts in the multivesicular bodies. *Plant Cell*, *18*, 2567–2581.

Ovca, A., Van Elteren, J. T., Falnoga, I., & Šelih, V. S. (2011). Speciation of zinc in pumpkin seeds (*Cucurbita pepo*) and degradation of its species in the human digestive tract. *Food Chemistry*, *128*(4), 839–846.

Ozdemir, I. S., Karaoglu, O., Dag, C., & Bekiroglu, S. (2018). Assessment of sesame oil fatty acid and sterol composition with FT-NIRspectroscopy and chemometrics. *Turkish Journal of Agriculture and Forestry*, *42*, 444–452.

Panyam, D., & Kilara, A. (1996). Enhancing the functionality of food proteins by enzymatic modification. *Trends in Food Science & Technology*, *7*(4), 120–125.

Patel, S. (2013). Pumpkin (*Cucurbita* sp.) seeds as nutraceutic: A review on status quo and scopes. *Mediterranean Journal of Nutrition and Metabolism*, *6*(3), 183–189.

Patil, U., & Benjakul, S. (2017). Characteristics of albumin and globulin from coconut meat and their role in emulsion stability without and with proteolysis. *Food Hydrocolloids*, *69*, 220–228.

Paul, I. M., Bartok, C. J., Downs, D. S., Stifter, C. A., Ventura, A. K., & Birch, L. L. (2009). Opportunities for the primary prevention of obesity during infancy. *Advances in Pediatrics*, *56*(1), 107.

Pauling, L., & Corey, R. B. (1951). Configurations of polypeptide chains with favored orientations around single bonds: Two new pleated sheets. *Proceedings of the National Academy of Sciences*, *37*(11), 729–740.

Peng, D., Ye, J., Jin, W., Yang, J., Geng, F., & Deng, Q. (2022). A review on the utilization of flaxseed protein as interfacial stabilizers for food applications. *Journal of the American Oil Chemists' Society*, *99*(9), 723–737.

Perkins, E. G. (1995). Composition of different oilseeds-based food products. In D. S. Kritchevsky, C. E. Bonfield, & E. G. Perkins (Eds.), *Nutritional and Health Benefits of Soy Proteins* (pp. 215–241). Champaign, IL: American Oil Chemists' Society.

Petraru, A., Ursachi, F., & Amariei, S. (2021). Nutritional characteristics assessment of sunflower seeds, oil and cake. Perspective of using sunflower oilcakes as a functional ingredient. *Plants*, *10*(11), 2487.

Przybylski, R. (2005). Canola oil: Physical and chemical properties. *Bailey's Industrial Oil and Fat Products, 6*, 1–14.

Rahim Monfared, M., & Nouri, L. (2020). Incorporation of Sesamum indicum protein and transglutaminase into gluten-free rice flour cake: Assessment of physico-chemical and sensory properties. *Journal of Food Processing and Preservation*, *44*(9), e14660.

Rahman, M. A., Miaruddin, M., Khan, M. H. H., Masud, M. A. T., & Begum, M. M. (2013). Effect of storage periods on postharvest quality of pumpkin. *Bangladesh Journal of Agricultural Research*, *38*(2), 247–255.

Rahman, M. J., de Camargo, A. C., & Shahidi, F. (2017). Phenolic and polyphenolic profiles of chia seeds and their in vitro biological activities. *Journal of Functional Foods*, *35*, 622–634.

Raikos, V., Neacsu, M., Duthie, G., Nicol, F., Reid, M., Cantlay, L. L., & Ranawana, V. (2017). Proteomic and glucosinolate profiling of rapeseed isolates from meals produced by different oil extraction processes. *Journal of Food Processing and Preservation*, *41*(4), 1–8. https://doi.org/10.1111/jfpp.13060

Ravindran, V., & Blair, R. (1992). Feed resources for poultry production in Asia and the Pacific. II. Plant protein sources. *World's Poultry Science Journal*, *48*, 205–231.

Reyes-Caudillo, E., Tecante, A., & Valdivia-Lopez, M. A. (2008). Dietary fibre content and antioxidant activity of phenolic compounds present in Mexican chia (*Salvia hispanica* L.) seeds. *Food Chemistry*, *107*, 656–663.

Richardson, J. S. (1981). The anatomy and taxonomy of protein structure. *Advances in Protein Chemistry*, *34*, 167–339.

Rico, M., Bruix, M., González, C., Monsalve, R. I., & Rodríguez, R. (1996). ^{1}H NMR assignment and global fold of napin BnIb, a representative 2S albumin seed protein. *Biochemistry*, *35*, 15672–15682.

Rodrıguez, M. L., Ortiz, L. T., Treviño, J., Rebole, A., Alzueta, C., & Centeno, C. (1998). Studies on the nutritive value of full-fat sunflower seed in broiler chick diets. *Animal Feed Science and Technology*, *71*, 341–349.

Russo, R., & Reggiani, R. (2013). Variability in antinutritional compounds in hempseed meal of Italian and French varieties. *Plant*, *1*, 25–29.

Sá, A. G. A., Pacheco, M. T. B., Moreno, Y. M. F., & Carciofi, B. A. M. (2022). Cold-pressed sesame seed meal as a protein source: Effect of processing on the protein digestibility, amino acid profile, and functional properties. *Journal of Food Composition and Analysis*, *111*, 104634.

Sammour, R. H., El-Shourbagy, M. N., Abo-Shady, A. M., & Abasary, A. M. (1995). Characterization of sunflower seed proteins by electrophoretic techniques. *Arab Gulf Journal of Scientific Research*, *13*, 591–601.

Sandoval-Oliveros, M. R., & Paredes-López, O. (2013). Isolation and characterization of proteins from chia seeds (*Salvia hispanica* L.). *Journal of Agricultural and Food Chemistry*, *61*(1), 193–201. https://doi.org/10.1021/jf3034978

Sarwar, G., Sosulski, F. W., & Bell, J. M. (1973). Nutritional evaluation of oilseed meals and protein isolates by mice. *Canadian Institute of Food Science and Technology Journal*, *6*(1), 17–21.

Schwartz, J. M., Solé, V., Guéguen, J., Ropers, M. H., Riaublanc, A., & Anton, M. (2015). Partial replacement of β-casein by napin, a rapeseed protein, as ingredient for processed foods: Thermoreversible aggregation. *LWT - Food Science and Technology*, *63* (1), 562–5680.

Schwenke, K. D., Paehtz, W., Linow, K. J., Raab, B., & Schultz, M. (1979). On seed proteins Part 11. Purification, chemical composition, and some physico-chemical properties of the 11 S globulin (Helianthinin) in sunflower seed. *Food/Nahrung*, *23*(3), 241–254.

Schwenke, K. D., Raab, B., Plietz, P., & Damaschun, G. (1983). The structure of the 12S globulin from rapeseed. *Nahrung*, *27*, 165–175.

Seow, C. C., & Gwee, C. N. (1997). Coconut milk: Chemistry and technology. *International Journal of Food Science and Technology*, *32*(3), 189–201.

Shahidi, F. (Ed.). (1990). *Canola and Rapeseed: Production, Chemistry, Nutrition, and Processing Technology*. New York: Springer Science & Business Media.

Shahidi, F., & Wanasundara, P. K. J. P. D. (1998). Effect of acylation on flax protein functionality. *ACS Symposium Series* (pp. 96–120).

Sharma, A., Thakur, A., & Nanda, V. (2025). Impact of bee pollen cell-wall disrupting techniques on the structural integrity, functional attributes, and nutritional quality of bee pollen protein isolates for food application. *International Journal of Biological Macromolecules*, *305*, 141179.

Sharma, G., & Lakhawat, S. (2017). Development, quality evaluation and acceptability of pumpkin seed flour incorporated in gravy. *Journal of Nutrition & Food Sciences*, *7*(4), 1000613.

Sharma, L., Singh, C., & Sharma, H. K. (2016). Assessment of functionality of sesame meal and sesame protein isolate from Indian cultivar. *Journal of Food Measurement and Characterization*, *10*, 520–526.

Sharma, S., Kaur, M., Goyal, R., & Gill, B. S. (2014). Physical characteristics and nutritional composition of some new soybean (*Glycine max* (L.) Merrill) genotypes. *Journal of Food Science and Technology*, *51*, 551–557.

Shewry, P. R., Napier, J. A., & Tatham, A. S. (1995). Seed storage proteins: Structures and biosynthesis. *The Plant Cell*, *7*(7), 945–956.

Shi, W., & Dumont, M.-J. (2014). Processing and physical properties of canola protein isolate-based films. *Industrial Crops and Products*, *52*, 269–277. https://doi.org/10.1016/j.indcrop.2013.10.037

SHIH, F. F. (1987). Deamidation of protein in a soy extract by ion exchange resin catalysis. *Journal of Food Science*, *52*(6), 1529–1531.

Siano, F., Moccia, S., Picariello, G., Russo, G. L., Sorrentino, G., Di Stasio, M., … & Volpe, M. G. (2019). Comparative study of chemical, biochemical characteristic and ATR-FTIR analysis of seeds, oil and flour of the edible fedora cultivar hemp (*Cannabis sativa* L.). *Molecules*, 24, 83. https://doi.org/10.3390/molecules24010083

Siddiqui, I. R., & Wood, P. J. (1977). Carbohydrates of rapeseed: A review. *Journal of the Science of Food and Agriculture*, *28*(6), 530–538.

Sim, S. Y. J., Srv, A., Chiang, J. H., & Henry, C. J. (2021). Plant proteins for future foods: A roadmap. *Foods*, *10*(8), 1967.

Simopoulos, A. P. (1999). New products from the agri-food industry: The return of n-3 fatty acids into the food supply. *Lipids*, *34*, S297–S301.

Singh, K. K., Mridula, D., Rehal, J., Barnwal, P. (2011). Flaxseed—A potential source of food, feed and fiber. *Critical Review in Food Science and Nutrition*, *51*, 210–222.

Singh, N. (2017). Pulses: An overview. *Journal of Food Science and Technology*, 54(4), 853–857. https://doi.org/10.1007/s13197-017-2537-4

Sjödahl, S., Rödin, J., & Rask, L. (1991). Characterization of the 12S globulin complex of *Brassica napus*: Evolutionary relationship to other 11–12S storage globulins. *European Journal of Biochemistry*, *196*, 617–621.

Slabi, S. A., Mathe, C., Basselin, M., Framboisier, X., Ndiaye, M., Galet, O., & Kapel, R. (2020). Multi-objective optimization of solid/liquid extraction of total sunflower proteins from cold press meal. *Food Chemistry*, *317*, 126423.

Solangi, A. H., & Iqbal, M. Z. (2011). Chemical composition of meat (kernel) and nut water of major coconut (*Cocos nucifera* L.) cultivars at coastal area of Pakistan. *Pakistan Journal of Botany*, *43*(1), 357–363

Soni, R. P., Katoch, M., Kumar, A., & Verma, P. (2016). Flaxseed – Composition and its health benefits. *Research in Environment and Life Sciences*, *9*(3), 310–316. https://rels.comxa.com/fullpapers9/volume9paper92.pdf.

Srilatha, K., & Krishnakumari, K. (2003). Proximate composition and protein quality evaluation of recipes containing sunflower cake. *Plant Foods for Human Nutrition*, *58*(3), 1–11.

Staswick, P. E., Hermodson, M. A., & Nielsen, N. C. (1981). Identification of the acidic and basic subunit complexes of glycinin. *Journal of Biological Chemistry*, *256*(16), 8752–8755.

Steffolani, M. E., D'Inca, R., Carrá, C., & León, A. E. (2014). Incorporation of chia (*Salvia hispanica* L.) in bread: Influence on dough and bread characteristics. *LWT - Food Science and Technology, 58*(2), 484–490.

Sultan, M. T., Butt, M. S., Anjum, F. M., Jamil, A., Akhtar, S., & Nasir, M. (2009). Nutritional profile of indigenous cultivar of black cumin seeds and antioxidant potential of its fixed and essential oil. *Pakistan Journal of Botany*, *41*(3), 1321–1330.

Sun-Waterhouse, D., Zhao, M., & Waterhouse, G. I. (2014). Protein modification during ingredient preparation and food processing: Approaches to improve food processability and nutrition. *Food and Bioprocess Technology*, *7*(7), 1853–1893. https://doi.org/10.1007/s11947-014-1326-6

Taha, F. S., Ibrahim, M. A., & Ismail, A. (2002). Effect of partial enzymatic hydrolysis on the molecular weight of some oilseed protein. *Egyptian Journal of Food Science*, *30*(2), 247–268.

Tai, S. S., Lee, T. T., Tsai, C. C., Yiu, T. J., & Tzen, J. T. (2001). Expression pattern and deposition of three storage proteins, 11S globulin, 2S albumin and 7S globulin in maturing sesame seeds. *Plant Physiology and Biochemistry*, *39*(11), 981–992.

Tan, S. H., Mailer, R. J., Blanchard, C. L., & Agboola, S. O. (2014). Emulsifying properties of proteins extracted from Australian canola meal. *LWT - Food Science and Technology*, *57*, 376–382. https://doi.org/10.1016/j.lwt.2013.12.040

Tan, S. H., Mailer, R. J., Blanchard, C. L., & Agboola, S. O. (2011). Canola proteins for human consumption: Extraction, profile, and functional properties. *Journal of Food Science*, *76*(1), R16–R28.

Tandang-Silvas, M. R. G., Fukuda, T., Fukuda, C., Prak, K., Cabanos, C., Kimura, A., … & Maruyama, N. (2010). Conservation and divergence on plant seed 11S globulins based on crystal structures. *Biochimica et Biophysica Acta (BBA)-Proteins and Proteomics*, *1804*, 1432–1442.

Tang, C.-H., Ten, Z., Wang, X.-S., & Yang, X.-Q. (2006). Physicochemical and functional properties of hemp (*Cannabis sativa* L.) protein isolate. *Journal of Agricultural and Food Chemistry*, *54*, 8945–8950.

Tanwar, B., & Goyal, A. (Eds.). (2021). *Oilseeds: Health Attributes and Food Applications* (p. 20220004233). Singapore: Springer Singapore.

Tao, R., Sedman, J., & Ismail, A. (2022). Characterization and in vitro antimicrobial study of soy protein isolate films incorporating carvacrol. *Food Hydrocolloids*, *122*, 107091.

Tay, S. L., Kasapis, S., Perera, C. O., & Barlow, P. J. (2006). Functional and structural properties of 2S soy protein in relation to other molecular protein fractions. *Journal of Agricultural and Food Chemistry*, *54*(16), 6046–6053.

Thanh, V. H., & Shibasaki, K. (1979). Major proteins of soybean seeds. Reversible and irreversible dissociation of. beta.-conglycinin. *Journal of Agricultural and Food Chemistry*, *27*(4), 805–811.

U.S Department of Agriculture. (2019). *Agricultural Research Service*. FoodData Central. Available at: https://fdc.nal.usda.gov/ (accessed 29 July 2020).

Uppstrom, B. (1995). Seed chemistry. In *Brassica Oilseeds: Production and Utilization* (pp. 217–242).

USDA. (2008). Available at: www.nal.usda.gov/fnic/foodcomp/cgi-bin/list_nut_edit.pl (accessed 25 July 2009).

USDA. (2018). USDA National Nutrient Database for Standard Reference, Release 28. Available at: https://www.ars.usda.gov/ba/bhnrc/ndl (accessed on 3 May 2019).

Venktesh, A., & Prakash, V. (1993). Functional properties of the total proteins of sunflower (*Helianthus annus* L.) seed-effect of physical and chemical treatments. *Journal of Agricultural and Food Chemistry*, *41*, 18–23.

Vieira, F., Santana, H. E., Jesus, M., Santos, J., Pires, P., Vaz-Velho, M., … & Ruzene, D. S. (2024). Coconut waste: Discovering sustainable approaches to advance a circular economy. *Sustainability*, *16*(7), 3066.

Von Der Haar, D., Müller, K., Bader-Mittermaier, S., & Eisner, P. (2014). Rapeseed proteins–Production methods and possible application ranges. *Ocl*, *21*(1), D104.

Wan, Z. L., Guo, J., & Yang, X. Q. (2015). Plant protein-based delivery systems for bioactive ingredients in foods. *Food & Function*, *6*(9), 2876–2889.

Wanasundara, J. P. (2011). Proteins of Brassicaceae oilseeds and their potential as a plant protein source. *Critical Reviews in Food Science and Nutrition*, *51*, 635–677.

Wanasundara, J. P., Kapel, R., & Albe-Slabi, S. (2024). Proteins from canola/rapeseed: Current status. In Nadathur, S., Wanasundara, D. J. P. D., & Scanlin, L. (Eds.), *Sustainable Protein Sources*, 1st edition (pp. 285–304). Cambridge, MA: Academic Press.

Wang, C., Wu, J., & Bernard, G. M. (2014a). Preparation and characterization of canola protein isolate-poly(glycidyl methacrylate) conjugates: A bio-based adhesive. *Industrial Crops and Products*, *57*, 124–31. https://doi.org/10.1016/j.indcrop.2014.03.024

Wang, L., Liu, H., Liu, L., Wang, Q., Li, Q., Du, Y., & Zhang, J. (2014b). Protein contents in different peanut varieties and their relationship to gel property. *International Journal of Food Properties*, *17*(7), 1560–1576.

WCRF (World Cancer Research Fund/American Institute for Cancer Research). (1997). *Food, Nutrition and Prevention of Cancer: A Global Perspective*. Washington, DC: American Institute for Cancer Research.

Wickramasuriya, S. S., Yi, Y. J., Yoo, J., Kang, N. K., & Heo, J. M. (2015). A review of canola meal as an alternative feed ingredient for ducks. *Journal of Animal Science and Technology*, *57*(1), 29.

Wildermuth, S. R., Young, E. E., & Were, L. M. (2016). Chlorogenic acid oxidation and its reaction with sunflower proteins to form green-colored complexes. *Comprehensive Reviews in Food Science and Food Safety*, *15*(5), 829–843.

Withana-Gamage, T. S. (2013). *Structure and Properties of Cruciferin: Investigation of Homohexameric Cruciferin Expressed in Arabidopsis*, Ph.D. Thesis. Saskatoon, SK, Canada: University of Saskatchewan.

Wolever, T. M., Jenkins, D. J., & Jenkins, A. L. (2011). Flaxseed as a functional food for improving glycemic control. *Nutrition Reviews, 69*(8), 459–468.

Wu, G., & Meininger, C. J. (2002). Regulation of nitric oxide synthesis by dietary factors. *Annual Review of Nutrition*, *22*(1), 61–86.

Wu, J., & Muir, A. D. (2008). Comparative structural, emulsifying, and biological properties of 2 major canola proteins, cruciferin and napin. *Journal of Food Science*, *73*(3), C210–C216.

Wu, S., Wang, X., Qi, W., & Guo, Q. (2019). Bioactive protein/peptides of flaxseed: A review. *Trends in Food Science & Technology*, *92*, 184–193.

Xu, B., Chang, S. K., Liu, Z., Yuan, S., Zou, Y., & Tan, Y. (2010). Comparative studies on the chemical and cell-based antioxidant activities and antitumor cell proliferation properties of soy milk manufactured by conventional and commercial UHT methods. *Journal of Agricultural and Food Chemistry*, *58*(6), 3558–3566. https://doi.org/10.1021/jf903796c

Yagasaki, K., Kousaka, F., & Kitamura, K. (2000). Potential improvement of soymilk gelation properties by using soybeans with modified protein subunit compositions. *Breeding Science*, *50*(2), 101–107.

Yan, S., Xu, J., Zhang, S., & Li, Y. (2021). Effects of flexibility and surface hydrophobicity on emulsifying properties: Ultrasound-treated soybean protein isolate. *LWT - Food Science and Technology*, *142*, 110881. https://doi.org/10.1016/j.lwt.2021.110881

Yong, J. W., Ge, L., Ng, Y. F., & Tan, S. N. (2009). The chemical composition and biological properties of coconut (*Cocos nucifera* L.) water. *Molecules*, *14*(12), 5144–5164.

Yoshie-Stark, Y., Wada, Y., & Wäsche, A. (2008). Chemical composition, functional properties, and bioactivities of rapeseed protein isolates. *Food Chemistry*, *107*, 32–39. https://doi.org/10.1016/j.foodchem.2007.07.061

Youle, R. J., & Huang, A. H. (1981). Occurrence of low molecular weight and high cysteine containing albumin storage proteins in oilseeds of diverse species. *American Journal of Botany*, *68*(1), 44–48.

Zdunczyk, Z., & Jankowski, J. (2013). Poultry meat as functional food: Modification of the fatty acid profile – a review. *Annals of Animal Science, 13*(3), 463–480. https://doi.org/10.2478/aoas-2013-0039

Zdunczyk, Z., Minakowski, D., Frejnagel, S., & Flis, M. (1999). Comparative study of the chemical composition and nutritional value of pumpkin seed cake, soybean meal and casein. *Food/Nahrung, 43*(6), 392–395.

Zhang, C., Wang, Z., Li, Y., Yang, Y., Ju, X., & He, R. (2019). The preparation and physiochemical characterization of rapeseed protein hydrolysate-chitosan composite films. *Food Chemistry, 272*, 694–701. https://doi.org/10.1016/j.foodchem.2018.08.097

Zhang, M., Wang, O., Cai, S., Zhao, L., & Zhao, L. (2023). Composition, functional properties, health benefits and applications of oilseed proteins: A systematic review. *Food Research International, 171*, 113061.

Zhang, Q., Li, Y., Chin, K. L., & Qi, Y. (2017). Vegetable soybean: Seed composition and production research. *Italian Journal of Agronomy, 12*(3), 872.

Zhang, S. B., Wang, X. H., Li, X., & Yan, D. Q. (2020). Effects of Tween 20 and transglutaminase modifications on the functional properties of peanut proteins. *Journal of the American Oil Chemists' Society, 97*(1), 93–103. https://doi.org/10.1002/aocs.12309

Zhang, Y., Liu, Q., & Rempel, C. (2018). Processing and characteristics of canola protein-based biodegradable packaging: A review. *Critical Reviews in Food Science and Nutrition, 58*(3), 475–485.

Zhu, X., Zeng, J., Sun, B., Regenstein, J. M., Zhao, J., Liu, L., … & Huang, Y. (2022). Extraction, conformation characteristics and functional properties of soybean lipophilic proteins. *Food Bioscience, 49*, 101907. https://doi.org/10.1016/j.fbio.2022.101907

Ziemichód, A., Wójcik, M., & Różyło, R. (2019). Seeds of *Plantago psyllium* and *Plantago ovata*: Mineral composition, grinding, and use for gluten-free bread as substitutes for hydrocolloids. *Journal of Food Process Engineering, 42*(1), e12931.

Zorzi, C. Z., Garske, R. P., Flôres, S. H., & Thys, R. C. S. (2020). Sunflower protein concentrate: A possible and beneficial ingredient for gluten-free bread. *Innovative Food Science & Emerging Technologies, 66*, 102539. https://doi.org/10.1016/j.ifset.2020.102539

5 Fruit Proteins

Amit Kotiyal, Aditi Thakur, Joshi Thoudam, Rohit Ghatuary, and Rupesh Kaushik

INTRODUCTION

Fruit proteins, derived from multiple fruit species, are a valuable but under-exploited group of plant-based proteins with immense potential to improve human nutrition and well-being. With the trend toward plant-based foods and sustainable food sources accelerating, the nutritional and functional value of fruit proteins is increasingly obvious. Present in different concentrations in the pulp, seeds, and fruit skin, these proteins are rich in essential amino acids, antioxidants, and bioactive compounds. They have many health benefits, such as antioxidant, anti-inflammatory, and antimicrobial activities (Bajramova & Spégel, 2022). Fruit proteins are also being seen as a sustainable substitute for animal proteins, with environmental advantages like lower greenhouse gas emissions and less land use (Wang et al., 2023).

One of the main issues with fruit proteins is that they have a relatively lower content of proteins relative to other plant foods. For example, tropical fruits like bananas, avocados, and papayas have moderate protein levels. Although these are less than in legumes and cereals, they still provide the necessary amino acids needed by the human body (Gupta et al., 2023). Moreover, fruit proteins, when appropriately processed, can be highly digestible, making them suitable for individuals with sensitive digestive systems(Loveday, 2022). Interestingly, papaya proteins are distinguished by their proteolytic enzyme activity, which facilitates effective protein digestion (Li et al., 2022).

Fruit protein bioactive peptides are becoming popular due to their health-promoting abilities. Research indicates that fruits such as kiwi, pomegranate, and watermelon provide antihypertensive and anti-inflammatory effects that aid cardiovascular health and immune function (Cordiano et al., 2024; Bahari et al., 2023; Stonehouse et al., 2023; Figueroa et al., 2014). Kiwi proteins, for instance, have been demonstrated to enhance immune function and decrease oxidative stress, leading to general health (Zhao et al., 2021). These results offer new opportunities to create functional foods that extend beyond minimal nutrition, like fortified food products and beverages, to enhance the health effects.

From a sustainability perspective, fruit protein production is a good way to minimize the environmental footprint. Fruits use fewer resources, such as water, land, and energy, than animal proteins, thus providing a greener alternative (Allaqaband et al., 2022). Studies have indicated that fruit proteins, such as bananas, have a much lower carbon footprint compared to conventional animal proteins (Arte et al., 2024). Fruit proteins can also be incorporated into circular economies, where waste products such as fruit peels and seeds are utilized for protein extraction (Campos et al., 2020).

DOI: 10.1201/9781003613909-5

Recent improvements in food processing technology have accelerated the production of fruit proteins. Technologies like cold pressing, enzymatic hydrolysis, and fermentation are more efficient in protein extraction while maintaining their bioactive properties (Fadimu et al., 2022). These technologies allow for the application of fruit proteins in various food products such as protein bars, smoothies, and vegetable-based meat substitutes, and the food industry becomes more accessible with sustainable and allergen-free protein sources (Moyo, 2024).

Despite the promise of fruit proteins, the problem of scaling up production and having uniform protein content in different fruit varieties has been a challenge. Research is ongoing to improve the extraction processes and increase protein yields from fruit sources, particularly from by-products like peels and seeds (Rifna et al., 2023). More research is required to determine the entire spectrum of bioactive activities of fruit proteins and maximize their utilization in food applications to address the increasing global demand for sustainable protein supplies.

COMPOSITION OF VARIOUS FRUITS

Fruits constitute a dense package of essential nutrients with an important role to play in maintaining human health. Fruits are especially valued for their low-calorie nature, high water content, and bioactive compound accessibility (Figure 5.1). Though nutrition in fruits is diverse to a great extent depending upon fruits, they consist mostly of a combined variety of macronutrients, micronutrients, and phytochemicals, and all of them play a critical role in promoting overall health and well-being (Kalcsits et al., 2020).

Carbohydrates

Carbohydrates are the principal macronutrient of fruits in the form of natural sugars such as glucose, fructose, sucrose, and dietary fiber. These carbohydrates are a convenient source of easily accessible and efficient energy for the body (Wahlqvist & Wattanapenpaiboon, 2020). The fruit sugars differ depending on the fruit type and ripeness, with fruits such as bananas and mangoes having a higher sugar content than berries and citrus fruits (Arumugam & Manikandan, 2011). In addition to sugars,

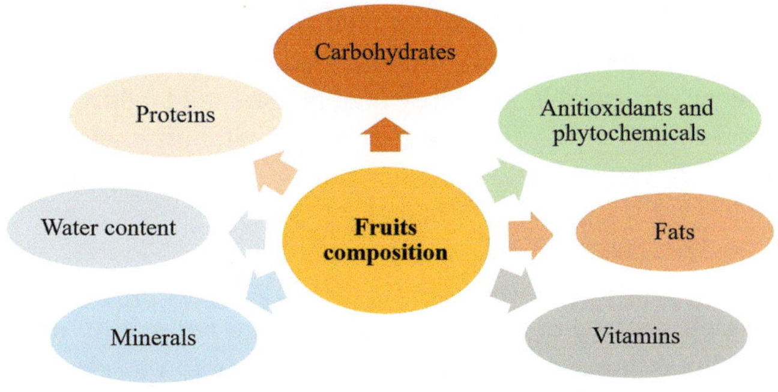

FIGURE 5.1 Composition of fruits.

fruits also consist of dietary fiber, particularly soluble fiber such as pectin, which regulates blood sugar levels and aids in digestive health. For instance, one 100 g serving of apples or pears provides about 3–5 g of fiber, which provides a feeling of satiety and a healthy gut (Smith and Tucker, 2011).

Proteins

While fruits are not significant sources of protein compared to other food groups like legumes or animal foods, fruits contain minute quantities of protein. The proteins are usually between 0.2 g and 2 g/100 g, and fruit species like guava, avocado, and jackfruit contain higher content (Swami et al., 2012). Fruity proteins will contain a mixture of amino acids, including essential amino acids like leucine, valine, and isoleucine, which are essential for muscle repair, immune function, and enzymes (Kalse et al., 2025). A good example is that, avocados have about 2 g of protein in each 100 g, which is one of the fruits with significant protein content (Li et al., 2022; Stephen and Radhakrishnan, 2022).

Fats

Although the majority of fruits contain low-fat content, there are some exceptions. Avocados, olives, and coconuts are rich in good fats, mostly monounsaturated fats like oleic acid. These fats contribute to heart well-being by reducing LDL cholesterol content (Stolp & Kodali, 2022). Fruits such as apples, grapes, and oranges have less than 1 g of fat content per 100 g serving. Low-fat versions are thus suitable for people seeking to regulate their calorie needs but still gain from the high content of micronutrients and bioactive substances found in them (Simmonds and Preedy, 2015).

Vitamins

Fruits are good sources of key vitamins, particularly vitamin C, vitamin A (in the form of provitamin A carotenoids), and the B-vitamins. Vitamin C is critical in the formation of collagen, immune system support, and resistance to oxidative stress. Fruits like oranges, lemons, and grapefruits are all rich in vitamin C, where a serving size of 100 g yields more than 50 mg of the vital nutrient (Patil et al., 2017). Vitamin A, responsible for eyesight, skin, and immune function, is abundant in fruits such as mangoes (Burton-Freeman et al., 2017). In addition, B-vitamins such as folate (vitamin B9) play a critical role in energy metabolism and healthy nervous systems. Bananas, avocados, and citrus fruits are rich in B-vitamin content (Ranjha et al., 2022; Richa et al., 2023).

Minerals

Fruits are also rich sources of essential minerals like potassium, magnesium, calcium, and iron. Potassium is required for heart function, muscle contraction, and fluid balance. Potassium is present in bananas, oranges, and melons, and one medium banana contains about 400 mg of potassium (Kandasamy and Shanmugapriya, 2015).

Magnesium, required for muscle and nerve function, is present in fruits like avocados and figs. Fruits like oranges and figs also contain small but significant amounts of calcium, required for bone health (Vincente et al., 2014). Iron, required for oxygen transport and energy metabolism, is present in moderate amounts in fruits like raisins, apricots, and dried figs (Carughi et al., 2020).

Antioxidants and Phytochemicals

Fruits are also high in bioactive compounds like antioxidants and phytochemicals that have protective effects against oxidative stress and inflammation. The flavonoids, phenolic acids, carotenoids, and anthocyanins are the major compounds responsible for their beneficial effects. Berries such as blueberries, strawberries, and raspberries are full of anthocyanins, flavonoids, and polyphenols that are highly active antioxidants and anti-inflammatory compounds (Gündeşli et al., 2019). Citrus fruits are also full of flavonoids such as hesperidin and naringenin that are linked with reduced inflammation and cardiovascular health (Alam et al., 2022). Another potent antioxidant, lycopene, found in tomatoes and watermelon, has been associated with a reduced risk of chronic diseases (Przybylska, 2020).

Water Content

One of the most distinguishing characteristics of most fruits is their high water content, which is extremely crucial for body hydration and fluid balance. Fruits such as watermelon, which contain more than 90% water, are particularly refreshing and hydrating (Sangeeta and Muzaffar, 2020). The high water content also makes fruits low in caloric density, making them an ideal choice for individuals who desire weight management or satiety. The water content also facilitates digestion and nutrient absorption. Fruits such as oranges, strawberries, and melons supply hydration and essential nutrients, a healthy and efficient means of staying hydrated all day (Jhade et al., 2015).

FRUIT PROTEIN CONTENT AND AMINO ACID PROFILE

Fruits are generally not good sources of protein compared to other foods like legumes, nuts, and animal foods. Fruits have a minute quantity of protein, and the amino acid composition can help assess their nutritional quality (Table 5.1). The protein content in fruits is typically low, but they still play a complementary role in overall protein intake, especially when consumed in combination with other protein-rich foods.

Protein Content in Fruits

Fruits, although not major protein providers, contribute to a varied and balanced diet with moderate amounts of protein. On average, the protein content of most fruits ranges from 0.5 g to 2.6 g/100 g serving. Tropical fruits, such as guava, avocado, and durian, tend to have higher protein concentrations than temperate fruits like

TABLE 5.1
Protein Content and Amino Acid Profile of Fruits

Fruit	Protein Content (g/100 g)	Amino Acids	References
Mango	0.8	Aspartic acid, glutamic acid, serine, alanine, valine	Alabi et al. (2024)
Avocado	2	Glutamic acid, alanine, arginine, leucine, valine	Stephen and Radhakrishnan (2022)
Peach	0.9	Asparate, pyruvate	Sun et al. (2022)
Banana	1.3	Glutamic acid, asparagine, alanine, leucine, valine	Zaini et al. (2022)
Orange	0.9	Glutamic acid, phenyl amine, glycine	Wistaff et al. (2021)
Kiwi	1.1	Glutamic acid, arginine	Xiong et al. (2020)
Grapes	0.6	Glutamic acid, alanine, serine, proline	Tangolar et al. (2019)
Blueberry	0.7	Glutamic acid, alanine, leucine, valine	Zhang et al. (2014)
Guava	2.6	Glutamic acid, aspartic acid, alanine, leucine, arginine	Raimbault et al. (2013)
Durian	2.5	Glutamic acid, asparagine, alanine, leucine, lysine, aspartic acid	Amid et al. (2012)
Jackfruit	1.7	Aspartic acid, glutamic acid, leucine, isoleucine, valine	Swami et al. (2012)

apples and oranges. The protein in fruits primarily consists of globulins, albumins, and other minor proteins, with a small amount of essential and non-essential amino acids (Yu et al., 2021). Fruits, such as bananas and kiwi, offer lower protein content, typically between 0.5 g and 1.5 g/100 g serving (Rice, 2022). Fruits also contain significant amounts of vitamins, minerals, and fiber, which support overall health and aid in the absorption and utilization of the proteins and amino acids present. While fruits cannot replace primary protein sources, they contribute to a well-rounded diet, especially in vegetarian or plant-based nutrition.

Amino Acid Profile of Fruits

The composition of amino acids in fruits varies significantly depending on the type and species. Fruits generally provide a mix of non-essential and essential amino acids (Figure 5.2), which serve as the building blocks of proteins and perform crucial functions in numerous physiological processes in the body. Certain fruits are particularly rich in specific amino acids like alanine, aspartic acid, and glutamic acid. These amino acids are vital for protein synthesis, brain function, and energy metabolism, highlighting the nutritional importance of including a variety of fruits in the diet.

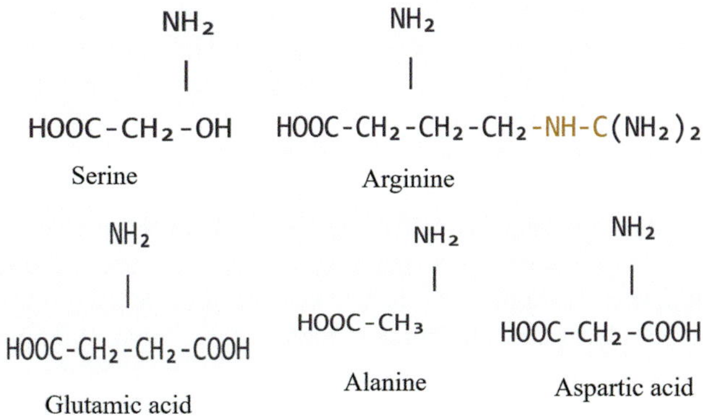

FIGURE 5.2 Chemical structure of different fruit protein.

Glutamic Acid: Found abundantly in fruits, glutamic acid plays a crucial role in protein synthesis and acts as a precursor to other amino acids (Trovato et al., 2021). It is also important for the brain's neurotransmitter functions.

Aspartic Acid: This amino acid is involved in the biosynthesis of other amino acids and plays a role in metabolic pathways, particularly in energy production. Many fruits like pineapples have moderate amounts of aspartic acid (Raimbault et al., 2013).

Alanine: Alanine is present in substantial amounts in several fruits and contributes to energy metabolism by converting into glucose when the body needs energy (Song et al., 2022).

Arginine: Arginine is an essential amino acid found in fruits like guava. It is involved in nitric oxide production, which is critical for vascular health and blood flow regulation (Almutairi et al., 2022).

COMPOSITION OF DIFFERENT FRUIT-BASED FOOD PRODUCTS

Fruits are widely known for their low protein content. Still, when processed into various food products, their protein levels can fluctuate depending on the type of fruit, processing techniques, and added ingredients. Fruit juices, for instance, generally contain minimal protein since the extraction process often removes much of the fruit's solid content, particularly the pulp and fiber. Fresh fruit juices, like those made from oranges or apples, typically provide around 0.2 to 1.0 g of protein/100 mL. The protein content in juices is largely dependent on the residual pulp remaining after extraction, making fruit juices a poor source of protein unless they are fortified (van Breda and de Kok, 2018).

Fruit jams and jellies, which are concentrated fruit preserves, retain more of the fruit's sugars but have limited protein content due to the processing methods. These foods contain substances that have a high sugar and pectin composition, thereby making the proteins low in content. Proteins in such jams are usually between 0.3 and 1.0 g for every 100 g of food, depending on the fruit used. Generally, food material usually employed in the production of jams, i.e., strawberries, raspberries, and apricots, contains minute percentages of protein. Their protein content, however, remains negligible in comparison to the amount of sugar present in them, making them a source of no proteins (Devi et al., 2015).

Conversely, fruit bars, typically made with dried fruits and mixed with nuts and seeds, have more protein. Additions of protein, like almonds, chia seeds, or hemp seeds, can contribute appreciably to the protein levels of these products. Fruit bars can have 1 to 3 g of protein/100 g, depending on the seeds and nuts added and their quantity. Fruit bars provide more protein than fruit juice or jam, as they tend to be more functional foods with greater concern for nutritional content. Some commercial fruit bars can include added protein isolates like pea or soy protein to boost their protein levels (Munir et al., 2016).

Fruit smoothies, which are usually made by mixing fruits with milk or plant milk alternatives, can be of moderate to high protein levels. Smoothies made using dairy products such as milk or yogurt can provide approximately 4 to 5 g of protein/100 g. Plant-based smoothies made using alternatives such as almond or soy milk also have some protein levels, although the amount is usually lower than in dairy-based smoothies. Protein levels can be increased further by adding ingredients such as nut butter or protein powder. Smoothies are a convenient way of having fruit with a protein element, and thus, they are a popular choice among those who wish to increase their daily protein levels without compromising the nutritional content of fruits (Tiwari, 2018; Bäuerle and Kühn, 2022).

Fruit ice creams, which are usually prepared with a blend of fruit purees and milk or milk alternatives, have protein based on the inclusion of protein-rich ingredients. Sorbets, which consist mostly of fruit juice and sugar, have zero to little protein content (Palka and Skotnicka, 2022). But, fruit ice cream prepared from milk or cream may contain as much as 3 g of protein/100 g, based on the dairy content concentration. Non-dairy versions of these ice creams made from plant-based milk alternatives can offer protein but in smaller quantities compared to their dairy counterparts. Protein content in fruit-based frozen desserts is often not a primary focus. However, it can still contribute to overall nutritional value, especially when the product includes ingredients like soy or almond milk, which are rich in protein (Pimentel et al., 2021).

CLASSIFICATION OF FRUIT PROTEINS

Fruit proteins are nature's diversity and have multiple functions in the plant. They are classified into some categories, such as storage proteins, structural proteins, enzymatic proteins, defense proteins, and miscellaneous proteins, each having a distinct function in the development, growth, and defense of the fruit. This categorization not only illustrates the biological importance of these proteins but also their possible uses in food technology, agriculture, and biotechnology.

Storage Proteins

Storage proteins are found primarily in seeds and fruits. During periods of germination and growth, these proteins serve as a reserve of amino acids and nitrogen.

Storage proteins are very crucial in fruit development, especially in seeds, which accumulate them for essential nutrients during seedling growth. Examples of storage proteins are globulins, albumins, and prolamins. For instance, apples contain small amounts of globulin proteins that also contribute to the total protein content in the fruit (Misran and Jaafar, 2019). Likewise, banana pulp may be low in total protein but has trace amounts of storage proteins, including chitinase (Toledo et al., 2012). Such storage proteins play a vital role in the early development of the fruit, but are available in much lower concentrations than in legumes or grains.

Structural Proteins

Structural proteins take part in the creation and sustenance of the fruit cell wall and the general structural integrity. Structural proteins enhance fruit texture, which evolves during ripening. Significant structural proteins in fruits are pectin methyl esterase, expansins, and xyloglucan endotransglucosylase. The proteins take part in altering the cell wall during fruit ripening, affecting traits like softness, texture, and firmness. For instance, the breakdown of pectin, a major cell wall component, is primarily due to pectinase, an enzyme found in fruits like tomatoes and pears, contributing to their softening during ripening (Wang et al., 2018). Similarly, mangoes contain expansins, which facilitate cell wall loosening and the softening of the fruit (Méndez-Yáñez et al., 2024).

Enzymatic Proteins

Enzymes are protein molecules that catalyze biochemical reactions, and in fruit, they perform essential functions in metabolic processes such as ripening, starch hydrolysis, and flavor compound synthesis. Key enzymes in fruit metabolism include amylase, polyphenol oxidase, and lipoxygenase. For instance, pineapple contains the protease enzyme bromelain, which is a significant component of protein degradation and also has medicinal value in that it is used for anti-inflammatory purposes (Varilla et al., 2021). Mangoes also release amylase, the enzyme responsible for starch degradation into sugars when a mango is ripe (Hossain et al., 2014). Enzymatic proteins, especially those related to the breakdown of starches, are critical in converting unripe, starchy fruits into sweeter, more palatable products.

Defense Proteins

Defense proteins are responsible for guarding fruits against pests, pathogens, and environmental stresses. They form part of the plant's defense system, and they comprise protease inhibitors, lectins, and chitinases with antifungal, antibacterial, and antiviral activities. For example, citrus fruits such as lemons and oranges yield chitinases, which protect the fruit against fungal pathogens (Liu et al., 2022). Likewise, papaya has protease inhibitors, which defend it against herbivores and insects by inhibiting their digestive enzymes (Macalood et al., 2014). The synthesis of these defense proteins is crucial for fruit survival to ensure that the fruit is intact during development and post-harvest storage.

Structure of Fruit Proteins

The protein structure of fruit is critical in determining its functional attributes, such as solubility, emulsification, and gelation (Table 5.2). Fruit proteins, just like other proteins, have their structure divided into four hierarchical categories: primary, secondary, tertiary, and quaternary structures. These are determined by factors such as the amino acid composition, the interactions of molecules, and the specific nature of the fruit source.

Primary Structure: The primary structure of fruit proteins is a linear arrangement of amino acids bonded together by peptide bonds. The sequence of the specific amino acids is governed by the genetic code present in the DNA of the fruit. Fruit proteins usually contain a wide variety of amino acids with different ratios of hydrophobic and hydrophilic residues. For instance, guava, banana, and papaya proteins contain more non-essential amino acids like glutamine and alanine (Raimbault et al., 2013; Zaini et al., 2022). The function of the protein and its interaction with other molecules, including water, lipids, and carbohydrates, are determined by the primary sequence.

Secondary Structure: The secondary structure is the folding of the protein chain into periodic structures like α-helices and β-sheets, stabilized by hydrogen bonds. In fruit proteins, the secondary structure is usually dependent on the environment of the protein, such as pH, temperature, and the presence of other solutes. For instance, citrus fruit proteins, like those in oranges, usually have a high percentage of α-helix structures, which account for their solubility and functional characteristics (Zhu et al., 2023). Secondary structure is crucial for the stability of the protein and its interaction with other biomolecules.

Tertiary Structure: The tertiary structure is the three-dimensional shape of a protein, formed when the secondary structure folds into a compact and functional shape. Numerous interactions, such as hydrophobic interactions, hydrogen bonding, disulfide bridges, and ionic bonds sustain this organization. For example, avocado seed proteins exhibit globular tertiary structures stabilized by hydrophobic interactions and disulfide bonds, which contribute to their emulsifying and functional properties in food systems (Wang et al., 2019). It is this level of organization that is important to the functional properties of fruit proteins, such as their enzymatic activity, emulsifying capacity, and gelation character.

Quaternary Structure: The quaternary structure describes the formation of two or more protein subunits into a bigger functional protein complex. Certain fruit proteins, specifically in fruits such as jackfruit and guava, have quaternary structures, wherein numerous polypeptide chains assemble into a functional entity. For example, guava proteins are known to create aggregates that increase their gel-forming capacity (Nuerjiang et al., 2023). Quaternary structure is important for the protein to carry out intricate functions like enzymatic activity or emulsion stabilization.

TABLE 5.2
Comparative Evaluation of Structural Traits in Proteins from Various Fruits

Protein Name	Source Fruit	Structural Traits	Reference
Thaumatin-like Protein (TLP)	Banana (*Musa acuminata*)	Two hundred-amino-acid single polypeptide chain; has fold structure similar to thaumatin; has electrostatically polarized surface.	Barre et al. (2000)
Calmodulin-like (CML) proteins	Passion Fruit (*Passiflora edulis*)	Contains conserved motifs and gene structures; involved in flower and fruit development.	Zhang et al. (2024)
Fruit bromelain	Pineapple (*Ananas comosus*)	Comprises I29 and PLCE domains; I29 domain is globular with three α-helices; exhibits proteolytic activity.	Pang et al. (2020)
Late embryogenesis abundant (LEA) proteins	Strawberry (*Fragaria × ananassa*)	Contains conserved domains; plays roles in fruit ripening and stress responses.	Banerjee and Roychoudhury (2016)
Arabinogalactan proteins (AGPs)	Various Fruits	Exhibits structural diversity due to protein domain and carbohydrate moiety; involved in fruit growth and development.	Seifert and Roberts (2007)
Pectate lyase (Pel)	Apricot (*Prunus armeniaca*)	Contains β-folding formation; conserved residues RXPXXR and WIDH; involved in cell wall degradation.	Marin-Rodriguez et al. (2002)
YABBY proteins	Star Fruit (*Averrhoa carambola*)	Contains α-helix, strand, and coil structures; involved in fruit development.	Li et al. (2022)
HVA22 proteins	Passion Fruit (*Passiflora edulis*)	Contains conserved TB2/DP1/HVA22 domain; involved in stress responses.	Chen et al. (2024)
Actinidin	Kiwifruit (*Actinidia deliciosa*)	Cysteine protease; similar to papain; active over a wide pH range; comprises up to 50% of the fruit's soluble protein content.	Chalabi et al. (2014)

(Continued)

TABLE 5.2 (*Continued*)
Comparative Evaluation of Structural Traits in Proteins from Various Fruits

Protein Name	Source Fruit	Structural Traits	Reference
Brazzein	Oubli (*Pentadiplandra brazzeana*)	Small, heat-stable sweet-tasting protein; consists of 54 amino acids; contains four disulfide bonds.	Zhao et al. (2021)
Miraculin	Miracle Fruit (*Synsepalum dulcificum*)	Glycoprotein; modifies taste perception by binding to sweetness receptors; consists of 191 amino acids.	Swamy et al. (2014)
Monellin	Serendipity Berry (*Dioscoreophyllum cumminsii*)	Sweet-tasting protein; composed of two noncovalently associated polypeptide chains; molecular weight of 10.7 kDa.	Ayobami et al. (2019)
Pentadin	Oubli (*Pentadiplandra brazzeana*)	Sweet-tasting protein; molecular weight of 12 kDa; resembles monellin in sweetness.	Wel et al. (1989)
Caricain	Papaya (*Carica papaya*)	Cysteine protease; synthesized as a preproenzyme; primary structure determined with 216 amino acids.	Sebastián et al. (2018)
Thaumatin	Katemfe Fruit (*Thaumatococcus daniellii*)	Sweet-tasting protein; consists of 207 amino acids; contains eight disulfide bonds.	An et al. (2024)
Cystatin	Tomato (*Solanum lycopersicum*)	Cysteine protease inhibitor; exhibits beta sandwich structure; involved in defense mechanisms.	Tremblay et al. (2019)
Papain	Papaya (*Carica papaya*)	Cysteine protease; single-chain globular protein; adaptation to different environments.	Liu et al. (2018)
Polyphenol oxidase	Apple (*Malus domestica*)	Copper-containing enzyme; dimeric structure; involved in browning of cut fruits.	Yoruk and Marshall (2003)
Alpha-amylase	Mango (*Mangifera indica*)	Enzyme with catalytic triad; hydrolyzes alpha-1,4 glycosidic bonds during fruit maturation.	Mishra and Patel (2017)
Dehydrin	Grape (*Vitis vinifera*)	Group II LEA protein; protects cells from dehydration during stress; contains K-segment.	Hanana et al. (2014)

Fruit proteins may be grouped into a number of families depending on their function and structure. These include albumins, globulins, prolamins, and glutelins. For instance, albumins are soluble in water and can be commonly found in fruits such as apples and citrus (Hutapea et al., 2023). Prolamins and glutelins are less predominant but may be found in fruits such as papaya (Olivares-Galván et al., 2020). Fruit proteins are involved in cell wall integrity and texture contribution. For instance, the protein matrix in bananas contributes to the structure and texture of the fruit upon ripening. Proteins like polygalacturonase in tomato and papaya fruits also contribute to the softening of fruit during ripening (Wang et al., 2023).

PROPERTIES OF FRUIT PROTEINS

Even though they occur in lower amounts than animal proteins, fruit proteins are crucial to the nutritional and functional attributes of fruit products. Their distinctive physical and chemical characteristics play a significant role in their use in the food, pharmaceutical, and cosmetic industries. Such factors as processing methods, the fruit type, and environmental factors determine these characteristics. This part explores the physical and chemical nature of fruit proteins (Figure 5.3).

Physical Properties

Solubility: Solubility is among the most important physical characteristics of fruit proteins, which affect their incorporation into liquid food products. Fruit proteins, like those from banana, guava, and papaya, are less soluble than animal proteins because they are more hydrophobic and have a different amino acid composition (Olunusi et al., 2024). Solubility, however, is pH, temperature, and fruit species dependent. For instance, guava

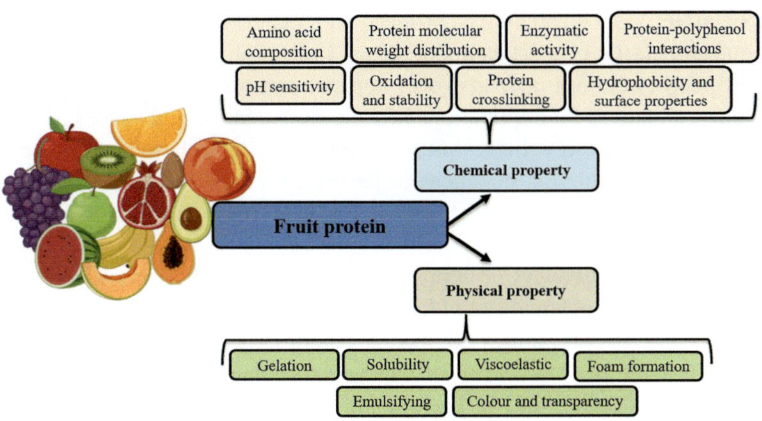

FIGURE 5.3 Fruit protein properties.

proteins exhibit greater solubility when in an acidic environment, and hence, they are utilized in drinks and fruit juice. Likewise, solubility is increased after treatments such as fermentation or enzymatic hydrolysis (Cruz-Casas et al., 2021).

Gelation: Fruit proteins, especially from pectin-containing fruits like apples, citrus fruits, and bananas, have gelation characteristics when exposed to heat and water. Protein-water and protein–protein interactions affect the process of gelation. For example, guava proteins have shown good gel-forming capacity, which can be utilized in gel-based foods such as jellies (Zhang et al., 2023). Papaya and avocado proteins have also been reported to produce gels with different levels of stability based on the level of hydration and the addition of other ingredients, such as acids and sugars (Trujillo-Cayado et al., 2024).

Emulsifying: Emulsifying properties pertain to the capacity of proteins to stabilize emulsions through decreasing surface tension between oil and water phases. Fruit proteins, such as papaya and avocado, are good emulsifiers that ensure the formation of stable emulsions for application in sauces, salad dressings, and mayonnaise. The presence of lipids in fruits such as avocado facilitates emulsification, which supports the overall functionality of such proteins in food (Flores et al., 2019). Other fruits, such as mango and citrus, display moderate emulsifying activity since they contain special protein and fat composition (Deng et al., 2020).

Formation of Foam: Foam development is another crucial physical characteristic of fruit proteins, which finds its application in whipped product manufacturing, bakery items, and drinks. Fruits like guava, banana, and jackfruit have shown the ability to form foam when mechanically whipped (Wadia et al., 2020; Watharkar et al., 2021; Wu et al., 2022). The capacity of these proteins to form stable foams is dependent on their molecular weight, amino acid composition, and surface-active characteristics. These proteins are generally used in plant-based foods, including vegan meringues or foamed desserts.

Viscoelastic: Viscoelasticity is a property of a material that can show both viscous and elastic properties under stress. Fruit proteins, such as citrus fruits and pectin-containing fruits (e.g., apples), exhibit viscoelasticity, making them well-suited for applications in fruit pastes, jams, and sauces. The protein molecules can achieve both elasticity (during gelation) and viscosity (during the production of sauces or pastes) (Xu et al., 2025). This bimodal behavior is particularly useful in food products where texture and mouthfeel are important.

Color and Transparency: Fruit protein color matters in both consumer acceptance and the visual qualities of food products. Fruit proteins such as oranges, lemons, and pomegranates are mostly colorless or pale yellow, while those of fruits such as dragon fruit or papaya can impart pale reddish or orange shades (Bashir et al., 2024). The color of proteins can affect product perception, particularly in the case of beverage and confectionery products.

Chemical Properties of Fruit Proteins

Composition of Amino Acid: The amino acid composition of fruit proteins is one of the key factors that determine their nutritional quality. Fruit proteins, e.g., those in guava, banana, and papaya, have a relatively higher percentage of non-essential amino acids (e.g., glutamine, alanine, and serine) but, sometimes, lower contents of certain essential amino acids (e.g., lysine and methionine, relative to animal proteins). However, some fruits, like guava, are relatively rich in essential amino acids, making their proteins more nutritionally complete for human consumption (Almutairi et al., 2022).

Weight Distribution of Protein Molecules: Fruit proteins generally possess a broad distribution of molecular weights, which determines their functionality, digestibility, and solubility. Fruits such as banana, avocado, and jackfruit have proteins that consist of a blend of globulins, albumins, and prolamins with different molecular weights (Kumar et al., 2023). Smaller proteins are more soluble and bioavailable, whereas larger proteins could provide functionalities like gelation or emulsification (Feng et al., 2021). Fruit proteins' molecular weight also impacts digestibility and uptake in the body.

Enzymatic Activity: Some fruits, specifically papaya (papain) and pineapple (bromelain), have intrinsic enzymic activities that are capable of breaking down protein molecules into peptide units or amino acids. Such proteases have specific functions during digestion and function in food manufacturing for the tenderization of meat as well as clearing juices (Oliver-Simancas et al., 2024). The enzymatic activity of fruit proteins is responsible for their functional characteristics in food systems, including texture improvement and protein degradation during fermentation or digestive processes (Toy et al., 2022).

Interactions of Protein–Polyphenol: Fruit proteins tend to interact with polyphenolic compounds, which are widely available in fruits such as pomegranate, blueberry, and grape. These interactions have the potential to influence protein solubility, stability, and functionality. For example, pomegranate polyphenols may bind to fruit proteins to create complexes that decrease protein solubility but increase their antioxidant activity (Trigueros et al., 2014). Protein–polyphenol interaction is very important in deciding the sensory and nutritional characteristics of fruit products, particularly in functional foods for offering health benefits.

pH Sensitivity: Fruit proteins are pH-sensitive, which can affect their solubility and structural integrity. The majority of fruit proteins, including citrus, guava, and mango proteins, show solubility at certain pH values. For example, guava proteins are most soluble under acidic pH, whereas the proteins of mango are best stabilized at neutral pH (Angulo-López et al., 2021). pH sensitivity becomes critical while developing food products with specific pH conditions for maximum protein functionality, for example, fruit juices or fermented foods.

Oxidation: Oxidation is one of the primary factors influencing the stability and functionality of fruit proteins. Proteins, particularly those derived from fruits such as apples and citrus, are prone to oxidation when subjected to light, heat, and oxygen. Oxidation of amino acid residues such as cysteine and methionine can result in protein degradation and loss of function (Ahmad et al., 2017). To maintain protein quality, antioxidants or controlled storage conditions are employed in most fruit protein-fortified foods to reduce oxidative damage and extend shelf life (Khan et al., 2021).

Crosslinking of Proteins: Fruit proteins, especially those of high-molecular-weight fruits such as jackfruit and avocado, can be crosslinked when exposed to certain reagents, heat, or pressure. Crosslinking improves the stability and texture of protein gels and emulsions. Crosslinking stiffens the protein structures, rendering them more stable against degradation and enhancing their functionality in foods (Sulaiman et al., 2022).

Hydrophobicity and surface properties: Fruit protein hydrophobicity is the major factor affecting their emulsifying and foaming properties. Hydrophobic amino acid residues in fruit proteins, like those from papaya and avocado, would be able to bind to lipids and improve the stabilization of emulsions and foams. The surface characteristics of fruit proteins are important in deciding their ability to interact with other food system components (Amagliani et al., 2021).

PROCESSING IMPACT ON FRUIT PROTEINS

Fruit proteins are naturally rich in nutrients but can be subjected to drastic changes during processing. Processing techniques, including heat treatment, fermentation, drying, and mechanical treatments, can influence the structure, solubility, digestibility, and bioactivity of proteins.

Effect of Heat Treatment on Fruit Proteins

Heat processing, including pasteurization or boiling, results in protein denaturation. This results in the unwinding of the protein structure, impacting its functionality. In mango, papaya, and guava fruits, heat treatment may decrease protein solubility and alter their texture, rendering the proteins less bioavailable (Barba et al., 2015). Mild heat treatments, however, may increase the release of bioactive peptides from fruit proteins, enhancing their antioxidant and anti-inflammatory activities (Fadimu et al., 2022). Although heat processing can lower fruit protein digestibility by changing its molecular structure, it can inactivate antinutritional elements, including protease inhibitors that could otherwise restrict protein absorption. Papaya proteins, for instance, have enzymes (such as papain) that can be inactivated through heat, enhancing the digestibility of the protein (Ahmad et al., 2018).

Effect of Drying Process on Fruit Proteins

Drying processes, including freeze-drying or air-drying, may lead to protein aggregation, which influences the solubility and bioavailability of proteins. Freeze-drying, for instance, is more likely to maintain the native protein structure compared to hot-air-drying (Omolola et al., 2017). Drying, however, can decrease protein solubility, which makes it less desirable for uses such as protein powders in foods and beverages and food fortification. Drying processes can degrade the emulsifying, foaming, and gel-forming capability of fruit proteins, which are vital for food product formulations. For instance, protein from fruits such as bananas, jackfruit, and guava applied in emulsified foods such as sauce could lose functionality upon drying (Siddiqui et al., 2024). Although there is some loss of solubility and functionality, dried fruits such as guava still contain high amounts of bioactive compounds, including polyphenols and flavonoids, which improve the health value of fruit proteins (Vijaya Anand et al., 2020).

Impact of Fermentation on Fruit Proteins

Fermentation is among the most useful operations for enhancing the digestibility of fruit proteins. Microbial enzymes during fermentation degrade larger protein complexes into peptides and amino acids, improving their bioavailability (Wang et al., 2022). Fermented fruit proteins, especially from mango (Elechı et al., 2022) and banana (Coutinho et al., 2023), exhibit enhanced digestibility and elevated functional properties, such as enhanced antioxidant activity and gut health promotion. Fermentation may also improve the bioactive nature of fruit proteins. Fermented fruit proteins have also been found to possess greater antioxidant and antihypertensive activities because microbial fermentation releases bioactive peptides. For example, fermented guava proteins possess significant anti-inflammatory and antioxidant activities that can be utilized for health uses (Mohan et al., 2024). Fermentation reduces antinutritional components like phytates and tannins found in raw fruits that can impair protein absorption. Fermentation lowers these compounds, thereby enhancing the nutritional value of fermented fruit protein products (Samtiya et al., 2020).

Mechanical Processing Effect

Mechanical processing methods like grinding, crushing, and blending are commonly used to extract fruit proteins for various food products. These methods break down fruit tissues and help release soluble proteins. However, excessive mechanical processing can cause physical damage to proteins, leading to a decrease in their structural integrity and functionality (Raseetha et al., 2022). In some cases, mechanical processing can enhance the bioavailability of fruit proteins by making them more accessible for digestion. For instance, mechanical disruption of the cell wall in fruits like bananas and apples facilitates the release of proteins that are otherwise bound within the cell structure, thus enhancing their digestibility (Holland et al., 2020).

Ultrasonication and High-Pressure Processing Effect

Ultrasonication is a non-thermal processing method employing sound waves to break the protein structure, improving protein solubility and emulsifying capability. Ultrasonication is employed to enhance the extraction of proteins from fruits such as citrus and pomegranate without imposing severe heat damage (Barbhuiya et al., 2021).

High-pressure processing (HPP) is a novel technique for fruit protein extraction, especially from sensitive fruits like berries and mangoes. HPP preserves the bioactive peptides and minimizes the risk of protein denaturation relative to conventional thermal processing techniques. HPP can improve the functional properties of fruit proteins without altering their native structure and bioactivity (Houška et al., 2022).

Storage Impact on Fruit Proteins

Fruit proteins may degrade during storage through oxidation, enzymatic hydrolysis, or microbial spoilage. The oxidation of amino acids, especially cysteine and methionine, may lead to loss of nutritional value and functionality of proteins. Storage under proper conditions, like refrigeration or controlled atmosphere storage, reduces protein degradation (Bodbodak & Moshfeghifar, 2016). Freezing of fruits maintains their protein content for a longer time, but it can influence the protein solubility as well as the texture. Although freezing has little influence on the amino acid composition, the structure of the cells could be changed, and the solubility of proteins might be influenced when the fruit is defrosted (Bonat et al., 2016).

APPLICATION OF FRUIT PROTEINS

Fruit proteins, while normally present in limited amounts relative to other protein sources, are receiving consideration for their beneficial health aspects, functional uses, and environmental appeal.

Nutritional and Health Advantages

Fruits like guava and mulberries contain proteins that are high in essential amino acids, such as lysine and tryptophan, which are both significant for human well-being. Both of these amino acids are necessary for the preservation of muscle tissue and overall growth. Fruit proteins are also lower in fat and sometimes have fewer calories compared to animal proteins, and hence, they are also good for weight control. Fruit proteins, especially from guava and pomegranate, possess antioxidant peptides that are involved in lowering oxidative stress and inflammation, which are major contributors to aging and several chronic diseases (Bhardwaj et al., 2014). The bioactive molecules in fruit proteins can play an important role in maintaining heart health and lowering the risk of diseases like cancer. Some fruit proteins, like those from papaya and avocado, have shown anti-inflammatory activity. This renders them beneficial in disease control, such as that of arthritis, as well as promoting joint health (Jeon et al., 2022; Marra et al., 2024). The protein complement, along with other bioactive compounds, helps boost the immune system.

Applications in the Food Industry

As more people embrace vegan and vegetarian diets, fruit proteins have become a substitute for conventional animal proteins. Specifically, guava protein has been seen as a good addition to protein powders and smoothies (Mareguddikar et al., 2024).

Protein from bananas and jackfruits are also utilized to make plant-based protein bars and meal replacement products. Fruits like banana, jackfruit, and avocado are used in enriching products such as bakery products, snack foods, and breakfast cereals. Fruit proteins are added to elevate the nutritional composition of these items, making them healthier for fitness-oriented consumers (Kumoro et al., 2020). Fruit proteins, particularly those from avocado, possess good emulsifying attributes. This renders them useful in the manufacture of salad dressings, sauces, and mayonnaise, where a stable emulsification process is essential for texture and consistency (Wang et al., 2019). Fruits like pomegranate and raspberry are becoming commonly used in functional beverages, where their protein content not only enhances the nutritional value but also flavor and texture. These drinks are regularly promoted as giving additional health benefits, including enhancing skin health or supporting digestion (Gupta et al., 2023).

Products of Personal Care and Beauty

Fruit proteins are employed in skincare products to enhance skin health. For instance, pomegranate, avocado, and guava proteins can help ensure skin elasticity and hydration. The proteins have antioxidants that shield the skin against oxidative stress and premature aging (Kumar et al., 2023). These proteins have become popular in creams, masks, and serums in the skincare industry. Avocado and coconut proteins can make hair follicles stronger, support hair growth, and enhance the texture of the hair. Avocado and coconut proteins are found to be increasingly applied in conditioners, shampoos, and hair masks because these proteins provide moisturization and defense against environmental harm (Barve and Dighe, 2016; Bhore et al., 2021).

Applications in the Medical and Pharmaceutical Industry

Proteins in fruits such as guava, pomegranate, apples, blueberries, and grapes have bioactive compounds like flavonoids and polyphenols that have been researched for their anticancer effects. These compounds neutralize free radicals and inhibit inflammation, both of which are linked to cancer (Jaglan et al., 2022). Proteins in fruits such as papaya have digestive enzymes like papain, which can be used to digest proteins and enhance digestion. These enzymes are also utilized in dietary supplements intended to alleviate bloating and support nutrient uptake (Kostiuchenko et al., 2022). Mangoes, as well as other tropical fruits, have been found to possess potential in the regulation of blood sugar levels and insulin sensitivity. Their protein content, along with other active compounds, could provide a natural method of diabetes management (Evans et al., 2014).

Application in Agriculture and Industries

The fruit residue-derived protein, for example, from banana and jackfruit, is gaining applications in animal feed. The fruit proteins are affordable, plant-based substitutes for conventional animal feed and provide a sustainable option in the poultry and livestock sectors (Tusé et al., 2024). Fruits like citrus are being researched in the production of biodegradable packaging materials. These proteins can be broken down into films and coatings that are not just biodegradable but also non-toxic, providing a sustainable alternative to plastic packaging (Dubey et al., 2023).

Green and Zero-Waste Innovations

Fruits like mangoes, papayas, and citrus contain considerable quantities of protein in their seeds, peels, and other by-products. Scientists are in the process of extracting these proteins from fruit waste to produce sustainable food ingredients or include them in functional products (Lau et al., 2021). This helps to reduce food waste and develop circular economies. Fruit proteins, especially citrus fruit proteins, are being explored for their capacity to create biodegradable plastics. These products are biodegradable and environmentally friendly, aiding in the prevention of plastic waste (Yaradoddi et al., 2022).

Biotechnological Applications

Biotechnological advancements have enabled the genetic modification of fruit crops, including bananas, to enhance their protein content. This may render fruits such as bananas and papayas more feasible as protein sources of food, particularly where protein malnutrition is a common issue (Ghag and Ganapathi, 2017). The fermentative enhancement of fruit proteins' bioavailability is becoming increasingly important. Fermentation may also degrade fruit proteins into bioactive peptides, which provide added health advantages in the form of better gut health and increased immunity (Septembre-Malaterre et al., 2018).

REFERENCES

Ahmad, I. Z., Tabassum, H., Ahmad, A., & Kuddus, M. (2018). Food enzymes in pharmaceutical industry: Perspectives and limitations. In *Enzymes in Food Technology: Improvements and Innovations* (pp. 41–62). https://doi.org/10.1007/978-981-13-1933-4_3

Ahmad, S., Khan, H., Shahab, U., Rehman, S., Rafi, Z., Khan, M. Y., Ansari, A., Siddiqui, R., Ashraf, J. M., Abdullah, S. M. S., & Uddin, M. (2017). Protein oxidation: An overview of metabolism of sulphur containing amino acid, cysteine. *Frontiers in Bioscience (Scholar Edition)*, 9(1), 71–87.

Alabi, M. A., Aremu, M. O., & Akpomie, T. M. (2024). Phytochemical, antioxidant, and nutritional evaluation of kernel and pulp of Bush mango (*Irvingia gabonesis*). *Human, Health & Halal Metrics*, 5(1). https://doi.org/10.30502/jhhhm.2024.450732.1087

Alam, F., Mohammadin, K., Shafique, Z., Amjad, S. T., & Asad, M. H. H. B. (2022). Citrus flavonoids as potential therapeutic agents: A review. *Phytotherapy Research*, 36(4), 1417–1441. https://doi.org/10.1002/ptr.7261

Allaqaband, S., Dar, A. H., Patel, U., Kumar, N., Nayik, G. A., Khan, S. A., Ansari, M., Alabdallah, N. M., Kumar, P., Pandey, V. K., Kovács, B., & Shaikh, A. M. (2022). Utilization of fruit seed-based bioactive compounds for formulating the nutraceuticals and functional food: A review. *Frontiers in Nutrition*, 9, 902554. https://doi.org/10.3389/fnut.2022.902554

Almutairi, K. F., Saleh, A. A., Ali, M. M., Sas-Paszt, L., Abada, H. S., & Mosa, W. F. (2022). Growth performance of guava trees after the exogenous application of amino acids glutamic acid, arginine, and glycine. *Horticulturae*, 8(12), 1110. https://doi.org/10.3390/horticulturae8121110

Amagliani, L., Silva, J. V., Saffon, M., & Dombrowski, J. (2021). On the foaming properties of plant proteins: Current status and future opportunities. *Trends in Food Science & Technology*, 118, 261–272. https://doi.org/10.1016/j.tifs.2021.10.001

Amid, B. T., Mirhosseini, H. & Kostadinović, S (2012). Chemical composition and molecular structure of polysaccharide-protein biopolymer from *Durio zibethinus* seed: Extraction and purification process. *Chemistry Central Journal* 6, 117. https://doi.org/10.1186/1752-153X-6-117

An, J. P., Wang, Y., Munger, S. D., & Tang, X. (2024). A review on natural sweeteners, sweet taste modulators and bitter masking compounds: Structure-activity strategies for the discovery of novel taste molecules. *Critical Reviews in Food Science and Nutrition*, 1–24. https://doi.org/10.1080/10408398.2024.2326012

Angulo-López, J. E., Flores-Gallegos, A. C., Torres-León, C., Ramírez-Guzmán, K. N., Martínez, G. A., & Aguilar, C. N. (2021). Guava (*Psidium guajava* L.) fruit and valorization of industrialization by-products. *Processes*, 9(6), 1075. https://doi.org/10.3390/pr9061075

Arte, P., Das, R. S., Tungare, K., Bhori, M., Kadam, A., & Kadam, D. (2024). Utilization of fruit waste as protein sources. In *Adding Value to Fruit Wastes* (pp. 107–130). Academic Press. https://doi.org/10.1016/B978-0-443-13842-3.00004-6

Arumugam, R., & Manikandan, M. (2011). Fermentation of pretreated hydrolyzates of banana and mango fruit wastes for ethanol production. *Asian Journal of Experimental Biological Sciences*, 2(2), 246–256.

Ayobami, O., Nwakego, A. O. H., Ndigwe, E. V., & Adunni, A. O. (2019). Chemical and sensory evaluation of Kunun-zaki sweetened with serendipity berry (*Dioscoresphyllum cumminsii*) and enriched with defatted moringa seed flour. *Annals of Food Science and Technology*, 20(2), 380–386.

Bahari, H., Rafiei, H., Goudarzi, K., Omidian, K., Asbaghi, O., Hosseini Kolbadi, K.S., Naderian, M., & Hosseini, A. (2023). The effects of pomegranate consumption on inflammatory and oxidative stress biomarkers in adults: a systematic review and meta-analysis. *Inflammopharmacology*, 31, 2283–2301. https://doi.org/10.1007/s10787-023-01294-x

Bajramova, A., & Spégel, P. (2022). A comparative study of the fatty acid profile of common fruits and fruits claimed to confer health benefits. *Journal of Food Composition and Analysis*, 112, 104657. https://doi.org/10.1016/j.jfca.2022.104657

Banerjee, A., & Roychoudhury, A. (2016). Group II late embryogenesis abundant (LEA) proteins: Structural and functional aspects in plant abiotic stress. *Plant Growth Regulation*, 79, 1–17. https://doi.org/10.1007/s10725-015-0113-3

Barba, F. J., Terefe, N. S., Buckow, R., Knorr, D., & Orlien, V. (2015). New opportunities and perspectives of high pressure treatment to improve health and safety attributes of foods. A review. *Food Research International*, 77, 725–742.

Barbhuiya, R. I., Singha, P., & Singh, S. K. (2021). A comprehensive review on impact of non-thermal processing on the structural changes of food components. *Food Research International*, 149, 110647. https://doi.org/10.1016/j.foodres.2021.110647

Barre, A., Hervé, C., Lescure, B., & Rouge, P. (2000). Purification and structural analysis of an abundant thaumatin-like protein from ripe banana fruit (*Musa acuminata*). *Planta*, 211(1), 82–89. https://doi.org/10.1007/s004250000354

Barve, K., & Dighe, A. (2016). *The Chemistry and Applications of Sustainable Natural Hair Products*. Springer International Publishing. https://doi.org/10.1007/978-3-319-29419-3

Bashir, I., Pandey, V. K., Dar, A. H., Dash, K. K., Shams, R., Mir, S. A., & Zahoor, I. (2024). Exploring sources, extraction techniques and food applications: A review on biocolors as next-generation colorants. *Phytochemistry Reviews*, 1–26. https://doi.org/10.1007/s11101-023-09908-6

Bäuerle, L., & Kühn, S. (2022). Development of a protein supplemented fruit smoothie using pea protein isolate as a plant-based protein alternative. *Future Foods*, 5, 100145. https://doi.org/10.1016/j.fufo.2022.100145

Bhardwaj, R. L., Nandal, U., Pal, A., & Jain, S. (2014). Bioactive compounds and medicinal properties of fruit juices. *Fruits*, 69(5), 391–412. https://doi.org/10.1051/fruits/2014027

Bhore, S. J., Ochoa, D. S., Houssari, A. A., Zelaya, A. L., Yang, R., Chen, Z., Deeya, S. S., Sens, S. C. D. S., Schumann, M., Zhang, Z., & Eltantawy, E. (2021). The avocado (*Persea americana* Mill.): A review and sustainability perspectives. *Preprints*. https://doi.org/10.20944/preprints202112.0523.v1

Bodbodak, S., & Moshfeghifar, M. (2016). Advances in controlled atmosphere storage of fruits and vegetables. In *Eco-Friendly Technology for Postharvest Produce Quality* (pp. 39–76). Academic Press. https://doi.org/10.1016/B978-0-12-804313-4.00002-5

Bonat Celli, G., Ghanem, A., & Su-Ling Brooks, M. (2016). Influence of freezing process and frozen storage on the quality of fruits and fruit products. *Food Reviews International*, 32(3), 280–304. https://doi.org/10.1080/87559129.2015.1075212

Burton-Freeman, B. M., Sandhu, A. K., & Edirisinghe, I. (2017). Mangos and their bioactive components: Adding variety to the fruit plate for health. *Food & Function*, 8(9), 3010–3032. https://doi.org/10.1039/C7FO00190H

Campos, D. A., Gómez-García, R., Vilas-Boas, A. A., Madureira, A. R., & Pintado, M. M. (2020). Management of fruit industrial by-products—A case study on circular economy approach. *Molecules*, 25(2), 320. https://doi.org/10.3390/molecules25020320

Carughi, A., Gallaher, D., & Mandalari, G. (2020). Bioavailability of nutrients and phytochemicals from dried fruits. In Preedy, V. R. (Ed.), *Health Benefits of Nuts and Dried Fruits* (pp. 369–396). Boca Raton, FL: CRC Press.

Chalabi, M., Khademi, F., Yarani, R., & Mostafaie, A. (2014). Proteolytic activities of kiwifruit actinidin (*Actinidia deliciosa* cv. Hayward) on different fibrous and globular proteins: A comparative study of actinidin with papain. *Applied Biochemistry and Biotechnology*, 172, 4025–4037. https://doi.org/10.1007/s12010-014-0812-7

Chen, Q., Huang, L., Li, X., Ma, Y., Wang, Z., Zhang, C., Lin, F., & Liu, C. (2024). Comprehensive genome-wide analysis and expression profiling of the *HVA22* gene family unveils their potential roles in soybean responses to abiotic stresses. *Journal of Plant Growth Regulation*, 44, 2122–2138. https://doi.org/10.1007/s00344-024-11531-4

Cordiano, R., Gammeri, L., Di Salvo, E., Gangemi, S., & Minciullo, P.L. (2024). Pomegranate (Punica granatum L.) Extract Effects on Inflammaging. *Molecules*, 29(17), 4174. https://doi.org/10.3390/molecules29174174

Coutinho, G. S. M., Ribeiro, A. E. C., Prado, P. M. C., Oliveira, É. R., Careli-Gondim, Í., Oliveira, A. R., & Vilas Boas, E. V. D. B. (2023). New plant-based fermented beverage made of baru nut enriched with probiotics and green banana: Composition, physicochemical and sensory properties. *Journal of Food Science and Technology*, 60(10), 2607–2618. https://doi.org/10.1007/s13197-023-05781-5

Cruz-Casas, D. E., Aguilar, C. N., Ascacio-Valdés, J. A., Rodríguez-Herrera, R., Chávez-González, M. L., & Flores-Gallegos, A. C. (2021). Enzymatic hydrolysis and microbial fermentation: The most favorable biotechnological methods for the release of bioactive peptides. *Food Chemistry: Molecular Sciences*, 3, 100047. https://doi.org/10.1016/j.fochms.2021.100047

Deng, Z., Pan, Y., Chen, W., Chen, W., Yun, Y., Zhong, Q., Zhang, W., & Chen, H. (2020). Effects of cultivar and growth region on the structural, emulsifying and rheological characteristic of mango peel pectin. *Food Hydrocolloids*, 103, 105707. https://doi.org/10.1016/j.foodhyd.2020.105707

Devi, M. P., Bhowmick, N., Bhanusree, M. R., & Ghosh, S. K. (2015). Preparation of value-added products through preservation. In *Value Addition of Horticultural Crops: Recent Trends And Future Directions* (pp. 13–41). https://doi.org/10.1007/978-81-322-2262-0_2

Dubey, P., Tripathi, G., Mir, S. S., & Yousuf, O. (2023). Current scenario and global perspectives of citrus fruit waste as a valuable resource for the development of food packaging film. *Trends in Food Science & Technology*, 141, 104190. https://doi.org/10.1016/j.tifs.2023.104190

Elechı, J. O. G., Nwiyi, I. U., & Oboh, E. J. (2022). Fermentation and diet diversity: Biochemical and functional properties of fermented mango (*Mangifera indica* L) pulp flour. *European Food Science and Engineering*, 3(2), 44–51. https://doi.org/10.55147/efse.1181022

Evans, S. F., Meister, M., Mahmood, M., Eldoumi, H., Peterson, S., Perkins-Veazie, P., Clarke, S. L., Payton, M., Smith, B. J., & Lucas, E. A. (2014). Mango supplementation improves blood glucose in obese individuals. *Nutrition and Metabolic Insights*, 7, NMI-S17028. https://doi.org/10.4137/NMI.S17028

Fadimu, G. J., Le, T. T., Gill, H., Farahnaky, A., Olatunde, O. O., & Truong, T. (2022). Enhancing the biological activities of food protein-derived peptides using non-thermal technologies: A review. *Foods*, 11(13), 1823. https://doi.org/10.3390/foods11131823

Feng, X., Wang, J., Zhang, Y., & Li, S. (2021). Improved solubility and interface properties of pigskin protein by microwave-assisted enzymatic hydrolysis. *Food Hydrocolloids*, 118, Article 106779. https://doi.org/10.1016/j.foodhyd.2020.106779

Figueroa, A., Wong, A., Kalfon, R., & Paolillo, J. (2014). Effects of watermelon supplementation on aortic blood pressure and wave reflection in individuals with prehypertension: a pilot study. *American Journal of Hypertension*, 27(7), 899–906. https://doi.org/10.1093/ajh/hpt155

Flores, M., Saravia, C., Vergara, C. E., Avila, F., Valdés, H., & Ortiz-Viedma, J. (2019). Avocado oil: Characteristics, properties, and applications. *Molecules*, 24(11), 2172. https://doi.org/10.3390/molecules24112172

Ghag, S. B., & Ganapathi, T. R. (2017). Genetically modified bananas: To mitigate food security concerns. *Scientia Horticulturae*, 214, 91–98. https://doi.org/10.1016/j.scienta.2016.11.023

Gündeşli, M. A., Korkmaz, N., & Okatan, V. (2019). Polyphenol content and antioxidant capacity of berries: A review. *International Journal of Agriculture Forestry and Life Sciences*, 3(2), 350–361.

Gupta, A., Sanwal, N., Bareen, M. A., Barua, S., Sharma, N., Olatunji, O. J., Nirmal, N. P., & Sahu, J. K. (2023). Trends in functional beverages: Functional ingredients, processing technologies, stability, health benefits, and consumer perspective. *Food Research International*, 170, 113046. https://doi.org/10.1016/j.foodres.2023.113046

Hanana, M., Daldoul, S., Fouquet, R., Deluc, L., Leon, C., Hoefer, M., Barrieu, F., & Ghorbel, A. (2014). Identification and characterization of a seed-specific grapevine dehydrin involved in abiotic stress response within tolerant varieties. *Turkish Journal of Botany*, 38(6), 1157–1168. https://doi.org/10.3906/bot-1405-35

Holland, C., Ryden, P., Edwards, C. H., & Grundy, M. M. L. (2020). Plant cell walls: Impact on nutrient bioaccessibility and digestibility. *Foods*, 9(2), 201. https://doi.org/10.3390/foods9020201

Hossain, M. A., Rana, M. M., Kimura, Y., & Roslan, H. A. (2014). Changes in biochemical characteristics and activities of ripening associated enzymes in mango fruit during the storage at different temperatures. *BioMed Research International*, 2014(1), 232969. https://doi.org/10.1155/2014/232969

Houška, M., Silva, F. V. M., Evelyn, Buckow, R., Terefe, N. S., & Tonello, C. (2022). High pressure processing applications in plant foods. *Foods*, 11(2), 223. https://doi.org/10.3390/foods11020223

Hutapea, T. P. H., Madurani, K. A., Syahputra, M. Y., Hudha, M. N., Asriana, A. N., & Kurniawan, F. (2023). Albumin: Source, preparation, determination, applications, and prospects. *Journal of Science: Advanced Materials and Devices*, 8(2), 100549. https://doi.org/10.1016/j.jsamd.2023.100549

Jaglan, P., Buttar, H. S., Al-bawareed, O. A., & Chibisov, S. (2022). Potential health benefits of selected fruits: Apples, blueberries, grapes, guavas, mangos, pomegranates, and tomatoes. In *Functional foods and Nutraceuticals in Metabolic and non-Communicable Diseases* (pp. 359–370). Academic Press. https://doi.org/10.1016/B978-0-12-819815-5.00026-4

Jeon, Y. A., Chung, S. W., Kim, S. C., & Lee, Y. J. (2022). Comprehensive assessment of antioxidant and anti-inflammatory properties of papaya extracts. *Foods*, 11(20), 3211. https://doi.org/10.3390/foods11203211

Jhade, R. K., Jain, P. K., & Verma, A. K. (2015). Fruits and their Processing for Nutritional Security-A Review. *Trends in Biosciences*, 8(17), 4439–4449.

Kalcsits, L., Lotze, E., Tagliavini, M., Hannam, K. D., Mimmo, T., Neilsen, D., … & Valverdi, N. A. (2020). Recent achievements and new research opportunities for optimizing macronutrient availability, acquisition, and distribution for perennial fruit crops. *Agronomy*, 10(11), 1738. https://doi.org/10.3390/agronomy10111738

Kalse, S. B., Khapre, L. S., Ranveer, R. C., & Rathod, N. B. (2025). Fruit and vegetable protein—Functions and potential health benefits. In *Plant-Based Proteins* (pp. 403–416). Academic Press. https://doi.org/10.1016/B978-0-443-13370-1.00014-4

Kandasamy, P., & Shanmugapriya, C. (2015). Medicinal and nutritional characteristics of fruits in human health. *Journal of Medicinal Plants Studies*, 4(4), 124–131.

Khan, M. R., Di Giuseppe, F. A., Torrieri, E., & Sadiq, M. B. (2021). Recent advances in biopolymeric antioxidant films and coatings for preservation of nutritional quality of minimally processed fruits and vegetables. *Food Packaging and Shelf Life*, 30, 100752. https://doi.org/10.1016/j.fpsl.2021.100752

Kostiuchenko, O., Kravchenko, N., Markus, J., Burleigh, S., Fedkiv, O., Cao, L., Letasiova, S., Skibo, G., Fåk Hållenius, F. and Prykhodko, O., (2022). Effects of proteases from pineapple and papaya on protein digestive capacity and gut microbiota in healthy C57BL/6 mice and dose-manner response on mucosal permeability in human reconstructed intestinal 3D tissue model. *Metabolites*, 12(11), 1027. https://doi.org/10.3390/metabo12111027

Kumar, P., Karam, N., & Choudhury, D. (2023). Assessment of the Phytochemical Constituents and Metabolites in the Medicinal Plants and Herbal Medicine Used in the Treatment and Management of Skin Diseases. In *Herbal Medicine Phytochemistry: Applications and Trends* (pp. 1–41). Cham: Springer International Publishing. https://doi.org/10.1007/978-3-031-21973-3_21-1

Kumoro, A. C., Alhanif, M., & Wardhani, D. H. (2020). A critical review on tropical fruits seeds as prospective sources of nutritional and bioactive compounds for functional foods development: A case of Indonesian exotic fruits. *International Journal of Food Science*, 2020(1), 4051475. https://doi.org/10.1155/2020/4051475

Lau, K. Q., Sabran, M. R., & Shafie, S. R. (2021). Utilization of vegetable and fruit by-products as functional ingredient and food. *Frontiers in Nutrition*, 8, 661693. https://doi.org/10.3389/fnut.2021.661693

Li, C., Dong, N., Shen, L., Lu, M., Zhai, J., Zhao, Y., Chen, L., Wan, Z., Liu, Z., Ren, H., & Wu, S. (2022). Genome-wide identification and expression profile of YABBY genes in *Averrhoa carambola*. *PeerJ*, 10, e12558. https://doi.org/10.7717/peerj.12558

Liu, J., Sharma, A., Niewiara, M. J., Singh, R., Ming, R., & Yu, Q. (2018). Papain-like cysteine proteases in *Carica papaya*: Lineage-specific gene duplication and expansion. *BMC Genomics*, 19, 1–12. https://doi.org/10.1186/s12864-017-4394-y

Liu, S., Du, Y., Zhang, D., Yang, F., He, X., & Long, C. (2022). Aluminum sulfate inhibits green mold by inducing chitinase activity of *Penicillium digitatum* and enzyme activity of citrus fruit. *Food Control*, 136, 108854. https://doi.org/10.1016/j.foodcont.2022.108854

Loveday, S. M. (2022). Protein digestion and absorption: The influence of food processing. *Nutrition Research Reviews*, 36(2), 177–196. https://doi.org/10.1017/S0954422422000245

Macalood, J. S., Vicente, H. J., Gorospe, J. G., Boniao, R. D., & Roa, E. C. (2014). Revisiting *Carica papaya* L. latex potentials may resolve agricultural infestation problems. *International Journal of Scientific & Technology Research*, 3(1), 95–8.

Mareguddikar, S. C., Siddiqua, A., Khayum, A., Sharma, S., Divya, C., Sahu, S., & Jigile, S (2024). Nature's protein bounty: Diving into fruit-derived proteins. *Plant Archives*, 24(2), 116–126. https://doi.org/10.51470/PLANTARCHIVES.2024.v24.no.2.015

Marin-Rodriguez, M. C., Orchard, J., & Seymour, G. B. (2002). Pectate lyases, cell wall degradation, and fruit softening. *Journal of Experimental Botany*, 53(377), 2115–2119. https://doi.org/10.1093/jxb/erf089

Marra, A., Manousakis, V., Zervas, G. P., Koutis, N., Finos, M. A., Adamantidi, T., Panoutsopoulou, E., Ofrydopoulou, A. and Tsoupras, A., 2024. Avocado and its by-products as natural sources of valuable anti-inflammatory and antioxidant bioactives for functional foods and cosmetics with health-promoting properties. *Applied Sciences*, 14(14), 5978. https://doi.org/10.3390/app14145978

Méndez-Yáñez, A., Carrasco-Orellana, C., Ramos, P., & Morales-Quintana, L. (2024). Alpha-expansins: More than three decades of wall creep and loosening in fruits. *Plant Molecular Biology*, 114(4), 84. https://doi.org/10.1007/s11103-024-01481-6

Mishra, A., & Patel, V. H. (2017). Evaluation of alpha amylase, alpha glucosidase and protein glycation inhibitory activities of different elements of mango (*Mangifera indica*). *PRAJNA-Journal of Pure and Applied Science*, 24–25, 174–181.

Misran, A., & Jaafar, A. H. (2019). Protein. In *Postharvest Physiology and Biochemistry of Fruits and Vegetables* (pp. 315–334). Woodhead Publishing. https://doi.org/10.1016/B978-0-12-813278-4.00015-4

Mohan, H., Sathya, P. M., Lee, S. W., Jeong, H. J., Ryu, S., Lee, A. H., & Oh, B. T. (2024). Enhanced health benefits of *Psidium guajava* peel fermented with *Lactiplantibacillus* from Myeolchi-jeot: Antioxidant, anti-inflammatory, and anti-cancer properties. *Food Bioscience*, 59, 104211. https://doi.org/10.1016/j.fbio.2024.104211

Moyo, H. N. (2024). Innovations in plant-based proteins. *Nutritional Value and Consumer Acceptance*, 8(20), 412–434.

Munir, M., Nadeem, M., Qureshi, T. M., Jabbar, S., Atif, F. A., & Zeng, X. (2016). Effect of protein addition on the physicochemical and sensory properties of fruit bars. *Journal of Food Processing and Preservation*, 40(3), 559–566. https://doi.org/10.1111/jfpp.12635

Nuerjiang, M., Li, Y., Yue, X., Kong, B., Liu, H., Wu, K., & Xia, X. (2023). Analysis of inhibition of guava (*Psidium guajava* l.) leaf polyphenol on the protein oxidative aggregation of frozen chicken meatballs based on structural changes. *Food Research International*, 164, 112433. https://doi.org/10.1016/j.foodres.2022.112433

Olivares-Galván, S., Marina, M. L., & García, M. C. (2020). Extraction and characterization of antioxidant peptides from fruit residues. *Foods*, 9(8), 1018. https://doi.org/10.3390/foods9081018

Oliver-Simancas, R., Labrador-Fernández, L., Abellán-Diéguez, C., García-Villegas, A., Del Caro, A., Leyva-Jimenez, F. J., & Alañón, M. E. (2024). Valorization applications of pineapple and papaya byproducts in food industry. *Comprehensive Reviews in Food Science and Food Safety*, 23(3), e13359. https://doi.org/10.1111/1541-4337.13359

Olunusi, S. O., Ramli, N. H., Fatmawati, A., Ismail, A. F., & Okwuwa, C. C. (2024). Revolutionizing tropical fruits preservation: Emerging edible coating technologies. *International Journal of Biological Macromolecules*, 130682. https://doi.org/10.1016/j.ijbiomac.2024.130682

Omolola, A. O., Jideani, A. I., & Kapila, P. F. (2017). Quality properties of fruits as affected by drying operation. *Critical Reviews in Food Science and Nutrition*, 57(1), 95–108. https://doi.org/10.1080/10408398.2013.859563

Palka, A., & Skotnicka, M. (2022). The health-promoting and sensory properties of tropical fruit sorbets with inulin. *Molecules*, 27(13), 4239. https://doi.org/10.3390/molecules27134239

Pang, W. C., Ramli, A. N. M., & Hamid, A. A. A. (2020). Comparative modelling studies of fruit bromelain using molecular dynamics simulation. *Journal of Molecular Modeling*, 26, 1–18. https://doi.org/10.1007/s00894-020-04398-1

Patil, B. S., Jayaprakasha, G. K., & Murthy, K. N. C. (2017). Beyond vitamin C: The diverse, complex health-promoting properties of citrus fruits. *Citrus Research & Technology*, 38(1), 107–121. https://dx. doi.org/10.4322/crt.ICC063

Pimentel, T. C., de Oliveira, L. I. G., de Souza, R. C., & Magnani, M. (2021). Probiotic non-dairy frozen dessert: Technological and sensory aspects and industrial challenges. *Trends in Food Science & Technology*, 107, 381–388. https://doi.org/10.1016/j.tifs.2020.11.008

Przybylska, S. (2020). Lycopene – A bioactive carotenoid offering multiple health benefits: A review. *International Journal of Food Science & Technology*, 55(1), 11–32. https://doi.org/10.1111/ijfs.14260

Raimbault, A. K., Zuily-Fodil, Y., Soler, A., & de Carvalho, M. H. C. (2013). A novel aspartic acid protease gene from pineapple fruit (*Ananas comosus*): Cloning, characterization and relation to postharvest chilling stress resistance. *Journal of Plant Physiology*, 170(17), 1536–1540.

Ranjha, M. M. A. N., Irfan, S., Nadeem, M., & Mahmood, S. (2022). A comprehensive review on nutritional value, medicinal uses, and processing of banana. *Food Reviews International*, 38(2), 199–225. https://doi.org/10.1080/87559129.2020.1725890

Raseetha, S., Aida, F. M. N. A., Chompoorat, P., Murtini, E. S., Fuggate, P., Roslan, N. F. A., & Nur-Diana, S. A. (2022). Disintegration with considerable changes in form: Cutting/dicing, crushing and grinding, shredding, sheeting, and pulping. In *Postharvest and Postmortem Processing of Raw Food Materials* (pp. 181–240). Woodhead Publishing. https://doi.org/10.1016/B978-0-12-818572-8.00004-8

Rice, S. (2022). *Supercharge Your Diet: Ten Easy Ways to Get Everything You Need from Your Food*. London: Hachette UK.

Richa, R., Kohli, D., Vishwakarma, D., Mishra, A., Kabdal, B., Kothakota, A., … & Naik, B. (2023). Citrus fruit: Classification, value addition, nutritional and medicinal values, and relation with pandemic and hidden hunger. *Journal of Agriculture and Food Research*, 14, 100718. https://doi.org/10.1016/j.jafr.2023.100718

Rifna, E. J., Misra, N. N., & Dwivedi, M. (2023). Recent advances in extraction technologies for recovery of bioactive compounds derived from fruit and vegetable waste peels: A review. *Critical Reviews in Food Science and Nutrition*, 63(6), 719–752. https://doi.org/10.1080/10408398.2021.1952923

Samtiya, M., Aluko, R. E., & Dhewa, T. (2020). Plant food anti-nutritional factors and their reduction strategies: An overview. *Food Production, Processing and Nutrition*, 2, 1–14. https://doi.org/10.1186/s43014-020-0020-5

Sangeeta, Nayik, G. A., & Muzaffar, K. (2020). Watermelon. In Raghavendra, M. S., Rao, A. R., & Nayik, K. (Eds.), *Antioxidants in Fruits: Properties and Health Benefits* (pp. 333–364). Singapore: Springer. https://doi.org/10.1007/978-981-15-7285-2_17

Sebastián, D. I., Guevara, M. G., Rocío, T. F., & Virginia, T. C. (2018). An overview of plant Proteolytic enzymes. In *Biotechnological Applications of Plant Proteolytic Enzymes* (pp. 1–19). https://doi.org/10.1007/978-3-319-97132-2_1

Seifert, G. J., & Roberts, K. (2007). The biology of arabinogalactan proteins. *Annual Review of Plant Biology*, 58, 137–161. https://doi.org/10.1146/annurev.arplant.58.032806.103801

Septembre-Malaterre, A., Remize, F., & Poucheret, P. (2018). Fruits and vegetables, as a source of nutritional compounds and phytochemicals: Changes in bioactive compounds during lactic fermentation. *Food Research International*, 104, 86–99. https://doi.org/10.1016/j.foodres.2017.09.031

Siddiqui, S. A., Ucak, İ., Jain, S., Elsheikh, W., Ali Redha, A., Kurt, A., & Toker, O. S. (2024). Impact of drying on techno-functional and nutritional properties of food proteins and carbohydrates-A comprehensive review. *Drying Technology*, 42(4), 592–611. https://doi.org/10.1080/07373937.2024.2303580

Simmonds, M., & Preedy, V. R. (Eds.). (2015). *Nutritional Composition of Fruit Cultivars*. London: Academic Press.

Smith, C. E., & Tucker, K. L. (2011).Health benefits of cereal fibre: A review of clinical trials. *Nutrition Research Reviews*, 24(1), 118–131. https://doi.org/10.1017/S0954422411000023

Song, X., Ren, Z., & Wang, X. (2022). Alanine metabolism and its role in glucose homeostasis: A comprehensive review. *Food Science & Human Wellness*, 11(4), 1258–1266. https://doi.org/10.1016/j.fshw.2022.01.012

Stephen, J., & Radhakrishnan, M. (2022). Avocado (*Persea americana* Mill.) fruit: Nutritional value, handling and processing techniques, and health benefits. *Journal of Food Processing and Preservation*, 46(12), e17207. https://doi.org/10.1111/jfpp.17207

Stolp, L. J., & Kodali, D. R. (2022). Naturally occurring high-oleic oils: Avocado, macadamia, and olive oils. In *High Oleic Oils* (pp. 7–52). AOCS Press. https://doi.org/10.1016/B978-0-12-822912-5.00003-4

Stonehouse, W., Gammon, C.S., Beck, K.L., Conlon, C.A., von Hurst, P.R., & Kruger, R. (2013). Kiwifruit: our daily prescription for health. *Canadian Journal of Physiology and Pharmacology,* 91(6), 442–447. https://doi.org/10.1139/cjpp-2012-0303

Sulaiman, N. S., Sintang, M. D., Zaini, H. M., Munsu, E., Matanjun, P., & Pindi, W. (2022). Applications of protein crosslinking in food products. *International Food Research Journal*, 29(4), 723–739. https://doi.org/10.47836/ifrj.29.4.01

Sun, M., Zhao, B., Cai, Z., Yan, J., Ma, R., & Yu, M. (2022). Amino acid profiles in peach (*Prunus persica* L.) fruit. *Foods*, 11(12), 1718. https://doi.org/10.3390/foods11121718

Swami, S. B., Thakor, N. J., Haldankar, P. M., & Kalse, S. B. (2012). Jackfruit and its many functional components as related to human health: A review. *Comprehensive Reviews in Food Science and Food Safety*, 11(6), 565–576. https://doi.org/10.1111/j.1541-4337.2012.00210.x

Swamy, K. B., Hadi, S. A., Sekaran, M., & Pichika, M. R. (2014). The clinical effects of *Synsepalum dulcificum*: A review. *Journal of Medicinal Food*, 17(11), 1165–1169. https://doi.org/10.1089/jmf.2013.3084

Tangolar, S., Turan, M., Tangolar, S., & Ateş, F. (2019). Evaluation of amino acid contents and enzyme activities of seeds from Semillon and Carignane wine grape cultivars grown under different irrigation conditions. *Scientia Horticulturae*, 251, 181–188.

Tiwari, U. (2018). Production of fruit-based smoothies. In Rajauria, G. & Tiwari, B. K. (Eds.), *Fruit Juices* (pp. 261–278). London: Academic Press.

Toledo, T. T., Nogueira, S. B., Cordenunsi, B. R., Gozzo, F. C., Pilau, E. J., Lajolo, F. M., & do Nascimento, J. R. O. (2012). Proteomic analysis of banana fruit reveals proteins that are differentially accumulated during ripening. *Postharvest Biology and Technology*, 70, 51–58. https://doi.org/10.1016/j.postharvbio.2012.04.005

Toy, J. Y. H., Lu, Y., Huang, D., Matsumura, K., & Liu, S. Q. (2022). Enzymatic treatment, unfermented and fermented fruit-based products: Current state of knowledge. *Critical Reviews in Food Science and Nutrition*, 62(7), 1890–1911. https://doi.org/10.1080/10408398.2020.1848788

Tremblay, J., Goulet, M. C., & Michaud, D. (2019). Recombinant cystatins in plants. *Biochimie*, 166, 184–193. https://doi.org/10.1016/j.biochi.2019.06.006

Trigueros, L., Wojdyło, A., & Sendra, E. (2014). Antioxidant activity and protein–Polyphenol interactions in a pomegranate (*Punica granatum L.) yogurt. Journal of Agricultural and Food Chemistry*, 62(27), 6417–6425. https://doi.org/10.1021/jf501503h

Trovato, M., Funck, D., Forlani, G., Okumoto, S., & Amir, R. (2021). Amino acids in plants: Regulation and functions in development and stress defense. *Frontiers in Plant Science*, 12, 772810. https://doi.org/10.3389/fpls.2021.772810

Trujillo-Cayado, L. A., García-Domínguez, I., Rodríguez-Luna, A., Hurtado-Fernández, E., & Santos, J. (2024). Cricket protein as an innovative emulsifier for avocado oil: Formulation and characterization of sustainable emulsions. *Applied Sciences*, 14(4), 1674. https://doi.org/10.3390/app14041674

Tusé, D., McNulty, M., McDonald, K. A., & Buchman, L. W. (2024). A review and outlook on expression of animal proteins in plants. *Frontiers in Plant Science*, 15, 1426239. https://doi.org/10.3389/fpls.2024.1426239

van Breda, S. G., & de Kok, T. M. (2018). Smart combinations of bioactive compounds in fruits and vegetables may guide new strategies for personalized prevention of chronic diseases. *Molecular Nutrition & Food Research*, 62(1), 1700597. https://doi.org/10.1002/mnfr.201700597

Varilla, C., Marcone, M., Paiva, L., & Baptista, J. (2021). Bromelain, a group of pineapple proteolytic complex enzymes (*Ananas comosus*) and their possible therapeutic and clinical effects. A summary. *Foods*, 10(10), 2249. https://doi.org/10.3390/foods10102249

Vijaya Anand, A., Velayuthaprabhu, S., Rengarajan, R. L., Sampathkumar, P., & Radhakrishnan, R. (2020). Bioactive compounds of guava (*Psidium guajava* L.). *Bioactive Compounds in Underutilized Fruits and* Nuts, 503–527.

Vincente, A. R., Manganaris, G. A., Ortiz, C. M., Sozzi, G. O., & Crisosto, C. H. (2014). Nutritional quality of fruits and vegetables. In *Postharvest Handling* (pp. 69–122). Academic Press. https://doi.org/10.1016/B978-0-12-408137-6.00005-3

Wadia, C., Champawat, P. S., & Jain, S. K. (2020). Influence of foam thickness, whipping time and drying temperature on production of guava powder during foam mat drying. *Journal of Pharmacognosy and Phytochemistry*, 9(5S), 720–727.

Wahlqvist, M. L., & Wattanapenpaiboon, N. (2020). Macronutrients: Carbohydrates. In Wahlqvist, M. L. (Ed.), *Food and Nutrition* (pp. 208–224). London: Routledge.

Wang, D., Yeats, T. H., Uluisik, S., Rose, J. K., & Seymour, G. B. (2018). Fruit softening: Revisiting the role of pectin. *Trends in Plant Science*, 23(4), 302–310. https://doi.org/10.1016/j.tplants.2018.01.006

Wang, J. S., Wang, A. B., Zang, X. P., Tan, L., Xu, B. Y., Chen, H. H., Jin, Z. Q., & Ma, W. H. (2019). Physicochemical, functional and emulsion properties of edible protein from avocado (*Persea americana* Mill.) oil processing by-products. *Food Chemistry*, 288, 146–153. https://doi.org/10.1016/j.foodchem.2019.02.098

Wang, W., Yang, P., Rao, L., Zhao, L., Wu, X., Wang, Y., & Liao, X. (2022). Effect of high hydrostatic pressure processing on the structure, functionality, and nutritional properties of food proteins: A review. *Comprehensive Reviews in Food Science and Food Safety*, 21(6), 4640–4682. https://doi.org/10.1111/1541-4337.13033

Wang, Y., Fan, Z., Zhai, Y., Huang, H., Vainstein, A., & Ma, H. (2023). Polygalacturonase gene family analysis identifies FcPG12 as a key player in fig (*Ficus carica* L.) fruit softening. *BMC Plant Biology*, 23(1), 320. https://doi.org/10.1186/s12870-023-04315-7

Watharkar, R. B., Chakraborty, S., Srivastav, P. P., & Srivastava, B. (2021). Foaming and foam mat drying characteristics of ripe banana [*Musa balbisiana* (BB)] pulp. *Journal of Food Process Engineering*, 44(8), e13726. https://doi.org/10.1111/jfpe.13726

Wel, H. V. D., Larson, G., Hladik, A., Hladik, C. M., Hellekant, G., & Glaser, D. (1989). Isolation and characterization of pentadin, the sweet principle of *Pentadiplandra brazzeana* Baillon. *Chemical Senses*, 14(1), 75–79. https://doi.org/10.1093/chemse/14.1.75

Wistaff, E. A., Beller, S., Schmid, A., Neville, J. J., & Nietner, T. (2021). Chemometric analysis of amino acid profiles for detection of fruit juice adulterations–Application to verify authenticity of blood orange juice. *Food Chemistry*, 343, 128452. https://doi.org/10.1016/j.foodchem.2020.128452

Wu, J., Zhou, X., Zhou, L., Liu, W., Zhong, J., Zhang, Y., & Liu, C. (2022). Physicochemical, structural, and functional properties of protein fractions and protein isolate from jackfruit seeds. *Journal of Food Science*, 87(4), 1540–1551. https://doi.org/10.1111/1750-3841.16104

Xiong, Y., Yan, P., Du, K., Li, M., Xie, Y., & Gao, P. (2020). Nutritional component analyses of kiwifruit in different development stages by metabolomic and transcriptomic approaches. *Journal of the Science of Food and Agriculture*, 100(6), 2399–2409. https://doi.org/10.1002/jsfa.10251

Xu, C., Zhang, P., Huang, X., Liu, J., & Li, Y. (2025). Characterization and evaluation of the viscoelastic models for normal and bruised apple pulp. *Computers and Electronics in Agriculture*, 229, 109756. https://doi.org/10.1016/j.compag.2024.109756

Yaradoddi, J. S., Banapurmath, N. R., Ganachari, S. V., Soudagar, M. E. M., Sajjan, A. M., Kamat, S., Mujtba, M. A., Shettar, A. S., Anqi, A. E., Safaei, M. R., Elfasakhany, A., Siddiqui, M. I. H., & Ali, M. A. (2022). Bio-based material from fruit waste of orange peel for industrial applications. *Journal of Materials Research and Technology*, 17, 3186–3197. https://doi.org/10.1016/j.jmrt.2021.09.016

Yoruk, R., & Marshall, M. R. (2003). Physicochemical properties and function of plant polyphenol oxidase: A review 1. *Journal of Food Biochemistry*, 27(5), 361–422. https://doi.org/10.1111/j.1745-4514.2003.tb00289.x

Yu, X. Y., Zou, Y., Zheng, Q. W., Lu, F. X., Li, D. H., Guo, L. Q., & Lin, J. F. (2021). Physicochemical, functional and structural properties of the major protein fractions extracted from *Cordyceps militaris* fruit body. *Food Research International*, 142, 110211. https://doi.org/10.1016/j.foodres.2021.110211

Zaini, H. M., Roslan, J., Saallah, S., Munsu, E., Sulaiman, N. S., & Pindi, W. (2022). Banana peels as a bioactive ingredient and its potential application in the food industry. *Journal of Functional Foods*, 92, 105054. https://doi.org/10.1016/j.jff.2022.105054

Zhang, H., Wang, Z. Y., Yang, X., Zhao, H. T., Zhang, Y. C., Dong, A. J., Jing, J., & Wang, J. (2014). Determination of free amino acids and 18 elements in freeze-dried strawberry and blueberry fruit using an Amino Acid Analyzer and ICP-MS with micro-wave digestion. *Food Chemistry*, 147, 189–194. https://doi.org/10.1016/j.foodchem.2013.09.118

Zhang, H., Wang, Z., & Li, Y. (2024). Genome-wide identification and expression analysis of calmodulin and calmodulin-like genes in passion fruit (*Passiflora edulis*) and their involvement in flower and fruit development. *BMC Plant Biology*, 24, 52. https://doi.org/10.1186/s12870-024-05295-y

Zhang, M., Wu, W., Li, C., & Chen, H. (2023). Structural characteristics and gelation behavior of polysaccharides extracted from guava (*Psidium guajava*) fruit. *Food Hydrocolloids*, 140, 108587. https://doi.org/10.1016/j.foodhyd.2023.108587

Zhao, X., Wang, C., Zheng, Y., & Liu, B. (2021). New insight into the structure-activity relationship of sweet-tasting proteins: Protein sector and its role for sweet properties. *Frontiers in Nutrition*, 8, 691368. https://doi.org/10.3389/fnut.2021.691368

Zhu, C., Zhang, S., Zhu, N., Wu, Q., Du, M., He, X., Bai, Y., & Wang, S. (2023). Effects of citrus fiber on the emulsifying properties and molecular structure of mutton myofibrillar protein: An underlying mechanisms study. *Food Bioscience*, 51, 102304. https://doi.org/10.1016/j.fbio.2022.102304

6 Vegetable Proteins

Tehreem Ashraf, Rabia Naseem, and Noor Ul Ain Malhotrah

INTRODUCTION

Plant portions that humans and other animals eat are called vegetables. Edible portion of plant, including stems, fruits, leaves, flowers, seeds, and roots, is referred to by their actual name, which is still commonly used today. Another definition of the term is used rather haphazardly, frequently about cultural and culinary customs. The food items made up of particular plants, such as cereal grains, nuts, flowers and fruits can be excluded but they consist of savory fruits like tomatoes, crops like broccoli, and grains like pulses. During the emergence of a new agricultural living, most likely between 10,000 and 7,000 BC, Hunter-gatherers initially collected vegetables from the wild before they were cultivated worldwide (Aryee et al., 2018). Locally grown plants were initially grown, but as time passed, trade brought both common and exotic crops from other areas to increase domestic diversity. Most vegetables are cultivated throughout the globe these days, as long as the climate allows; however plants can be cultivated in protected areas in less desirable places. China is the world's greatest producer of vegetables, and international commerce in agricultural goods allows people to purchase vegetables grown in overseas countries.

Production ranges from Agribusinesses that grow single-product crops on vast areas to subsistence farmers providing for their families' food needs. According to the variety of vegetable, grading, storage, processing, and marketplace sales happen following crop harvest. A key component of human nutrition, vegetables can be eaten fresh or cooked and are mostly low in fat and carbohydrates but high in vitamins, minerals, and dietary fiber. It is highly advised by dietitians that people should consume five or more servings of fresh fruit and vegetables every single day to stay healthy.

Multiple U.N., French, and German records from the last century demonstrate that acute malnutrition was more prevalent in hospitalized children in Germany, UK, America, and France. According to the survey, 24% of children were underweight (90th percentile weight for height) in a large government hospital in Germany, with mild malnutrition accounts for 17.9%, moderate malnutrition for 4.4%, and severe malnutrition for 1%. Forty percent of patients with chondrogenesis, 34% of patients with infection, 33% of patients with cystic fibrosis, 38% of patients with heart disease, 27% of patients receiving cancer treatment, and 23% of patients with gastrointestinal disorders were found to be undernourished. For healthy growth and to prevent malnutrition, protein is essential (Sakthivel and Guruvayoorappan, 2015).

By 2063, there will be 13 billion people on the planet, double the current population of 6.5 billion to make sure there is enough protein present for the population's consumption (Nolden and Forde, 2023). Protein is important since it is one of three macronutrients in

DOI: 10.1201/9781003613909-6

our diet, with the remaining two being fats and carbohydrates. A range of amino acid residues that are necessary for replenishment, tissue growth, and repair make up proteins. Because of societal changes like growing incomes, population growth, and improved living standards, as well as increasing knowledge about the significance of protein for aging well and a balanced diet, protein prices are on the rise. Significant population dynamics and nutritional status are being produced by economic urban development, especially in moderate and high-income countries. Most of the global increase in the demand for animal-based diets are occurring in developing nations.

COMPOSITION OF DIFFERENT VEGETABLES

Vegetables and their products, including salads, pickled, dried, frozen, marinated, and preserved vegetables, as well as fermented and non-fermented pickles and prepared sauces, include compounds that have pharmacological action. Since vegetables consist of many bio-products necessary for the existence of animal kingdom, humans and vegetables are inextricably linked. Vegetables' complex chemical composition, which is vital to the human body and includes organic substances (carbohydrates, proteins, lipids, and organic acids), phytoncides and antimicrobial substances, a high concentration of minerals (Ca, P, Fe, K, Mg, S, Cl, Zn and Cu), and a high concentration of vitamins (A, B complex, C, E, F, K) (Butnariu and Butu, 2015).

Phytochemicals and antimicrobials, organic substances (carbohydrates, proteins, lipids, and organic acids), a high concentration of minerals (Ca, P, Fe, K, Mg, S, Cl, Zn and Cu), and a high concentration of vitamins (A, B complex, C, E, F, K, P and PP) are among the complex chemicals that make vegetables important from the perspective of the food industry. The complex chemicals found in vegetables are beneficial to the humans because they have a high nutritional and energy value, regulate metabolism overall, improve the enzymatic activity, internal glands, skeletal and muscular systems, and hydration due to higher content of water. One of the characteristics of vegetables is their high nutritional density.

Vegetables are a vital component of the basic diet because they include a range of biological phytonutrients and have a significant nutritional density with minimal calorie input (Table 6.1). Vegetables should be eaten fresh when they have the highest

TABLE 6.1
Vitamin and Mineral Composition of Some High-Protein Vegetables per 100 g

Vegetable	Ca(mg)	Fe(mg)	P(mg)	Na(mg)	K(mg)	Vit A(µg)	Vit B1(mg)	Vit B2(mg)	Niacin(mg)	Vit C(mg)
Brussels Sprouts	4`	0.8	51	5	308	0	0.01	0.03	0.30	3.0
Broccoli	47	0.7	66	33	316	623	0.07	0.12	0.64	89.2
Asparagus	24	2.1	52	2	202	756	0.14	0.14	0.98	5.6
Artichoke	28.0	0.6	72.3	3	561	–	0.07	0.06	1.53	6
Green peas	25	1.5	108	5	244	765	0.27	0.13	2.09	40.0
Mushroom	15	3.5	57	9	506	–	0.02	0.22	4.09	–

nutritional value as some processing techniques and storage methods destroy some water-soluble vitamins and nutrients. Vegetables use solar energy to create the essential substances to stay alive and numerous organic phytochemicals can be obtained as raw materials with important applications in medicine, medical dermatology, beauty products, and science and technology.

A healthy body depends on vegetables and their products, which have reduced number of calories and increased vitamin and mineral content. Among the 1,200 plant species classified as "vegetable plants," only 250 are harvested. Most vegetables are 75%–95% water, despite their nutrient content. The dry matter of plants is composed of minerals, vitamins, phytonutrients, and biomolecules. Minerals, various flavors, water-soluble and liposoluble vitamins, and intestinal transit-facilitating fibers like cellulose and pectin are all abundant in vegetables. Because they are 80%–90% water, they also give energy and keeps body hydrated. Vegetables and plant-based foods with various uses are also a vital source of essential nutrients.

Depending on the species, a stem, roots, fruits, seeds, bulb, flowers, leaves, or entire plant may or may not be edible. Vegetables provide varying amounts of nutrients depending on the part of the plant consumed. While tubers have a high carbohydrate content but are low in vitamins, roots and stems are rich in cellulose and hemicellulose fiber but low in vitamin C; green vegetables are abundant in vitamin C, magnesium, carotenoids, and chlorophyll; and blossoms, which are not generally eaten intact (like cauliflower), are rich in vitamin B and K-complex vitamins. Cellulose makes up the majority of carbohydrates, while the composition varies throughout plant kinds. Many plants include sucrose, starch, and cellulose, including lettuce, tomatoes, etc.

Mono- and disaccharide carbohydrates like cellulose, pectin, and starch are commonly found in vegetables. It has been discovered that glucose is a free component of carrots, even though it is rarely ingested as a single carbohydrate. In general, veggies have between 1% and 7% fructose. About 3% of the weight of dried veggies is composed of fructose, the sweetest carbohydrate. Inulin-type oligocarbohydrates, which are naturally occurring fructose polymers that store carbohydrates, are found in onions and asparagus. They offer 1 kcal/g since they are not proximally digested. Some vegetables are high in fructo-oligocarbohydrates include tomatoes and garlic. Large amounts of sucrose are present in unripe vegetables, but as the fruit ripens, the amount of sucrose diminishes.

The complex insoluble molecules known as starches are found in vegetables like carrots, dry legumes, and potatoes like beans and peas. The starch granules in potatoes and raw grains are encased in tough cellulose walls that keep digestive enzymes from reaching them, making them challenging to digest. Granules swell, starch gelatinizes and cell walls relax and decompose when heated, making digestion easier. Large levels of starch are present in plants before they reach full maturity; as the fruit ages, this starch is converted into simpler sugars, which causes the amount to decrease. Since it's safe to eat, raw carrot starch is frequently suggested for babies. For people who have certain digestive issues, vegetable cellulose can be helpful as it has a finer structure than grains. Vegetable cellulose and pectin compounds combine to form pectin-cellulose complexes, which may facilitate digestive secretions and gut motility. The cellulose content of vegetables can vary, ranging from 3.6% in parsnips to 0.3% in zucchini.

Hemicelluloses, which bind cations and help the intestines retain water, are found in vegetables including carrots, beets, and zucchini. This indigestible poly-carbohydrate is related to lignin and cellulose, other than high-fiber crops including radishes, cabbages, and eggplants. Pectin, which is found in significant amounts in beets and carrots, is used to treat digestive issues, including enteritis, and to create absorbent gels and antiseptics. Vegetables generally contain little fat, except oilseeds. Peanuts and almonds have trace levels of saturated fatty acids such as stearic, myristic, and palmitic acids, while peas, beans, and spinach contain these same fatty acids. While spinach contains palmitoleic acid, which accounts for 23% of monounsaturated fatty acids, peas include oleic acid. Omega-3 is a polyunsaturated fatty acid found in beans and soybeans.

Most vegetables have a protein content of 0.5% to 1.5%, but green legumes have higher levels (5%–6% in beans and peas) and dried legumes can have 20%–34% in dried beans, lentils, and soybeans (Van Den Berg et al., 2022). Legumelin in peas, Phaseolin in beans, and Glycine in soybeans, and lentils are examples of globulin-type proteins found in these legumes. From a nutritional perspective, proteins in soybean are considered insufficient because they have all types of essential amino acids, but in unequal amounts, particularly lysine and methionine. A protein is classified as incomplete if it is completely lacking in one or more essential amino acids, or if it has all the required amino acids but not in equal proportions, as in dried legumes that have reduced methionine levels.

Vegetables are rich in minerals, particularly potassium, calcium, magnesium, phosphorus and iron. Because they also gather trace elements (chromium, copper, iodine, fluorine, zinc, magnesium, molybdenum and selenium) from the soil in addition to water, their amounts can change. Few vegetables, including cauliflower, cabbage, parsley, broccoli, beans, peas and onions, have calcium absorption rates of more than 50%. Two significant calcium sources are soybeans and beans. Spinach and tomatoes contain oxalic acid, which decreases the bioavailability of minerals by making ions of insoluble calcium oxalate. The optimal calcium-to-phosphorus ratios found in onions, lettuce and cabbage promote absorption. Vegetables provide less than 30% of the body's phosphorus requirements.

The insoluble complexes formed by phytotic acid, a form of phosphorus present in dry legumes, hinder the absorption of iron, calcium, zinc, and magnesium. Potassium is abundant in vegetables in the form of carbonate and organic acid salts that dissolve in water and digestive juices. Vegetables usually have a diuretic effect due to high potassium and water content. Significant sources of potassium include dry beans (1,500 mg/100 g), potatoes (500 mg/100 g), and spinach (700 mg/100 g). Dried legumes and dark green vegetables like nettles, green onions, spinach, and lettuce are good sources of magnesium, which is a necessary component of chlorophyll. After meat, vegetables, especially lentils and dried beans, are an excellent source of iron.

Iron absorption is enhanced by the high vitamin C content of even lower iron-containing vegetables, such as sweet peppers, cauliflower, broccoli, potatoes, tomatoes, and cauliflower. Cooking methods such as boiling and over-refining can lead to fragmentation, which exacerbates iron loss. The primary foods that contain sulfur (S) are dry beans, broccoli and cauliflower. Spinach, dried beans, lentils, and peas are examples of vegetables that are high in zinc (Zn) and typically have a high chlorophyll content.

Vegetables like carrots, cucumbers, and zucchini that have high levels of ascorbic oxidase are rare sources of vitamin C. Horseradish, parsley, dill and red pepper are all excellent sources of dehydroascorbic acid, a form of vitamin C. Oxidation and the addition of sodium bicarbonate to cooked vegetables to preserve and enhance their color deplete vitamin C. It can be effectively preserved by freezing and refrigeration, but cooking can lower its concentration by up to 45% and 52%, respectively. Deficiencies in vitamin C and iron (Fe) can affect the body's folic acid levels. Legumes such dry beans, leafy greens (spinach and lettuce), cabbage, and potatoes are good sources of vitamin B1. 0.2 to 0.3 mg of pulses (like lentils) and green leafy vegetables (spinach, broccoli, and parsley) contain vitamin B2. Dried legumes and a number of vegetables, such as beans, potatoes, spinach, eggplant, green peas, cauliflower, and tomatoes, contain nicotinic acid which is a type of niacin. Broccoli, tomatoes, and potatoes are the main plant sources of vitamin B5. Vitamin B6 is found in pulses like beans, lentils, and soybeans as well as vegetables like potatoes, cauliflower, and spinach.

Vegetables have a lower bioavailability of vitamin B6 than animal sources due to factors such as prolonged heat treatment, freezing, and long-term storage. Vegetables like tomatoes, beans, peas, spinach, cabbage, and carrots are good sources of vitamin B7 (H). Folic acid, which is also called vitamin B9, is abundant in vegetables like lettuce, spinach, and broccoli and is also present in legumes and potatoes. When vegetables are boiled, their B9 content might be lowered and oxidized between 50% and 90%. The existence or absence of inhibitors, chelating compounds, and B9-conjugated metal chelates affects the bioavailability of B9, and it has an effect on a nutritional and health status of an individual. Absorption of B9 in lettuce is half of that of bananas or yeasts.

Vegetables contain vitamin D, with spinach and cabbage having the highest concentrations. Plants are the only source of tocopherols and tocotrienols, with plant oils being the most abundant. 80%–95% of soybean oil is made up of a-tocopherol/g-tocopherol. Vitamin E is found in celery, seeds, spinach, peas, lettuce, cabbage, and legumes, and is known for its antioxidant qualities. Leafy greens including cabbage, parsley, broccoli, turnips, lettuce, and spinach, are rich in vitamin K, making up 40%–50% of the total amount consumed (Moure et al., 2006). Vitamin K1's bioavailability varies by food (for example, broccoli has a higher bioavailability than spinach) and by dietary composition (for example, adding lipids improves absorption).

Vegetables contain carotenoids, which include carotenoid hydrocarbons and their oxygenated derivatives, either bound or in free form to carbohydrates and holoproteins to create carotenoid glycosides and carotenoproteins. Lycopene, β-carotene, α-carotene, and γ-carotene -are the most significant hydrocarbon carotenoids containing 40 carbon atoms. A-carotene can change into b-carotene when heated. Commonly present in plants, b-carotene is a key precursor of vitamin A and is typically linked to chlorophyll. B-carotene is hydrolyzed by oxidative enzymes to produce two molecular units of vitamin A1. The estimated bioavailability of b-carotene in plants is 22%–24% for broccoli, 19%–34% for carrots, and 3%–6% for leafy greens.

Eating broccoli and peas can raise plasma levels of retinol and b-carotene more than eating spinach.

According to research, b-carotene from certain veggies can raise plasma b-carotene levels 2.6–6 times more effectively than that from green leafy vegetables (Baroni, 2010). The differences result from the position of carotenoids within cells; they are located in chromoplasts in other goods and chloroplasts in vegetables. Carrots are the main source of g-carotene, a relatively uncommon plant compound that resembles red crystals with blue accents. By preventing detrimental destabilizing processes, phytonutrients, including anthocyanins and carotenoids found in carrots, spinach, green onions, tomatoes, and lettuce act as powerful cellular protectors.

In addition to being cholagogues (like artichokes), depurative (like radishes, asparagus, lettuce, dandelion greens, leeks, watercress, chicory, and celery), diuretics (like dandelions, asparagus, leeks, watercress, and celery), emollients (like leeks, spinach, sorrel, onions, lettuce, potatoes, and carrots), expectorants (like red cabbage), and vermifuges (like onion, garlic, and cabbage), vegetables can also be used as snacks like celery, peas, and chicory. Vegetables can be cooked to make purees and puddings or eaten raw in salads. Water-soluble vitamins (such as those in the B complex and vitamin C), minerals and other nutrients are present in the water used to cook vegetables.

PROTEIN CONTENT AND AMINO ACID PROFILE OF DIFFERENT VEGETABLES

Plant-based proteins have less amount of essential amino acids, i.e., about 26% ± 2% of total protein than animal-based proteins, which is 37% ± 2% and skeletal muscle protein, which is 38% of total protein. The WHO/FAO/UNU Expert Consultation states that the microalgae (23%), lupin (21%), hemp (23%), wheat (22%), and oat (21%), which are plant-based proteins, have necessary amino acid compositions (Gorissen et al., 2018). As a result, consuming merely one of these proteins would not satisfy the necessary human needs. To fulfill the protein needs of an average adult, 0.66 g/kg body weight protein is recommended per day.

Although corn and potatoes have high leucine content, the normal amount of leucine in plant-based proteins was lower (7.1% ± 0.8%) than that of animal-based proteins (8.8% ± 0.7%). Plant-based proteins contain considerably reduced amount of lysine and methionine (3.6% ± 0.6% and 1.0% ± 0.3%, respectively) than animal-based proteins (7.0% ± 0.6% and 2.5% ± 0.1%, respectively) and skeletal muscle protein (7.8% and 2.0%, respectively). However, there is significant variation among plant-based protein sources. The concentration of lysine (Table 6.2) in brown rice, corn, lupin, hemp, wheat, and oats is significantly lower than the WHO/FAO/UNU requirements compared to potatoes, peas, microalgae, and soy (Zahari et al. 2020).

According to WHO/FAO/UNU, the methionine concentration (in Table 6.3) is lower in wheat (0.9%), peas (0.4%), lupin (0.3%), oats (0.2%), soy (0.4%), and microalgae (0.0%) as compared to brown rice (2.5%), hemp (2.0%), corn (1.7%), and potatoes (1.6%) (Magro et al., 2024)

TABLE 6.2
Concentration of Lysine

Plant-Based Protein Sources	Concentration of Lysine
Brown rice	2.4%
Corn	1.5%
Lupin	3.5%
Hemp	2.8%
Wheat	1.4%
Oats	2.1%
Potatoes	6.0%
Peas	5.9%
Microalgae	5.3%
Soy	4.6%

TABLE 6.3
Concentration of Methionine

Plant-Based Proteins	Concentration of Methionine
Wheat	0.9%
Peas	0.4%
Lupin	0.3%
Oats	0.2%
Soy	0.4%
Microalgae	0.0%
Brown rice	2.5%
Hemp	2.0%
Corn	1.7%
Potato	1.6%

There was minimal difference in the quantities of histidine, isoleucine, threonine, valine, and phenylalanine in proteins derived from plants and animals. With the exception of potato protein, plant-based proteins did not meet WHO/FAO/UNU requirements and contained lower quantities of the branched-chain amino acids valine and isoleucine than animal-based proteins (Montesano et al., 2020).

Concentration of protein in skeletal muscle of humans and from animal sources are compared to the concentration of amino acids in the several plant-based protein isolates, such as those from oats, lupins, hemp, wheat, soy, microalgae, peas, brown rice, corn, and potatoes. Leucine and other essential amino acids are slightly less abundant in plant-based proteins than in those that come from animals and human skeletal muscle. Furthermore, amounts of methionine and lysine are low in few plant-based protein isolates (Mir et al., 2020). Considering the wide variations in the amino acid composition of the different plant-based protein sources, a well-balanced combination of unique plant-based proteins may result in a higher-quality protein mixture.

Lower quantities of essential amino acids or a relative deficiency of leucine, methionine or lysine in plant-based proteins may be the cause of their decreased anabolic ability. But there are significant differences in the amino acid composition of different plant-based protein sources. The quantity of methionine, leucine, and lysine in different plant-based, animal-based, and human skeletal muscle proteins, as well as their protein composition and essential and non-essential amino acid levels. Among all the amino acids, essential amino acids seem to postprandially promote synthesis of muscle protein.

COMPOSITION OF DIFFERENT VEGETABLE-BASED FOOD PRODUCTS

Vegetable-based food products are made from plants and consist of whole vegetables and various plant-made substitutes for meat and dairy. The reason these products are becoming more popular is because they are good for your health, sustainable and fit for vegetarians and vegans. Composition of some vegetable-based food products is as follows:

Composition of Whole Vegetables Products

The varying nutritional content in whole vegetables is largely determined by their water content, which is typically between 70% and 95%. Between 3% and 20%, carbohydrates make up a portion of their composition and potatoes and sweet potatoes have more carbs compared to leafy greens (Table 6.4). In general, vegetables have a protein content of 1%–5%, except for peas and some beans. Vegetables have lesser amount of fat, so they are generally low in calories. The reason whole vegetables are healthy is that they provide lots of essential nutrients and phytochemicals. Dietary fiber makes up approximately 1%–4% of whole grains, providing benefit to both the digestive system and the hunger. Large amounts of vitamin A from beta-carotene in carrots and sweet potatoes, vitamin C in bell peppers and broccoli, vitamin K in

TABLE 6.4
Key Ingredients in Vegetable-Based Products

	Key Ingredients in Vegetable-Based Products
Soy	Soy products, such as tempeh and tofu, are a mainstay of plant-based diets and are high in vitamins, minerals, and protein. Soy milk is a common dairy substitute that offers both nutritional advantages and a velvety mouthfeel.
Peas	Because of their high protein content, peas are being used more and more in plant-based diets as a base for goods like meat alternatives and pea protein powder.
Oats	Oats are a common ingredient in plant-based drinks like oat milk. They are hypoallergenic and offer fiber and other nutrients.
Rice	Prized for its hypoallergenic qualities and ease of digestion, rice is frequently used in dairy substitutes like rice milk.
Nuts and seeds	Nut milk (like almond milk) and butter are made with cashews and almonds. They are abundant in vitamins, protein, and good fats.

leafy greens and folate in Brussels sprouts can be obtained. Potassium, magnesium and iron are all present in generous amounts in different varieties of whole vegetables. Vegetables, in addition, contain phytochemicals called carotenoids, flavonoids and glucosinolates, both of which can reduce inflammation and prevent oxidation.

COMPOSITION OF PROCESSED VEGETABLE PROTEIN-BASED PRODUCTS

Carbohydrates and moisture are decreased in processed vegetable protein products due to engineered process that concentrates plant protein sources. For example, tofu is a common soy-based product that contains at least 8% protein and tempeh, which has an increased protein due to fermentation, contains as much as 20%. The nutritional profiles of processed vegetable products, including dry, frozen, and canned vegetables, can differ.

The vitamin and mineral content of the processed vegetable products can be impacted by processing techniques, even though they still contain some of the original nutrients. While frozen vegetables typically retain more nutrition, canned vegetables can have extra sodium. Among the plant proteins, pea protein isolates are well known because they have a protein concentration of around 80% to 90%. When compared to other proteins, wheat gluten or seitan, has about 75%–80% protein, though it lacks essential amino acids such as lysine. Mycoprotein is gotten from fungi and offers 11%–15% protein (Jin et al., 2021). Since it is fibrous, it is a common ingredient in meat alternatives. Manufacturers generally texturize processed proteins to boost their feel in the mouth and commonly fortify them with nutrients.

Composition of Vegetable-Based Meat Alternatives

Vegetable-based meat alternatives are prepared to taste, feel and provide the same nutrition as slaughtered animal meats. Manufacturers achieve an alternative fat content by mixing protein isolates (for example, from soy, pea or wheat) with vegetable oils, usually coconut or canola. To obtain the desired consistency, they include binders such as methylcellulose and starches, with flavors from yeast extract and colors from beet juice to increase taste and make food more appealing. In terms of nutrition, the aim is to have protein concentrations equal to meat, with more than 15% and less than 25%. Often, vitamin B12, iron and zinc are added to overcome some nutrition deficiencies (Menal-Puey et al., 2016). Even though they are a sustainable and ethical meat alternative, their ingredients may differ depending on the product, which means some can be higher in sodium and saturated coconut fat.

CLASSIFICATION OF VEGETABLE PROTEINS

The vegetable proteins are categorized into:

a. **Leguminous proteins:** are obtained from legumes, which include peas, beans, and lentils. These proteins are frequently used as meat substitutes because they are high in critical amino acids.

b. **Oilseed proteins** are derived from oilseeds like sunflower seeds, peanuts, and soybeans. They are frequently found in goods like textured vegetable protein (TVP) and tofu.
c. **Cereal proteins** are present in cereals such as corn, rice, and wheat. These proteins are important sources of energy even though they frequently contain less lysine than leguminous proteins.
d. **Leaf protein** is a protein-rich extract made from the green leaves of different plants and are also called leaf protein concentrate, or LPC. They provide a wholesome and sustainable food supply that is especially helpful in alleviating worldwide protein challenges. Examples of plants that are used to produce LPC include sugar beet, spinach, alfalfa, and Moringa oleifera (Gharibzahedi and Smith, 2020).

FUNCTIONAL PROPERTIES

The functional characteristics of vegetable proteins, which dictate how they are used in food items, also have an impact on how they are classified:

a. **Emulsification:** Certain proteins can stabilize emulsions, which makes them beneficial in dressings and sauces.
b. **Foaming Ability:** Products like meringues or whipped toppings benefit from the ability of certain proteins to trap air and produce foams.
c. **Gel Formation:** Certain legume proteins or non-plant-based gelatin can generate gels that improve the texture of a variety of food products (Li et al., 2019).

STRUCTURE OF VEGETABLE PROTEINS

The majority of biological functions in cells are carried out by proteins, which are biological macromolecules that occasionally work in tandem with other biomolecules. Proteins are therefore biomolecules with inherent functional and structural characteristics, which are frequently difficult to replicate in the lab. Bacterial flagella, which are microscopic propellers, and spider silk proteins, which combine exceptional elasticity and strength, are two common proteins with extraordinary functions. Enzymes, which are proteins, catalyze reactions quickly and with high selectivity. Like proteins in general, plant proteins perform a wide range of enzymatic, structural, and functional roles, including photosynthesis, biosynthesis, transport, immunology, etc. They also act as storage mediums to satisfy the nutritional and growth needs of growing seedlings. These functions are carried out by proteins in their chemical composition and particular structures, such as folding, which can be defined and compact or unfolded and inherently disordered.

The American Committee on Protein Nomenclature gave the first systematic study on seed storage proteins by using the classification scheme (Caunii et al., 2010). The extractability and solubility of plant proteins were the basis for the first attempts to categorize them. Proteins are essentially divided into three categories by this nomenclature: simple, conjugated, and derived. There are four categories of proteins found in all plant tissues. These four plant protein types albumins, globulins, prolamins, and glutelins are primarily linked to seed storage proteins and are distinguished by Osborne fractionation with alkali, water, alcohol, and alkali. Subsequent attempts have been made to classify plant proteins in more intricate ways, based on factors like their chemical structure, biological function, mechanism or mechanisms of action, or location within the plant.

Subsequent attempts have been made to classify plant proteins in more intricate ways, based on factors like their mechanisms of action, chemical structure, location within the plant, and biological function. In spite of these efforts to adopt more contemporary Osborne classification is still the most popular system, especially for processes involving protein extraction and purification. But this classification has only been applied to seed storage proteins in practice; other plant proteins are typically classified in a more complex and occasionally unclear manner.

Globular proteins that coagulate when heated are known as albumins. The two most common albumins are egg white and serum albumin, which is the main protein in human blood. Leucine in barley, wheat, and other plant seeds is an example of the 2S albumin storage protein found in plant seeds and rye, as well as ricin in castor beans, phase-in kidney beans, and legume in peas, soybeans, and cowpeas. Even though they are water soluble, many of the proteins found in green plant tissue such as ribulose-1,5-bisphosphate carboxylase oxygenase (RuBisCO), an enzyme that catalyzes the first step of carbon fixation do not qualify as albumins. According to Osborne's definition, RuBisCO should be classified as an albumin protein since it is soluble in water and coagulable when heated. Likewise, most plant enzymatic proteins are soluble in water and coagulate by heat, but they have not been further classified as a type of protein beyond being identified as enzymes.

Similar to albumins, globulins are globular proteins with a larger molecular weight that dissolve in diluted salt solution but are not soluble in water. Serum gamma globulin, a significant human blood protein, is the most well-known globulin. Globulins are found in plants as the most prevalent class of storage proteins is found in both dicots and monocots. Plant storage globulins are essentially classified into two types based on the sedimentation coefficient: 7S vicilin-type, that is identified and thoroughly examined in peas, soybeans, etc., and 11S legumin-type, which is the main storage protein in the majority of legumes and dicots like brassica, rice, and oats (Lin et al., 2017). Certain plant leaf proteins are insoluble in water directly.

Prolamins and glutelin, the other two protein types, are specifically present as storage proteins in grass seeds (Triticale), where they predominate and account for up to 85% of total protein. Gliadins are the prolamins present in wheat, whereas The Latin names of the prolamins in other cereals hordein in barley, zein in maize, secalin in rye, etc. are used to determine their nomenclature. Although barley and rye also contain glutelin, wheat's glutenin is the most widely distributed type.

An increased percentage of proline and glutamine, as well as an increased percentage of repetitive motifs or sequences with non-repetitive domains in their N- and C-terminals, are among the similarities between prolamins and glutelins. Despite having different molar masses, the primary distinction between the two protein types is that prolamins and glutelins, respectively, form intra- and inter-molecular disulfide bonds in their native states, which accounts for the variations in their extractability.

It is evident from the preceding sections that the seed contains a variety of seed storage proteins, such as albumin, globulin, prolamin, and glutelin. In the plant cell, during processing and for different applications, these proteins may exhibit distinct properties, where the storage proteins make things work. But up until now, the primary method used to characterize the seed storage proteins has been their chemical performance during sedimentation, fractionation, etc.

Many different plant components are used to make vegetables. In reality, some are fruit, while others are storage organs, buds, leaves, stems, or roots. The location of the vegetable on the plant influences many of its sensory attributes as well as how well it preserves.

In general, vegetables fall into one of three categories: (1) roots, tubers, and bulbs; (2) buds, stems, leaves, and flowers; and (3) pods, seeds, and fruit. The best way to handle veggies after harvest depends on where they come from. Vegetables grown above ground, for instance, are more likely to develop waxy skins to prevent water loss. Since they lack this defense, roots, and tubers could need high humidity levels to prevent dehydration. Seeds have inevitably stored energy in the form of carbs or sugars. Because they are growing quickly and typically have limited reserves, buds and shoots require cautious handling to be preserved after harvest (Soria-Hernández et al., 2015).

The plants grow fruits for human consumption, and if they are not actively harvested, they will eventually fall off on their own. The majority of vegetables, in contrast to fruit, are not designed to survive without the entire plant. Vegetables are separated from their water and nutrient sources after harvest. They are typically taken out of the protection of soil or vegetation since they cannot photosynthesize without sunshine. Cucumber and eggplant are among the many fruiting vegetables that are harvested when they are still immature. The sugar reserves and/or protective skins of fruit like apples or bananas are absent from these "baby" fruits. Most vegetables are headed toward senescence and death as soon as they are harvested, except those that are mature fruit, bulbs, or storage organs (sweet potatoes, onions, pumpkin, etc.). Delaying deterioration and preserving freshness, nutrition, flavor, and texture until vegetables are consumed by the final customer is the goal of post-harvest management.

Vegetables are composed of cells, just like other living things. Even a basic vegetable contains a wide variety of cell types, each of which serves a distinct purpose for the plant. The cellulose, lignin's, and certain proteins that makeup plant cells' rather stiff cell walls. For cell walls to develop, calcium is essential. The growing tips of fruiting vegetables like zucchini and eggplants will lose structural integrity and may decompose if there is a calcium deficit (caused by low soil levels or development that exceeds the rate of transfer within the plant). Blossom end rot is caused by this.

Both water and solutes can pass through the cell wall. The plasmalemma functions as a liner inside the cell wall. The plasmalemma keeps the cell turgid by assisting in the maintenance of internal pressure. This turgidity is what maintains the crispness and firmness of veggies. A layer of pectin's holds adjacent cells together. The interchange of different chemicals between cells is made possible by a network of channels called plasmodesmata. The liquid "soup" inside the cell is termed cytoplasm. Numerous organelles and structures, each with a distinct function, float within the cytoplasm. The cell's power plant, the mitochondria, transforms breakdown products from carbohydrates into usable energy. The green portions of plants contain chloroplasts. They are in charge of photosynthesis and contain chlorophyll. When chlorophyll degrades, chromoplasts emerge from chloroplasts. Carotenoids, the red and yellow pigments that give some vegetables their color, are found in them. Amyloplasts, which are composed of starch grains.

Plant cells come in a variety of forms and serve a variety of purposes for the plant. Water loss and "keeping" quality are significantly impacted by the arrangement of plant cells and whether the vegetable has a waxy cuticle or a thick epidermis. Numerous cell types and air gaps can be found in plant tissues. These consist of veins, including: Xylem vessels, which move minerals and water from the roots to the leaves. Sugars are transported from leaves to roots by phloem channels. Lenticels and stomata are apertures that permit gas exchange between the inside and exterior of plant tissues. Guard cells actively regulate stomata; they can open to permit gas exchange or seal to minimize water loss. At harvest, they often close in reaction to a little decrease in the water pressure within the cells. Numerous stomata may be dispersed throughout the surface of leaves. Sunken holes that are unable to seal are called lenticels. They are common in fruit and vegetable storage organs, including sweet potatoes, cucumbers, and others. The skin of the plant, or epidermal cells. These can form a waxy cuticle that prevents water loss from the underlying tissue. A corky layer is formed by many layers of epidermal cells in certain storage organs, like potatoes.

The majority of the interior tissue is made up of parenchyma cells, which are non-specialized cells that perform tasks, including photosynthesis (mesophyll). Collagenous cells: these are elastic cells that are high in cellulose and pectin and contribute to the strength of the plant; for example, the majority of the cells in celery strings are collenchyma. The number of lenticels and stomata, as well as the structure of the epidermis, determine how rapidly a vegetable loses water. For instance, lenticels and stomata are absent from solanaceous fruiting vegetables, such as tomatoes, eggplant, and capsicum. This lowers the fruit's rate of water loss. Since many green vegetables have many stomata on their surface and cucumbers have many lenticels, both lose water quickly. Compared to a zucchini's robust epidermal layer, its thin epidermis provides minimal protection against water loss.

PROPERTIES OF VEGETABLE PROTEINS

Physical Properties

Vegetable proteins' solubility, water absorption capacity, texture, and heat stability are examples of their physical characteristics. Their usefulness in food applications is greatly impacted by these characteristics.

Solubility: Several variables, including pH and ionic strength, affect how soluble vegetable proteins are. For example, solubility might vary from 10% to 95% based on the processing conditions and protein source. In general, more solubility is preferred for uses such as sauces and beverages.

Water Absorption Capacity (WAC): The ability of a protein to absorb water is a key factor in food products' mouthfeel and texture. Vegetable proteins can differ greatly in WAC; for instance, pea protein has a higher water absorption rate than other proteins.

Texture: Processing techniques like extrusion can change the texture of vegetable proteins. This procedure can produce fibrous structures that resemble the textures of flesh, making them appropriate for use as meat substitutes. The main textural characteristics that can be changed during formulation and processing are cohesion, chewiness, and hardness.

Thermal Stability: When exposed to heat, vegetable proteins frequently denaturize, which alters their functional characteristics. Heating, for instance, can enhance emulsification but, if improperly managed, may decrease solubility (O'Sullivan et al., 2016).

Chemical Properties

The amino acid composition, reactivity with other substances, and presence of functional groups that affect how they behave in food systems are examples of the chemical characteristics of vegetable proteins.

Amino Acid Composition: Vegetable proteins have different amino acid composition from that of animal proteins. Legumes offer a better-balanced profile of essential amino acids, even though they frequently contain less of some of them (such as lysine in grains). The Protein Digestibility-Corrected Amino Acid Score (PDCAAS) is commonly utilized to evaluate the quality of protein.

Emulsification: The emulsifying characteristics of vegetable proteins enable them to stabilize water and oil mixes. For goods like sauces and salad dressings, this is essential.

Proteins differ in their ability to emulsify; soy protein, for instance, has been demonstrated to have exceptional emulsifying qualities.

Hydrolysis: To improve the flavor and solubility of vegetable proteins, chemical hydrolysis can break them down into smaller peptides or amino acids. Hydrolyzed vegetable protein (HVP) is frequently added to a variety of food items to enhance their flavor.

Reactivity with Other Ingredients: During processing, vegetable proteins may interact with lipids and carbohydrates to produce Maillard reactions that improve flavor and color but may also compromise nutritional value (Aletor et al., 2002).

EFFECT OF PROCESSING ON VEGETABLE PROTEINS

The digestibility, nutritional value, and functional characteristics of vegetable proteins are all greatly impacted by processing techniques. Depending on the kind of processing heat treatment, fermentation, and extraction methods these impacts can change. A thorough summary of the effects of processing on vegetable proteins may be found below.

Physical Changes

Denaturation

a. **Heat Treatment:** Denaturation, or the unfolding of the protein structure, can result from heating vegetable proteins. By exposing peptide bonds to digestive enzymes, this procedure can improve digestibility. However, too much heat can lead to aggregation, which lowers digestion and solubility.

b. **Extrusion:** This thermomechanical technique modifies proteins' molecular makeup to produce fibrous textures that resemble flesh. By dissolving cell walls and deactivating anti-nutritional factors (ANFs), such as trypsin inhibitors, it can increase the digestibility of proteins.
   - **Textural Changes:** Texture modification by altering the texture of vegetable proteins, processing techniques like fermentation and extrusion can make them more appetizing and appropriate for a range of culinary uses. For example, extrusion is used to manufacture textured vegetable protein (TVP), which has a texture similar to meat.

Nutritional Changes

a. **Protein Quality:**
   - **Amino Acid Composition:** Processing may change the profile of amino acids. For instance, the Maillard reaction brought on by heat treatment may make lysine, an important amino acid, less readily available. On the other hand, it has been demonstrated that cooking soybeans at high temperatures improves protein quality by increasing digestibility and deactivating ANFs.
   - **Digestibility:** Protein digestibility might vary significantly as a result of processing. Research has indicated that the Protein Digestibility-Corrected Amino Acid Scores (PDCAASs) produced by various processing techniques vary. For example, isolates of unprocessed soy protein exhibit greater PDCAAS values than isolates that have been treated with alkaline heat.

b. **Nutrient Loss:**
   - **Minerals and Vitamins:** Certain processing techniques may cause the minerals and vitamins included in vegetable proteins to deteriorate. For instance, delicate vitamins like vitamin C and other B vitamins may be harmed by high heat or lengthy cooking times.

c. **Properties of Function:**
   - **Emulsification and Foaming:** Vegetable proteins can be processed to improve their emulsifying qualities, which makes them appropriate for use in sauces and dressings. In a variety of food applications, the capacity to stabilize emulsions is essential for attaining desired textures.

- **Gelling Ability:** As a result of modifications to their molecular structure brought about by fermentation or heat treatment, certain processed vegetable proteins have better gelling qualities. This helps in the development of plant-based gels and meat substitutes (Sedaghat Doost et al., 2019).

APPLICATION OF VEGETABLE PROTEINS

The ability of leguminous crops to accumulate significant amounts of proteins and other beneficial, physiologically active organic and inorganic substances in fruits and seeds is what makes them so amazing. The use of compounds made from plant raw materials as food technology additives to enhance the mechanical and structural characteristics of the final product as well as their organoleptic qualities has a long history on the global scene. In addition to increasing the variety and quality of food products, the use of natural additives derived from plants also makes it possible to use local resources wisely. Consequently, natural additives made from vegetable materials enhance the biological value of the final product in addition to their designated technological purposes because they contain naturally occurring, biologically active substances.

Plant materials with surface-active qualities are used as emulsifying and foaming compounds. These materials are primarily natural compounds that have a greater ability for absorption in the body of the individual. Proteins and pectins are examples of surfactants found in the majority of these substances. There is also a noteworthy group of compounds that contain saponins in addition to proteins and pectin's. Like pectin's, saponins have anticancer qualities and lower blood cholesterol, which lowers the risk of atherosclerotic vascular lesions (Barbosa-Nuñez et al., 2025). Concerns about the human health and environment have led to a growing trend in recent years to substitute plant-based ingredients for eggs.

The use of plant-based ingredients, such as plant proteins or starch, which have the same foaming and emulsifying qualities as chicken eggs and contain not only proteins and pectin's but also saponins, has become more popular in recent years due to health and environmental concerns. Three primary factors dietary preferences, economic considerations, and health support the growing demand for egg-free products. The existence of phenylketonuria is one of the justifications for avoiding eggs in the diet. If left untreated, phenylketonuria, a disorder of amino acid metabolism, can result in irreversible mental health impairment (Fu et al., 2023).

Foods high in protein, such as seafood, meat, eggs, nuts, milk, and related products, should be avoided when treating phenylketonuria. An additional health concern is egg allergy, which is among the most frequent and common food allergy, which are ranked second only to the allergy to cow's milk protein. Egg whites are the primary source of major egg allergens, which can cause symptoms that are potentially fatal. Because cholesterol is linked to cardiovascular diseases, there are still concerns about the amount of cholesterol in eggs. The so-called "bird flu," an asymptomatic viral infection that causes worldwide drops in egg sales, is another consumer health concern. As a result, bird flu outbreaks have had a significant negative financial impact on the global poultry industry by affecting both egg production and consumption.

Various food needs and preferences, which are influenced by consumer attitudes and beliefs, can result in the development of new food products as well as affect consumption and purchasing power. Religion and certain beliefs can also influence consumer food choices by establishing dietary guidelines about what foods are allowed and what are not. The high price, storage challenges, and limited distribution of the refrigeration chain of fresh eggs, along with their short shelf life, are economic factors related to egg replacement that drive producers to look for egg substitutes.

Many studies on the partial or complete substitution of egg substitutes from different sources for eggs in food products have been published in recent years (Tan et al., 2023). Analysis has been done on the possible utilization of bovine plasma as an egg replacement in cake making. Moreover, whey, legumes, particularly soy, lentils, peas, and lupin, as well as mixtures of hydrocolloids like guar gum, xanthan gum, hydroxypropyl methylcellulose, carboxymethyl cellulose, arabic gum, and carrageenan, can be used in place of eggs when making cakes. The potential of using chia or banana seed gel as an egg substitute has been assessed by some researchers. Additionally, aquafaba, the cooking water of legumes, particularly, chickpeas, are becoming more and more popular as a thickening, emulsifying, and foaming agent.

Additionally, aquafaba, the cooking water of legumes, particularly chickpeas, is becoming more and more popular as an ingredient with strong foaming, emulsifying, and thickening properties because of its content of protein, carbohydrates, and saponins, at the proper pH and concentration of NaCl. As a result, aquafaba has gained recognition for transforming food waste into a useful ingredient for biology and nutrition. In this sense, aquafaba may offer a practical and affordable alternative to eggs in the manufacturing of food items like mayonnaise, ice cream, and candies. Few researchers have used aquafaba to make cakes, despite the fact that it is categorized as a functional ingredient for confectionery products.

Because of their high protein content, legumes can be utilized in replacement for meat in meals. Meat-like products like burgers, salami, and sausages are made by substituting either all or part of the meat with legume flour. But in Western nations, there is still a general lack of acceptance regarding cutting back on meat consumption and substituting it with other proteins. Customers must understand the significance of switching to plant-based foods in order to support animal rights and welfare organizations' campaigns calling for people to eat less meat and to address the increased greenhouse gas emissions that livestock farming causes, which are harmful to the environment.

First and foremost, consumer attitudes regarding the utilization of plant-based alternatives to obtain meat analogs are critical due to the production of nutritional analogs with enhanced nutritional status and high biological value. It is consistent with the encouragement of animal welfare, environmental sustainability, and a healthy lifestyle within the framework of a sustainable circular bioeconomy.

Vegetable proteins are increasingly being used in many different industries, most notably the food industry, due to their nutritional benefits, usefulness, and versatility. Some significant applications for vegetable proteins include the following:

Food and Beverages Industry

Meat Alternatives

- The texture and flavor of meat can be replicated by using Textured Vegetable Proteins (TVP), which are derived from soy, peas, or other legumes. Products that satisfy flexitarian and vegetarian diets include meatballs, sausages, and burgers
- **Hydrolyzed Vegetable Proteins:** These are added to processed meat products to improve texture and flavor. They are perfect for sauces and prepared foods because they stabilize emulsions and hold onto moisture (Xiao et al., 2023).

Dairy Alternatives

- **Plant-Based Kinds of Milk:** Soy milk, almond milk, and oat milk are well-liked lactose-free dairy alternatives that offer flavors and textures comparable to those of conventional dairy products.
- **Dairy Analogs:** Products like tofu, vegan cheeses, and whipped toppings contain vegetable proteins. These goods frequently have dairy-like flavors and textures while providing functional and nutritional benefits.

Goods from Bakeries

- By enhancing texture and protein content, vegetable proteins improve baked foods' nutritional profile. They can also help prolong shelf life and retain moisture.
- **Supplemental Nutrition:** Because of their high protein content and important amino acids, vegetable proteins are being used more and more in nutritional supplements. They serve bodybuilders, athletes, and health-conscious shoppers looking for plant-based protein sources. Protein powders derived from rice, peas, or soy are common types.
- **Nutritional Foods:** Dietetic diets that attempt to control weight or provide particular health advantages frequently contain vegetable proteins. They help items aimed at health-conscious consumers have a higher protein-to-fat ratio.
- **Useful Substances:** In food applications, vegetable proteins have a variety of useful functions.
- **Emulsification:** It stabilizes water-oil blends in sauces and dressings.
- **Foaming:** For items like whipped toppings or meringues, certain vegetable proteins can produce stable foams.
- **Gelling:** Proteins can gel to provide textures for desserts or meat substitutes.

Feed for Animals

Animal feed compositions use a large amount of the protein produced from vegetables. Compared to conventional feed sources, these proteins are more sustainable and give cattle the vital elements they need.

CONCLUSION

In conclusion, vegetable proteins are an essential part of contemporary diets, providing a host of health advantages and advancing sustainability objectives. They supply vital nutrients that are required to sustain general health from a variety of sources. Because of their structural characteristics, these proteins can be used in a variability of creative ways in the food industry to fulfill the increasing demand for plant-based substitutes. The role of vegetable proteins in nutrition and food technology is expected to grow as more research is done on their potential, making them essential for both industrial and dietary applications. People can meet their nutritional needs and support environmental sustainability by focusing on a balanced intake of different plant protein sources.

REFERENCES

Aletor, O., A. Oshodi and K.J.F.C. Ipinmoroti. 2002. Chemical composition of common leafy vegetables and functional properties of their leaf protein concentrates. *Food Chemistry*, 78:63–68.

Aryee, A.N.A., D. Agyei and C.C. Udenigwe. 2018. Impact of processing on the chemistry and functionality of food proteins. In Yada, R.Y. (ed), *Woodhead Publishing Series in Food Science, Technology and Nutrition, Proteins in Food Processing*, 2nd ed., pp. 27–45. Woodhead Publishing. https://doi.org/10.1016/B978-0-08-100722-8.00003-6

Barbosa-Nuñez, J.A., H. Espinosa-Andrews, A.A.V. Cardona and J.N. Haro-González. 2025. Polymer-based encapsulation in food products: A comprehensive review of applications and advancements. *Journal of Future Foods*, 5:36–49.

Baroni, L. 2010. VegPyramid: A proposal for a vegetarian food guide for Italian people. *Nutrition and Metabolism*, 3:71–80.

Butnariu, M. and A. Butu. 2015. Chemical composition of vegetables and their products. In Cheung, P. and B. Mehta (eds), *Handbook of Food Chemistry*. Springer, Berlin, Heidelberg. https://doi.org/10.1007/978-3-642-36605-5_17

Caunii, A., R. Cuciureanu, A.M. Zakar, E. Tonea and C. Giuchici. 2010. Chemical composition of common leafy vegetables. *Studia Universitatis "Vasile Goldis" Arad. Seria Stiintele Vietii (Life Sciences Series)*, 20:45.

Fu, Q., J. Zhao, S. Rong, Y. Han, F. Liu, Q. Chu, S. Wang and S. Chen. 2023. Research advances in plant protein-based products: Protein sources, processing technology, and food applications. *Journal of Agricultural and Food Chemistry*, 71:15429–15444.

Gharibzahedi, S.M.T. and B. Smith. 2020. The functional modification of legume proteins by ultrasonication: A review. *Trends in Food Science & Technology*, 98:107–116. https://doi.org/10.1016/j.tifs.2020.02.002

Gorissen, S.H., J.J. Crombag, J.M. Senden, W.H. Waterval, J. Bierau, L.B. Verdijk and L.J. Van Loon. 2018. Protein content and amino acid composition of commercially available plant-based protein isolates. *Amino Acids*, 50:1685–1695.

Jin, J., O.D. Okagu, A.E.A. Yagoub and C.C. Udenigwe. 2021. Effects of sonication on the in vitro digestibility and structural properties of buckwheat protein isolates. *Ultrasonics Sonochemistry*, 70:105348.

Li, R., Q. Cui, G. Wang, J. Liu, S. Chen, X. Wang, X. Wang and L. Jiang. 2019. Relationship between surface functional properties and flexibility of soy protein isolate-glucose conjugates. *Food Hydrocolloids*, 95:349–357. https://doi.org/10.1016/j.foodhyd.2019.04.030

Lin, D., W. Lu, A.L. Kelly, L. Zhang, B. Zheng and S. Miao. 2017. Interactions of vegetable proteins with other polymers: Structure-function relationships and applications in the food industry. *Trends in Food Science & Technology*, 68:130–144.

Magro, P.C., I. Maestro-Gaitán, M.R. Blázquez, J.M. Prieto, F.M.S. Iñiguez, V.C. Sobrado and M.J.R.J.F.C. Gómez. 2024. Determination of nutritional signatures of vegetable snacks formulated with quinoa, amaranth, or wheat flour. *Food Chemistry*, 433:137370.

Menal-Puey, S., M.M. Del Ruste and I. Marques-Lopes. 2016. Nutritional composition of common vegetarian food portions. *Nutrición Hospitalaria*, 33:386–394.

Mir, N.A., C.S. Riar and S. Singh. 2020. Structural modification in album (*Chenopodium album*) protein isolates due to controlled thermal modification and its relationship with protein digestibility and functionality. *Food Hydrocolloids*, 103:105708. https://doi.org/10.1016/j.foodhyd.2020.105708

Montesano, D., M. Gallo, F. Blasi and L. Cossignani. 2020. Biopeptides from vegetable proteins: New scientific evidence. *Current Opinion in Food Science*, 31:31–37.

Moure, A., J. Sineiro, H. Domínguez and J.C. Parajó. 2006. Functionality of oilseed protein products: A review. *Food Research International*, 39:945–963.

Nolden, A.A. and C.G. Forde. 2023. The nutritional quality of plant-based foods. *Sustainability*, 15:3324.

O'Sullivan, J., B. Murray, C. Flynn and I. Norton. 2016. The effect of ultrasound treatment on the structural, physical, and emulsifying properties of animal and vegetable proteins. *Food Hydrocolloids*, 53:141–154.

Sakthivel, K. M. and C. Guruvayoorappan. 2015. Acacia ferruginea inhibits cyclophosphamide-induced immunosuppression and urotoxicity by modulating cytokines in mice. *Journal of Immunotoxicology*, 12:154–163.

Sedaghat Doost, A., M. Nikbakht Nasrabadi, V. Kassozi, K. Dewettinck, C.V. Stevens and P. Van der Meeren. 2019. Pickering stabilization of thymol through green emulsification using a soluble fraction of almond gum—Whey protein isolate nano complexes. *Food Hydrocolloids*, 88:218–227

Soria-Hernández, C., S. Serna-Saldívar and C. Chuck-Hernández. 2015. Physicochemical and functional properties of vegetable and cereal proteins as potential sources of novel food ingredients. *Food Technology and Biotechnology*, 53:269–277.

Tan, M., M.A. Nawaz and R. Buckow. 2023. Functional and food application of plant proteins–A review. *Food Reviews International*, 39:2428–2456.

Van Den Berg, L.A., J.J. Mes, M. Mensink and A.J. Wanders. 2022. Protein quality of soy and the effect of processing: A quantitative review. *Frontiers in Nutrition*, 9:1004754.

Xiao, X., P.-R. Zou, F. Hu, W. Zhu and Z.J. Wei. 2023. Updates on plant-based protein products as an alternative to animal protein: Technology, properties, and their health benefits. *Molecules*, 28:4016.

Zahari I, F. Ferawati, A. Helstad, C. Ahlström, K. Östbring, M. Rayner and J.K. Purhagen. 2020. Development of high-moisture meat analogs with hemp and soy protein using extrusion cooking. *Food*, 9(6):77.

7 Meat Proteins

Pir Mohammad Junaid, Tariq Khan, Syed Shahid Reyaz Shah, Siddharth Thakur, and Mudasir Ahmad Malik

INTRODUCTION

Meat is a vital component of the human diet, providing a rich source of high-quality protein, essential amino acids, and significant levels of nutrients such as vitamin B12, zinc, phosphorus, and iron (Kumar et al., 2023). While low in carbohydrates and devoid of dietary fiber, meat is consistently accepted for its protein content and nutrient density, which remain relatively steady across different types and qualities of meat. As a complete protein source, meat contains essential amino acids that the body cannot synthesize on its own, making it indispensable for muscle growth, tissue repair, and the synthesis of hormones, enzymes, and immune molecules. Meat is also a primary source of heme iron, which is more bioavailable than non-heme iron from plant sources, as well as bioavailable B vitamins and zinc, which support nerve function, immune health, and metabolic processes (Lombardi-Boccia et al., 2004). Meat proteins, classified into myofibrillar, sarcoplasmic, and stromal proteins, play distinct functional roles in muscle contraction, enzymatic activity, and structural support, respectively. Understanding the composition, nutritional value, and functional attributes of meat proteins is vital for optimizing their use in food science and addressing the growing demand for protein diets.

COMPOSITION OF MEAT FROM DIFFERENT SOURCES

Fresh meats refer to uncooked animal products typically sold shortly after slaughter, including beef, pork, lamb, and poultry. The freshness of these meats is critical, as it directly influences their flavor, texture, and nutritional value. Fresh meats are generally characterized by their bright color, firm texture, and absence of off-putting odors. Proper handling and storage are essential to maintain their quality and safety, given their highly perishable nature. The composition of fresh meats is diverse, encompassing water, proteins, fats, vitamins, and minerals, all of which contribute to their nutritional and sensory properties. Water constitutes the largest proportion, typically ranging from 60% to 75% of the total weight, playing a crucial role in maintaining juiciness and texture (Barone, & Aschemann-Witzel, 2022). Proteins, the predominant macronutrient, provide essential amino acids for muscle repair, growth, and various metabolic functions. The protein content varies by meat type, with lean cuts generally containing higher protein levels and lower fat content. Fats in fresh meats are categorized into saturated and unsaturated fats, influencing flavor, tenderness, and palatability (Ren et al., 2022). The fat content varies significantly across species

DOI: 10.1201/9781003613909-7

and cuts, affecting both culinary applications and health considerations. Fresh meats are also rich in vitamins, particularly B vitamins such as B12, niacin, and riboflavin, which are essential for energy metabolism and overall health. Meat provide essential minerals like iron, zinc, and phosphorus, further enhancing their role in a balanced diet. The approximate composition of fresh meats includes water (70%–75%), protein (18%–22%), fat (2%–30%, varying by cut and type), minerals (1%–1.5%), and trace amounts of carbohydrates (e.g., glycogen) (Warriss, 2010). The protein content in fresh meats is a critical component of their nutritional value, providing essential amino acids essential for various bodily functions. The protein composition varies depending upon the type of meat and its cut. The composition of meat varies significantly across different sources, including red meat, poultry, and seafood, each offering distinct nutritional benefits.

Red Meats

Red meat is amongst the most important dietary sources of high-quality protein, making it crucial for human health and nutrition. Raw red muscle meat typically contains between 20 and 25 g of protein/100 g, while cooked red meat provides an even more concentrated source, with 28–36 g/100 g, due to water loss during cooking (Williams, 2007). The protein from red meat is highly digestible, boasting a digestibility rate of approximately 94%, notably higher than that of plant proteins like beans (78%) and whole wheat (86%). Red meat supplies essential amino acids required for human nutrition, including Lysine, Threonine, Methionine, Phenylalanine, Tryptophan, Leucine, Isoleucine, and Valine, without any limiting amino acids (Boateng et al., 2020). Its Protein Digestibility Corrected Amino Acid Score (PDCAAS) is around 0.9, very close to the maximum value of 1.0, underscoring its superior protein quality compared to plant-based foods, which generally score between 0.5 and 0.7. Glutamic acid/glutamine is the most abundant amino acids found in meat, followed by Arginine, Alanine, and Aspartic acid. Moreover, red meat provides additional bioactive peptides such as Carnosine, Anserine, Taurine, and Creatine, which further enhance muscle function, antioxidant defense, and energy metabolism. Carnosine, which is found in beef at around 365 mg/100 g and lamb at 400 mg/100 g, is a strong antioxidant that shields cells from oxidative damage. Taurine, abundant at 77 mg/100 g in beef and 110 mg/100 g in lamb, is vital for immune function and antioxidative processes. Creatine, found at approximately 350 mg/100 g, supports metabolism of muscle energy and athletic performance (Manhiani, 2010). Such substances are easily absorbed and have functional roles that go beyond basic protein supply. Additionally, lean red meat serves as an excellent protein source without excessive fat intake, particularly given the modern trend toward leaner cuts with less than 7% fat. Compared to vegetarian protein sources, red meat provides a far higher quantity and quality of protein with a broader amino acid profile and superior bioavailability. Even organ meats like liver, kidney, and heart, besides being rich in micronutrients, are substantial sources of highly digestible protein. The presence of bioavailable iron and zinc in red meat further enhances the efficiency of protein utilization in the body. During periods of high physiological demand such as growth, pregnancy, lactation, and athletic training, red meat offers a dense, efficient source of protein and related bioactive compounds critical for tissue repair, immune support, and muscle synthesis. The peptides and amino acids found in red flesh proteins contribute to the 'meat factor' that enhances the intake of minerals like iron and zinc from the digestive tract. Lean red meat remains one of the most comprehensive and superior natural sources of dietary protein, essential for maintaining optimal health, muscle function, and overall nutritional status across different life stages.

Poultry

Poultry meat including chicken and turkey, is a highly valued source of high-quality protein, essential for human nutrition and health. On average, raw poultry meat contains about 20 to 24 g of protein/100 g, and this concentration increases upon cooking due to water loss (Barbut, 2024). The protein found in poultry meat is highly digestible, with a biological value and net protein utilization close to that of eggs and dairy, ranking among the highest of all protein sources. Poultry protein provides essential amino acids in the correct proportions needed by humans, making it a complete protein source. Among the amino acids, poultry meat is particularly rich in lysine, leucine, isoleucine, and valine, which are vital for muscle maintenance, tissue repair, and enzyme production. Glutamic acid is the most abundant non-essential amino acid in poultry, contributing to the characteristic flavor profile of the meat (Maia et al., 2024). Meat from poultry is a significant source of bioactive peptides, such as carnosine, anserine, and taurine, which have antioxidant properties and support muscle function and cardiovascular health. The relatively low collagen content in poultry meat compared to red meats contributes to its tenderness and easy digestibility. Poultry proteins also enhance the absorption of important minerals such as iron and zinc from the diet, supporting overall nutrient bioavailability. Due to low fat and high protein content, poultry meat is often recommended in health-conscious and therapeutic diets. Although there are some differences depending on the age, sex, and food of the bird, the amino acid makeup of poultry meat is generally constant throughout breeds and raising settings. The superior digestibility, complete amino acid profile, and associated functional peptides make poultry meat an exceptional dietary protein source, suitable for populations of various age groups, including infants, athletes, and the elderly

Seafood

Seafood, including fish and shellfish, is another excellent source of high-quality protein, with protein content that usually ranges from 18% to 22% (Nurdiani et al., 2020). Fish protein is of high biological value, containing essential amino acids in optimal proportions, making it a complete protein source. This quality is particularly beneficial for human nutrition, as fish consumption is linked to the prevention of cardiovascular diseases and other health benefits. Fatty fish, such as salmon and mackerel, also provide beneficial omega-3 fatty acids, which promote heart health and have anti-inflammatory qualities (Patted et al., 2024). Lean fish, like cod and haddock, offer high protein content with minimal fat, making them ideal for low-calorie diets. Seafood, including fish and shellfish, is valued for its lean proteins (~18%–22%) and beneficial omega-3 fatty acids, which promote heart health. Salmon and other fatty fish and mackerel contain up to 20% fat, mostly in the form of healthy unsaturated fats, while lean fish like cod have less than 5% (Patted et al., 2024). In addition to vitamin D, seafood is a great source of important minerals, including iodine and selenium

Processed Meats

Processed meats, such as sausages and deli meats, generally have a protein content similar to fresh meats, ranging from 15% to 20% (Deshpande, 2025). However, the protein quality may be compromised due to the addition of fillers, preservatives, and high levels of sodium. Despite their convenience, processed meats are often less nutritious than fresh meats and should be consumed in moderation. Processed meats, such as sausages, hot dogs, and deli meats, present a different nutritional profile. While convenient and flavorful, they often contain added preservatives, sodium, and unhealthy fats, which can detract from their nutritional value. Processed meats, such as sausages and deli meats, generally contain 15%–20% protein by weight, though this can vary depending upon the type and processing methods. While they retain some essential amino acids found in fresh meats, their protein quality is often compromised due to the addition of preservatives, sodium, and non-meat fillers. Processed meats are convenient but are typically high in saturated fats and sodium, which can detract from their nutritional value. Regular consumption has been linked to increased health risks, such as cardiovascular diseases and certain cancers, highlighting the importance of moderation and opting for minimally processed alternatives (Iyiola & Bulus, 2024).

PROTEIN CONTENT AND AMINO ACID PROFILE OF MEAT FROM DIFFERENT SOURCES

Meat is a highly valued source of high-quality protein, providing essential amino acids required for human health. The protein content and Amino Acid Profile vary across different meat sources, including red meat, poultry, and seafood. Table 7.1 provides the details of the protein composition and amino acid profiles of these meat types.

Red Meats (Beef, Lamb, Pork)

Red meat is a complete source of dietary protein, providing Essential Amino Acids (EAAs) and Non-Essential Amino Acids (NEAAs) necessary for human health. The amino acid profile of a red meat varies depending on factors such as; species, cut, age, and sex of the animal. Beef and pork exhibit differences in both EAA and NEAA proportions, with beef containing higher levels of histidine (32% more), leucine (28% more), and isoleucine (21% more) compared to pork, while pork has higher concentrations of tyrosine (106% more) and glutamine (68% more) (Barr et al.,2025; Liao et al., 2024). Red meat is rich in branched-chain amino acids (BCAAs) leucine, isoleucine, and valine which play critical roles in muscle protein synthesis and repair. The digestibility of amino acids in red meat is high, with true fecal digestibility exceeding 94%, making it a highly bioavailable protein source. Red meat contains

TABLE 7.1
Comparative Analysis of Amino Acid Profiles and Protein Content

Meat Type	Protein Content (%)	Amino Acid Profile	Key Functions	References
Beef	20–22	Leucine, Lysine, Valine, Heme Iron	Muscle repair, iron absorption, energy metabolism, antioxidant activity	Alekseeva and Kolchina (2019)
Goat	18–20	Leucine, Lysine, Valine, Methionine	Lean protein source, supports muscle growth, rich in iron and zinc	Shan et al. (2007)
Lamb	19–21	Leucine, Lysine, Valine, Carnosine	High in omega-3 fatty acids (grass-fed), supports heart health and muscle repair	Yuldashbaev et al. (2021)
Camel	17–19	Leucine, Lysine, Valine, Heme Iron	Low-fat protein source, rich in iron and vitamin B12, supports immune function	Kadim et al. (2011)
Salmon	20–22	Lysine, Methionine, Taurine, Arginine	Rich in omega-3 fatty acids (EPA, DHA), supports heart and brain health	Espe et al. (2023)
Tuna	22–26	Lysine, Methionine, Taurine, Arginine	High protein content, supports muscle repair, rich in selenium and vitamin D	Peng et al. (2013)
Cod	18–20	Lysine, Methionine, Taurine, Arginine	Lean protein source, low in fat, supports weight management	Ouellet et al. (2017)
Mackerel	19–21	Lysine, Methionine, Taurine, Arginine	High in omega-3 fatty acids, anti-inflammatory, supports cardiovascular health	Salma and Nizar (2015)
Shellfish	18–22	Lysine, Methionine, Taurine, Arginine	Rich in zinc, iodine, and vitamin B12, supports immune and thyroid function	Asha et al. (2014)
Poultry	20–24	Leucine, Isoleucine, Valine, Lysine	Muscle growth, immune support, energy metabolism, collagen synthesis	Baker (2009)
Turkey	20–25	Threonine, Methionine	Lean protein source, supports weight management, promotes muscle recovery	Amirkhanov et al. (2017)

semi-essential amino acids like arginine and cysteine, which become conditionally essential during stress, pregnancy, or illness. These amino acids contribute to various physiological functions, including hormone secretion, nutrient metabolism, and immune support. The comprehensive amino acid profile of red meat underscores its nutritional value, particularly for muscle growth, recovery, and metabolic health. Red meats also contain creatine and carnosine, compounds that support muscle energy metabolism and act as antioxidants. However, their higher saturated fat content necessitates moderate consumption to avoid potential health risks.

Poultry (Chicken, Turkey)

The Amino Acid Profile of poultry is rich in branched-chain amino acids (BCAAs), such as leucine, isoleucine, and valine, which are essential for muscle recovery and growth. Poultry is also lower in saturated fats compared to red meats, making it a healthier alternative for those monitoring fat intakes. Its high digestibility and bioavailability of amino acids further enhance its nutritional value, particularly for athletes and individuals seeking to maintain muscle mass. Poultry meat, particularly duck, is rich source of high-quality protein, characterized by a balanced profile of essential (EAAs) and non-essential amino acids (NEAAs). In Peking and Muscovy ducks, glutamic acid, aspartic acid, lysine, and methionine are among the most abundant amino acids, with lysine and methionine being the predominant EAAs. Duck meat contains higher methionine levels compared to chicken and ostrich, enhancing its nutritional value. The total EAA content in Peking duck thighs is significantly higher than in breast cuts, while NEAAs are more concentrated in breast meat. Muscovy duck meat generally exhibits superior protein quality, as indicated by higher chemical and, amino acid scores compared to Peking duck, though the Essential Amino Acid Index (EAAI) remains comparable across species and cuts.

Fish

Fish muscle is rich source of high-quality protein, accounting for 15%–25% total protein content in fish. It is highly digestible and contains a well-balanced profile of essential amino acids, including lysine (8.8%), tryptophan (1.0%), histidine (2.0%), phenylalanine (3.9%), leucine (8.4%), isoleucine (6.0%), threonine (4.6%), and methionine-cystine (4.0%). Fish protein hydrolysates, produced through enzymatic hydrolysis, fermentation, or chemical methods, further enhance bioavailability and bioactivity. Enzymatic hydrolysis, using proteases like Alcalase, Flavourzyme, or Protamex, is preferred for its specificity and ability to preserve amino acid integrity. The degree of hydrolysis (DH), hydrolysis circumstances, and fish species all affect the amino acid content of hydrolysates. For instance, hydrophobic and aromatic amino acids (e.g., leucine, phenylalanine, tyrosine) contribute to antioxidant and antihypertensive properties, while glutamic acid and glycine enhance flavor. Studies highlight the potential of fish protein hydrolysates in functional foods and nutraceuticals, offering benefits such as ACE-inhibitory, antimicrobial, and anti-inflammatory effects. Seasonal, biological, and environmental factors also influence protein content, underscoring the importance of optimal processing methods to maximize nutritional value. Overall, fish proteins and their hydrolysates represent a sustainable and health-promoting alternative to terrestrial animal proteins.

COMPOSITION OF DIFFERENT MEAT-BASED PRODUCTS

The formulation of various meat-based products encompasses a diverse array of components and ingredients. These products can include processed meats, sausages, and ready-to-eat meals, each characterized by distinct formulations influencing their flavor, texture, and nutritional profile (Hathwar et al., 2012). Meat, the main component, can come from a variety of sources, including beef, hog, chicken, or game. It is frequently enhanced with chemicals like binders, taste enhancers, and preservatives. Additionally, spices, herbs, and other flavoring agents play a vital role in defining the sensory attributes of the final product. The composition may also vary based on dietary considerations, with options for reduced fat, low sodium, or gluten-free formulations catering to specific consumer preferences (Teixeira et al., 2020). Understanding the intricate balance of these components is essential for the development of high-quality meat-based products that meet both safety standards and consumer expectations.

Processed Meats

Meat items that have been altered by curing, smoking, or adding preservatives to improve flavor and prolong shelf life are referred to as processed meats. Among these products include items like sausages, hot dogs, bacon, and deli meats (Ahmad & Badpa, 2014). The processing often involves the addition of various chemicals, including nitrates and nitrites, which can contribute to the preservation of the meat but also have been linked to health concerns. Consumption of processed meats has been associated with an increased risk of certain health issues, like cardiovascular diseases and certain types of cancer. As a result, dietary guidelines often recommend limiting the intake of these products in favor of fresh, unprocessed meats and plant-based protein sources (Boada et al., 2016). Processed meats are modified through curing, smoking, or other preservation techniques.

Cured meats refer to various types of meat that have undergone preservation processes involving the addition of salt, nitrates, or other curing agents (Wolk, 2017). These methods not only enhance the flavor and texture of meat but also increases its shelf life by inhibiting the growth of spoilage-causing microorganisms. The curing process can involve techniques such as dry curing, wet curing, or smoking, each imparting distinct characteristic to the final product. Common examples of cured meats include prosciutto, salami, and bacon, which are often enjoyed in a variety of culinary applications, from charcuterie boards to sandwiches (Gómez et al., 2020). The cultural significance of cured meats varies globally, with each region showcasing unique recipes and traditions associated with their preparation and consumption.

Cured meats are products that undergo a preservation process involving the addition of salt, nitrates, or other curing agents, which inhibit the growth of bacteria and enhance flavor. The composition of cured meats typically includes various types of meat, such as pork, beef, or poultry, along with a blend of spices, sugars, and curing (Safa et al., 2017). The curing process can be either dry or wet, with dry curing involving the application of salt and spices directly to the meat, while wet curing utilizes a brine solution. Additionally, the composition may vary based on regional practices and specific recipes, leading to a diverse array of flavors and textures in cured meat products. Nutritionally, cured meats are often high in protein but it may also contain significant levels of sodium and preservatives, which are important considerations for health-conscious consumers (Xiong & Mikel, 2001).

Smoked meats refer to various types of meat that have undergone a preservation and flavor-enhancing process involving exposure to smoke. This technique not only imparts a distinctive taste but also helps in extending the shelf life of the meat (Ren et al., 2024). The process typically involves curing the meat with salt and other seasonings before it is subjected to smoke from burning wood or other materials. Different types of wood can be used, each contributing unique flavors to the final product. Common examples of smoked meats include brisket, ribs, and sausages, which are often enjoyed in various culinary traditions around the world (Nam et al., 2010). The smoking process can vary in duration and temperature, influencing the texture and flavor profile of the meat.

Composition: The composition of smoked meats encompasses various elements that contribute to their flavor, texture, and preservation. Primarily, smoked meats are derived from various types of animal proteins, such as beef, pork, poultry, and fish. The smoking process involves the application of heat and smoke, which not only imparts a distinctive taste but also aids in the preservation of the meat by inhibiting microbial growth. The composition also includes the seasoning and curing agents used before smoking (Di Santo, 2016).Common ingredients in the curing process may consist of salt, sugar, and various spices, which enhance the flavor profile and contribute to the overall quality of the final product. Additionally, the type of wood used for smoking such as hickory, mesquite, or Applewood can significantly influence the aroma and taste of the meat. Furthermore, the fat content within the meat plays a vital role in determining the juiciness and mouthfeel of the smoked product. The balance of lean and fatty portions can vary widely depending on the cut of meat selected. Overall, the intricate interplay of these components results in a diverse array of smoked meat products, each with unique characteristics and culinary applications.

Sausages

Sausages are a type of processed meat product that typically consists of ground meat, fat, and various seasonings, all encased in a natural or synthetic casing (Owusu-Apenten & Vieira, 2023). They can be made from a variety of meat, including pork, beef, chicken, and even plant-based alternatives. The preparation methods for sausages vary widely, encompassing techniques such as smoking, curing, and drying, which contribute to their distinct flavors and textures. Sausages are often categorized into fresh, cooked, or cured varieties, each with unique characteristics and culinary applications (Munekata et al., 2023). Additionally, they hold cultural significance in many cuisines around the world, often being featured in traditional dishes or served at social gatherings, like Hot dogs, bratwurst, salami.

Composition: The formulation of sausages involves a careful selection of ingredients and a precise method of preparation. Typically, sausages are composed of ground meat, which may include pork, beef, poultry, or a combination thereof. The meat is often supplemented with fat to enhance flavor and texture. Additionally, various seasonings, herbs, and spices are incorporated to impart distinct tastes, while fillers such as breadcrumbs or rice may be added to modify the consistency and volume (Dévi et al., 2020). The mixture is then encased in natural or synthetic casings, which serve to maintain the shape and integrity of the sausage during cooking. The final product can be subjected to various cooking methods, including grilling, frying, or smoking, each contributing to the overall flavor profile and culinary experience. Water: 45%–60%, Protein: 10%–20% (from meat and added fillers like soy protein), Fat: 10%–30%, Carbohydrates: 2%–10% (added starches, flours), Additives: Spices, preservatives, binders.

Meat Patties and Burgers

Meat patties and burgers are popular culinary items that consist primarily of ground meat, typically beef, shaped into a flat, round form. These products are often seasoned and cooked by grilling, frying, or baking (Bakhsh et al., 2021).The versatility of meat patties allows them to be incorporated into various dishes, with burgers being a prominent example. A burger typically consists of a meat patty placed within a sliced bun, accompanied by a variety of toppings such as lettuce, tomato, cheese, and condiments. The preparation and presentation of meat patties and burgers vary significantly across different cultures, reflecting regional tastes and preferences (Petrat-Melin, & Dam, 2023). Additionally, the rise of plant-based alternatives has led to the development of vegetarian and vegan patties, expanding the appeal of this food category to a broader audience. Examples: Beef burgers, and chicken patties.

Composition: The formulation of meat patties and burgers involves a careful selection of ingredients and a precise method of preparation. The main ingredient is usually ground meat, which can be any kind of meat—beef, hog, chicken, or a mix of these. Because it affects the final product's juiciness and flavor, the meat's fat level is very important; a higher fat content often makes the patty more succulent. To improve taste and texture, several seasonings and additives are added in addition to the meat. Common seasonings include salt, pepper, garlic powder, and onion powder, while binders such as breadcrumbs or eggs may be added to improve cohesion (Karimidastjerd et al., 2024). Some recipes also incorporate fillers like vegetables or grains to modify the nutritional profile or to cater to specific dietary preferences. The composition of burgers and beef patties is also greatly influenced by the cooking technique. Grilling, frying, or baking can yield different flavor profiles and textures, affecting the overall sensory experience. Furthermore, the choice of toppings and condiments, such as lettuce, tomato, cheese, and sauces, contributes to the complexity of flavors and the overall appeal of the dish. Water: 50%–60%, Protein: 15–25%, Fat: 10%–30% (higher in some formulations), Carbohydrates: 1%–5% (from fillers or breadcrumbs)

Fermented Meats

Products that go through a regulated fermentation process, involving the activity of microorganisms such bacteria, yeasts, and molds, are known as fermented meats (Ashaolu et al., 2023). By reducing the pH and preventing the formation of spoilage organisms, this procedure not only improves the meat's flavor and texture but also helps to preserve it. Common examples of fermented meats include salami, pepperoni, and prosciutto. The fermentation process typically involves the addition of salt, spices, and curing agents, which together create an environment conducive to the growth of beneficial microorganisms (Sharma et al., 2020). The resulting products are characterized by their distinctive taste, aroma, and shelf stability, making them popular in various culinary traditions worldwide. Furthermore, depending on the particular product and fermentation circumstances, fermented meats may have additional health advantages, such as better digestion and the presence of bacteria.

Composition: The formulation of fermented meats involves a complex interplay of ingredients and processes that contribute to their unique flavors, textures, and preservation qualities. Usually, the main ingredients include of different kinds of meat, including beef, pig, or chicken, which are frequently mixed with certain curing agents like nitrites and salt (Mao et al., 2024).These ingredients not only enhance flavor but also inhibit the growth of harmful bacteria. In addition to the meat and curing agents, fermentation relies on the introduction of starter cultures, which are specific strains of bacteria or molds. These microbes are essential to the fermentation process because they break down carbohydrates and create lactic acid, which gives the food its distinctively tangy flavor and prolongs its shelf life. The fermentation environment, including temperature and humidity, is meticulously controlled to ensure optimal growth of the starter cultures while preventing spoilage (Xiao-Hui et al., 2023). Furthermore, various spices, herbs, and flavoring agents are often incorporated to enhance the sensory attributes of the final product. The combination of these elements, along with the fermentation process, results in a diverse array of fermented meat products, each with distinct characteristics shaped by regional traditions and consumer preferences.

Meat Emulsions

Meat emulsions are complex mixtures formed by the combination of meat proteins and fats, resulting in a stable emulsion that is essential in various processed meat products (Guo et al., 2023). These emulsions are created through mechanical processes that involve the dispersion of fat droplets within a continuous protein matrix, often facilitated by the addition of emulsifying agents. The stability and texture of meat emulsions are influenced by several factors, including the type of meat used, the fat content, the presence of water, and the specific processing techniques employed (Zhang et al., 2022). Understanding the physicochemical properties of meat emulsions is vital for optimizing product quality, enhancing flavor, and improving mouthfeel in processed meats such as sausages and patties. Examples: Bologna, frankfurters.

Composition: The formulation of meat emulsions involves a complex interplay of various components that contribute to their texture, flavor, and stability. Typically, meat emulsions are created by finely grinding meat and combining it with fat, water, and other ingredients such as salt, spices, and binders. The emulsification process is crucial, as it allows for the uniform distribution of fat within the meat matrix, resulting in a cohesive (Gómez et al., 2020). The choice of meat type, fat content, and additional ingredients significantly influences the final characteristics of the emulsion, including its mouthfeel and sensory attributes. Understanding the biochemical interactions among these components is essential for optimizing the quality and functionality of meat emulsions in food applications.

Canned or Shelf-Stable Meats

A variety of meat products that are maintained in air-tight containers and may be kept at room temperature for lengthy periods of time without refrigeration are referred to as canned or shelf-stable meats. These products undergo a preservation process that typically involves cooking the meat and sealing it in a can or jar, which is then subjected to heat treatment to eliminate bacteria and other microorganisms (Maheswarappa et al., 2022). This technique preserves the meat's taste and nutritional content while also extending its shelf life. Common examples include canned chicken, beef, and tuna, which are often utilized in a variety of culinary applications, from quick meals to emergency food supplies. Customers looking for useful and adaptable food alternatives frequently choose these goods due to their lifespan and ease of use. Examples: Corned beef, potted meats, spam.

Composition: The formulation of canned or shelf-stable meats involves a careful selection of ingredients and preservation methods to ensure safety, flavor, and longevity. Typically, these products consist of various types of meat, such as beef, pork, chicken, or turkey, which are processed and cooked before being sealed in air-tight containers (Ramesh, 2020). The inclusion of preservatives, such as salt, sugar, or sodium nitrite, is common to inhibit microbial growth and enhance shelf life. Additionally, the use of broth or sauces can contribute to the overall taste and moisture content of the final product. The packaging process is critical, as it must maintain a sterile environment to prevent contamination and spoilage. Overall, the composition of these meats is designed to provide convenience and extended storage without compromising quality.

Meat-Based Snacks

Meat-based snacks are a class of food items that are mostly made of different kinds of meat and are frequently packed and processed for ease of use (Ramesh, 2020). These snacks can include jerky, meat sticks, and other dried meat products. They are typically high in protein and can serve as a satiating option for individuals seeking quick and portable sources of nutrition. The production of meat-based snacks often involves methods such as dehydration, smoking, or curing, which not only enhance flavor but also extend shelf life. To accommodate a range of consumer tastes, these snacks can also be seasoned with different spices and flavorings (Awanish & Sit, 2022). Innovations in tastes, textures, and packaging have resulted from the growth of the meat-based snack sector and the ongoing need for high-protein snacks.

Composition: The formulation of meat-based snacks encompasses a variety of components that contribute to their flavor, texture, and nutritional profile. Typically, these snacks are primarily composed of different types of meat, such as beef, pork, or poultry, which serve as the main ingredient. The meat is often processed through methods such as drying, smoking, or curing to enhance its preservation and flavor. In addition to the meat, various seasonings, and flavoring agents are incorporated to create distinct taste profiles (Ohlau et al., 2022). Frequently used additions that improve flavor and aid in food preservation include salt, spices, and occasionally sugars. Furthermore, binders and fillers, such as starches or proteins, may be included to improve texture and cohesiveness. Nutritional considerations are also paramount in the composition of these snacks. Manufacturers often aim to balance protein content with fat levels, catering to consumer preferences for healthier options. As a result, some meat-based snacks may be formulated to be lower in fat or enriched with additional nutrients, such as vitamins or minerals. Overall, the composition of meat-based snacks is a complex interplay of ingredients designed to satisfy customer needs for convenience, flavor, and nutritional content.

CLASSIFICATION OF MEAT PROTEINS

Muscle tissue predominantly comprises water, accounting for approximately 75% of its total mass. The remaining composition includes around 20% protein, 3% fat, and 2% soluble non-protein substances. Within this 2% of non-protein components, various essential elements are present: metals and vitamins make up 3%, non-protein nitrogen-containing compounds constitute 45%, carbohydrates represent 34%, and inorganic substances account for 18%. Proteins within muscle can be categorized into three primary types: myofibrillar proteins, sarcoplasmic proteins, and connective tissue proteins. The classification of meat proteins is typically based on their solubility, functional roles, and structural contributions to the overall architecture of the muscle. Sarcoplasmic proteins, such as myoglobin and enzymes, are involved in metabolism and oxygen storage, whereas myofibrillar proteins, such as actin and myosin, are necessary for muscle contraction. Stromal proteins such as collagen, elastin, and reticulin form connective tissue and contribute to muscle toughness. I'll make sure to explain in depth how these proteins interact and contribute to meat structure.

Myofibrillar Proteins

The most prevalent protein in muscle, myofibrillar proteins, are principally in charge of contraction and structural integrity. These proteins are salt-soluble and contribute to the water-holding capacity and emulsification properties of meat. Myosin and actin are the two principal proteins in this category. The motor protein myosin is made up of an α-helical coiled-coil tail that interacts with other myosin molecules to produce thick filaments and a globular head domain that hydrolyzes ATP (Squire, 2019). The myosin head binds to actin during muscle contraction, facilitated by ATP hydrolysis. Actin exists in two forms: G-actin (globular actin), a monomeric form, and F-actin (filamentous actin), which polymerizes to form a long, thin filaments that serve as the backbone of the contractile apparatus. Another myofibrillar protein that runs the length of the actin filament and controls its interactions with myosin is called tropomyosin. It is a two-stranded α-helical coiled-coil protein (Hitchcock-DeGregori & Barua, 2017). This regulation is controlled by the troponin complex, which consists of troponin C (TnC), responsible for binding calcium ions; troponin I (TnI), which inhibits actin-myosin interaction; and troponin T (TnT), which anchors the troponin complex to tropomyosin. The structural arrangement of these proteins forms the characteristic sarcomere, the functional unit of muscle contraction.

Myofibrillar proteins represent a significant portion of the total protein composition in muscle tissue, comprising approximately 50% to 55%. In contrast, sarcoplasmic proteins contribute around 30% to 34% to this total. Connective tissue proteins make up the remaining portion, which is believed to be between 10% and 15%. There are many subclasses of myofibrillar proteins. The structural stability of myofibrils depends on the myofilament fibrous proteins, which comprise actin and myosin. Regulatory proteins such as the tropomyosin-troponin complex, α- and β-actinin, M-protein, and C-protein play crucial roles in muscle contraction regulation. Furthermore, scaffold proteins, including titin, nebulin, desmin, vimentin, and synemin, provide essential support to the overall myofibrillar architecture (Xu et al., 2024). Titin is an exceptionally large protein, with molecular weight nearing 1 million Daltons. These myofibrillar proteins are integral to the process of muscle contraction and are predominantly found within the myofibrils of muscle cells. They exhibit solubility in saline solutions. Overall, myofibrillar proteins are a vital category of muscle tissue proteins, significantly influencing the structure, functionality, and quality of muscle tissue, as they are closely related with the myofibrils, which serve as the contractile units of muscle fibers.

Major Components

Myosin: Myosin is a substantial and intricate protein characterized by its distinct structural regions: the head, neck, and tail. The head region is crucial as it engages with actin filaments and catalyzes the hydrolysis of adenosine triphosphate (ATP), thereby producing the energy necessary for muscle contraction. Myosin is fundamentally responsible for muscle contraction through the formation of cross-bridges with actin, facilitating the sliding filament mechanism (Linari et al., 2020). Additionally, myosin significantly influences the texture and functional properties of processed meat products. It is essential for establishing these goods' ability to retain water and for creating a stable gel matrix, especially in products like sausages (Herz et al., 2023).

Actin: Actin is present in two distinct forms: globular actin (G-actin), which is the unpolymerized state, and filamentous actin (F-actin), which occurs when G-actin monomers polymerize within muscle fibers (Bai et al., 2023; Chen et al., 2025; Fu et al., 2020). This protein plays a vital role in conjunction with myosin to produce contractile force, essential for muscle contraction. Additionally, actin contributes to the stabilization of the myofibrillar architecture, thereby indirectly affecting the tenderness of muscle tissue. Its structural and functional properties are vital for maintaining the integrity and performance of muscle fibers during contraction and relaxation cycles (Lim & Frontera, 2022).

Tropomyosin: A filamentous protein that extends parallel to the actin filaments plays a vital role in modulating the interaction between actin and myosin (Szikora et al., 2022). This regulation occurs through the protein's ability to either shield or reveal the binding sites on the actin filaments. Furthermore, this protein has a major effect on the denaturation process of proteins during cooking, which affects the food's texture and structural qualities.

Troponin: A structural assembly comprising three distinct subunits Troponin C, which can bind calcium ions; Troponin I, which serves an inhibitory function; and Troponin T, which interacts with tropomyosin plays a vital role in the regulation of muscle contraction (Tobacman & Cammarato 2021). This complex modulates the positioning of tropomyosin in response to fluctuations in calcium ion concentrations, thereby influencing the contraction and relaxation cycles of muscle fibers (Rasmussen & Jin, 2021). Furthermore, this troponin complex's dynamics are important when discussing muscle relaxation and the rigor mortis phenomena, both of which have a big impact on the texture and quality of meat (Wang & Raunser, 2023).

Sarcoplasmic Proteins

Sarcoplasmic proteins are distinguished by their water solubility and important roles in a number of muscle cell enzymatic and metabolic activities (Sultana et al., 2022). These proteins reside in the sarcoplasm, which is the cytoplasmic matrix of muscle fibers. Although they are present in lower quantities compared to myofibrillar proteins, their contributions are essential for several physiological functions, including the regulation of meat color, the facilitation of metabolic pathways, and the enhancement of flavor profiles in meat products (Cao et al., 2022). Myoglobin, a tiny globular protein that stores oxygen in muscle cells, is the most well-known of them.

Myoglobin is made up of a single polypeptide chain with a heme group that binds oxygen that is folded into a compact α-helical shape. The color of meat is largely influenced by the oxidation state of the heme iron in myoglobin, with oxymyoglobin (bright red), deoxymyoglobin (purple), and metmyoglobin (brown) being the three primary states. Additionally, hemoglobin, which is structurally similar to myoglobin but tetrameric, may be present in small quantities in residual blood. Other sarcoplasmic proteins include enzymes involved in glycolysis, such as lactate dehydrogenase, phosphofructokinase, and creatine kinase, which play a role in postmortem muscle metabolism and influence meat quality parameters like pH and tenderness.

Major Components

Myoglobin: A globular protein called myoglobin, which includes heme, is essential for storing oxygen in muscle tissues (Pujol et al., 2023). It is structured from a globin protein component and a heme group, which includes an iron atom at its core (Matarneh et al., 2023). This protein is essential for the transport and retention of oxygen in muscle cells, thereby supporting aerobic metabolism. Additionally, myoglobin significantly influences the coloration of meat, with variations in color arising from the oxidation state of the myoglobin molecule. These changes in oxidation state can lead to different hues, reflecting the protein's interaction with oxygen and its environment.

Deoxy Myoglobin: Often observed in newly sliced meat that hasn't been exposed to air, it is distinguished by a purple-red color. Because myoglobin is present in its deoxygenated state, this hue reflects the freshness and absence of oxidative processes in the meat (Singh et al., 2022)

Oxymyoglobin: Oxymyoglobin's intense red coloration is a direct consequence of its interaction with oxygen. This pigment plays a vital role in the transport and storage of oxygen within muscle tissues, and its bright coloration is indicative of the presence of oxygen bound to the myoglobin molecule (King et al., 2023). The intensity of this red color can serve as a visual marker for the freshness and quality of meat products, as it reflects the degree of oxygenation that the myoglobin has undergone.

Metmyoglobin: The brownish-gray color of metmyoglobin, which denotes oxidation, is frequently linked to the aging or spoiling of meat. The process of cooking induces heat denaturation of myoglobin, resulting in the transformation of its color to a brown shade, which is commonly observed in cooked meat products. This color change is a critical indicator of the meat's thermal treatment and its subsequent quality (Gagaoua et al., 2023).

Stromal Proteins

Stromal proteins represent a category of structural proteins found within connective tissue, characterized by their insolubility in both salt and water (Obernikhina et al., 2020). These proteins play a vital role in providing support and a structural framework for muscular tissues. In the context of meat, stromal proteins are integral to the connective tissue matrix, impacting both the texture and tenderness of the final product while also contributing to its structural integrity. Despite their significant influence on the overall quality of meat, stromal proteins are generally insoluble and, as a result, are less advantageous in the formulation of processed meat products. Their properties necessitate careful consideration in meat processing and culinary applications, as they can affect the sensory attributes and palatability of the meat Stromal proteins provide structural support and are largely insoluble in both water and salt solutions. The most abundant of these is collagen, a fibrous protein composed of three polypeptide chains (α-chains) that form a triple-helix structure, stabilized by hydrogen bonds and covalent cross-links. Collagen is a key determinant of meat toughness, as it is highly resistant to enzymatic breakdown and forms a rigid network in connective tissues. The amount and type of collagen present in meat influence the texture, with type I collagen being the most common in muscles, while type III collagen is associated with more delicate connective tissue structures. Another important stromal protein is elastin, which, unlike collagen, provides elasticity due to its high hydrophobic amino acid composition and unique cross-linking involving desmosine and isodesmosine bonds. Elastin is found in ligaments and tendons, where flexibility is required. Reticulin, a lesser-known stromal protein, forms fine networks that support muscle fibers and blood vessels.

Major Components

Collagen: The predominant stromal protein constitutes 1%–10% of muscle protein and is composed of a triple helix coiled around three polypeptide chains arranged into fibrils (Gosline et al., 2002). These fibrils provide connective tissues with strength and give muscles their characteristic tensile strength. Higher amounts of collagen substance is found in tougher beef portions such as shank and brisket. High heat and moisture during cooking cause collagen to change into gelatin, which improves softness. Meat hardness is influenced by the amount of collagen and the degree of cross-linking (older animals have more cross-linked, tougher collagen) (Zając et al., 2021).

Elastin: An elastic protein found in connective tissues like ligaments and tendons is up of an extremely interconnected network of amino acids which gives tissues flexibility and durability much less common than collagen (Gosline et al., 2002). The toughness of some cuts attributed to elastin, which doesn't deteriorate when cooked as collagen does.

Reticulin: In connective tissues, this fine, fibrous protein creates a supporting structure, composed of collagen-woven tiny fibrils, which helps keep tissues' structural integrity intact less important than elastin and collagen; however, it might affect texture in elderly animals (Mathews, 2012).

Enzymes

Lactate dehydrogenase, Pyruvate kinase, and enolase are key enzymes involved in the metabolic pathways of energy production, specifically glycolysis and oxidative phosphorylation (Nusairat et al., 2022). Additionally, cytochrome c and succinate dehydrogenase play significant roles in these processes. Following death, these enzymes still have an impact on how adenosine triphosphate (ATP) is broken down and how glycogen is converted to lactic acid, which subsequently affects the pH levels in muscle tissue. The activity of these enzymes post-mortem is critical, as it determines both the rate and the extent of pH decrease, which in turn has profound implications for the tenderness, water-holding capacity, and color stability of the meat. While myoglobin is well-known for its role in meat color, a similar protein found in the blood also contributes to this aspect, albeit to a lesser extent. The residual blood in the meat can impart color, but its contribution is minor when compared of myoglobin (Faustman et al., 2023). Thus, the interplay of these enzymes and proteins is essential for understanding the biochemical changes that occur in meat post-mortem and their effects on quality attributes.

Other Proteins

A variety of transport proteins, albumin, and ferritin all have vital functions in the body. The movement and storage of vital nutrients are the main functions of these proteins. Furthermore, they have an indirect impact on metabolic functions, which helps to regulate metabolism in general. Their roles are essential for preserving homeostasis and guaranteeing the availability of nutrients for a range of physiological requirements.

STRUCTURE OF MEAT PROTEINS

Peptide bonds bind the 20 distinct amino acids that make up a protein in various proportions. The specific amino acid sequence of proteins determines their unique three-dimensional structure, which in turn controls their biological activities. Each protein is different in size and structure. A healthy diet must include meat since it contains critical nutrients that promote human health. Meat proteins are the main structural elements of meat products, making up around 20% of a muscle's weight. In terms of nutrition, meat is prized for its high-quality protein, which includes all of the essential amino acids, as well as its profusion of vitamins and minerals that are accessible.

The structures of meat proteins can be explained at four levels: primary, secondary, tertiary, and quaternary. The primary structure of a protein is the linear arrangement of amino acids in a polypeptide chain, connected by peptide bonds. The genetic code determines this sequence, which is essential since it determines the ultimate form and function of the protein. The protein's ability to fold into secondary, tertiary, and quaternary structures is influenced by the arrangement of amino acids in the chain. The characteristics and biological activity of a protein can be greatly affected by even a single alteration in the amino acid sequence. The entire protein function is constructed on top of the fundamental structure. The secondary structure of proteins is formed through hydrogen bonding between the backbone functional groups, specifically between the oxygen of the C=O group of one amino acid and the hydrogen of the N-H group of another. This bonding results in structures like the α-helix, where the backbone coils, and the β-pleated sheet, where segments fold in the same plane. The formation of these structures depends on the primary sequence of the protein, with certain amino acids favoring one structure over the other. The tertiary structure of a protein is its unique three-dimensional shape, formed by interactions between residues as the polypeptide chain bends and folds in space (Sun et al., 2004). These interactions often involve residues that are far apart in the primary sequence. Unlike secondary structures, which rely on hydrogen bonds between the backbones, tertiary structures involve a variety of bonds, including hydrogen bonding, hydrophobic interactions, van der Waals forces, covalent disulfide bonds, and ionic interactions. Hydrophobic side chains cluster at the protein's core, away from water, while polar side chains interact with each other and the backbone. These interactions collectively stabilize the protein's structure and contribute to its functional shape. The arrangement and interactions of many polypeptide chains, or subunits that combine to produce a functional protein are referred to as the quaternary structure of a protein. The quaternary structure is determined by the interactions between the primary, secondary, and tertiary structures of each component. Although some proteins may also have covalent bonds, such as disulfide bridges, between subunits, the majority of the time, the subunits are bound together by non-covalent interactions such as hydrogen bonds, ionic bonds, and hydrophobic interactions.

PROPERTIES OF MEAT PROTEINS

Myofibrillar proteins, particularly myosin, play a vital role in water binding and retention within muscle tissue. Structural change in these proteins during rigor mortis and post-mortem storage significantly influence water-holding capacity (WHC), directly affecting meat juiciness (Ayaz et al., 2024). Both myosin and actin are essential for the gelation process, which is vital for the quality of processed meat products. Heat-induced denaturation of these proteins facilitates the formation of a gel network that entraps water and fat, enhancing texture and mouthfeel (Hussain et al., 2024). Post-mortem proteolysis, driven by enzymes such as calpains, further contributes to tenderness by breaking down myofibrillar proteins. Insufficient degradation can lead to tougher meat, often observed in older or poorly aged cuts. Additionally, myofibrillar proteins stabilize emulsions by binding fat and water, a property crucial for products like sausages and meat batters (Bourouis et al., 2023).

The solubility and functionality of proteins are heavily influenced by pH levels. A rapid post-mortem pH decline can denature proteins, reducing WHC and resulting in pale, soft, and exudative (PSE) meat (Mohamed et al., 2022). Heat application during cooking denatures myosin and actin, altering meat texture and appearance. Formation of actomyosin cross-bridges during this process determines meat rigidity and firmness (Matarneh et al., 2023). Sarcoplasmic proteins, particularly myoglobin, are key determinants of meat color. Myoglobin concentration, higher in red meat than in white, dictates color based on oxygen availability and oxidation state. Cooking transforms the red hue to brown, while sarcoplasmic proteins and their by-products, such as peptides and amino acids, contribute to flavor precursors through interactions with the lipid oxidation and the Maillard reaction. Although sarcoplasmic proteins do not directly support structural integrity, they indirectly influence tenderness and WHC by regulating pH during post-mortem metabolism (Fu et al., 2020). Glycolysis, facilitated by sarcoplasmic enzymes, converts glycogen into Lactic acid post-slaughter, leading to a pH drop that profoundly impacts meat quality (López-Bote & Calvo, 2023). This pH decline affects protein functionality, influencing texture, flavor, and juiciness. Myofibrillar and sarcoplasmic proteins are heat-sensitive, denaturing at relatively low temperatures and altering the meat's visual and textural properties. Antioxidants like ascorbic acid and rosemary extracts stabilize myoglobin and inhibit oxidation, while modified atmosphere packaging (MAP) enhances color stability. Quantifying sarcoplasmic proteins, including myoglobin, aids in classifying meat varieties and optimizing cooking methods to improve flavor profiles (Zduńczyk et al., 2024).

Collagen significantly impacts meat texture, with extensive cross-linking in older animals or heavily exercised muscles resulting in tougher meat. Moist heat or slow cooking methods are necessary to break down collagen into gelatin, enhancing tenderness. In contrast, meat from younger animals contains less cross-linked, more soluble collagen, yielding a naturally tender texture. Elastin, another connective tissue protein, imparts chewiness and remains tough even after prolonged cooking, affecting mouthfeel (Mohd Azmi et al., 2023). Unlike myofibrillar proteins, stromal proteins like collagen and elastin do not directly influence WHC due to their insolubility. Excessive connective tissue can hinder muscle development, reducing WHC. Collagen and gelatin are rich in glycine, proline, and hydroxyproline, beneficial for skin and joint health, but stromal proteins are nutritionally incomplete due to a lack of essential amino acids. As animals age, collagen cross-linking increases, reducing solubility and increasing toughness. Cooking collagen above 70°C in moist conditions converts it to gelatin, enhancing tenderness in slow-cooked meats, while elastin remains resistant to heat, maintaining toughness (Bourouis et al., 2023)

Cooking techniques such as low and slow methods are ideal for collagen-rich cuts like brisket and shank, as they break down collagen into gelatin, improving tenderness and moisture. Grilling and roasting, while popular for flavor, are less effective for tough cuts due to insufficient collagen softening. Meat tenderization methods, including mechanical pounding, acidic marinades, and proteolytic enzymes like papain and bromelain, disrupt connective tissue, enhancing tenderness. Aging, whether wet or dry, primarily breaks down muscle proteins, with limited effects on stromal proteins. Collagen extraction for gelatin production is widely used in food products like jellies and sausages, while additives may be necessary to improve texture and flavor in low-quality, collagen-rich meats (Mohd et al., 2023). Non-protein nitrogen (NPN) compounds, include free amino acids, creatine, and peptides, participate in Maillard reactions during cooking, generating flavor and aroma compounds essential for the "meaty" taste. Nucleotides like inosine monophosphate (IMP) enhance umami flavor, often synergizing with glutamate (Ma et al., 2024). Creatine supports energy metabolism, while carnitine aids fat metabolism, and free amino acids and peptides are readily absorbed, contributing to physiological functions (Bonilla et al., 2021). Post-mortem degradation of NPN compounds influences meat pH, affecting tenderness, WHC, and microbial stability (Gómez et al., 2020). Post-slaughter, ATP degradation to IMP and hypoxanthine shapes meat flavor during aging and cooking. Enzymatic hydrolysis release free amino acids and peptides, enhancing taste, while prolonged aging increases NPN compounds, enriching flavor complexity but potentially compromising microbial stability (Kathera and Kim 2024). Controlled cooking techniques optimize Maillard reactions and other chemical processes involving NPN compounds, further enhancing flavor. In processed meats, NPN compounds like IMP and glutamate are often added to improve taste (Nowacka et al., 2023.)

Physical

Texture and Tenderness: Meat tenderness is a crucial quality attribute that significantly influences consumer satisfaction and culinary applications. Several factors determine tenderness, including the type and amount of connective tissue, composition of muscle fibers, and the activity of Proteolytic enzymes (Kemp et al., 2010). Connective tissue, primarily composed of collagen, can affect tenderness; meat with higher levels of cross-linked collagen, typically found in older animals, tends to be tougher. Conversely, younger animals generally have less cross-linked collagen, resulting in more tender meat. Postmortem aging is another vital process that enhances tenderness. During these times, proteolytic enzymes naturally present in the muscle break down myofibrillar proteins, leading to increased tenderness (Lonergan et al., 2010). This enzymatic activity is crucial for improving the mouthfeel and overall quality of the meat. Additionally, the muscle fiber composition, including the ratio of fast-twitch to slow-twitch fibers, plays a role in texture; fast-twitch fibers are typically tenderer. Understanding these factors allows meat producers and chefs to select appropriate cuts and aging methods to achieve the desired tenderness in their products.

Meat Color and Appearance: The color of meat is an important factor influencing consumer perception and purchasing decisions, primarily determined by the oxidation state of myoglobin, the main pigment protein in muscle. Myoglobin exists in various forms, each contributing to the meat's visual appeal (Young & West, 2001). Deoxymyoglobin, the purple-red form, occurs in vacuum-packed meat where oxygen is depleted, often leading consumers to perceive it as less fresh. When exposed to oxygen, myoglobin converts to oxymyoglobin, resulting in a bright red color that is more visually appealing and signals freshness to consumers. Conversely, metmyoglobin, which appears brown due to oxidation, may indicate spoilage and can deter consumers. Meat color stability may be preserved in a number of ways, such as by using antioxidants and efficient packing methods that reduce oxygen exposure. These techniques aid in avoiding oxidative reactions, which may result in unintended hue shifts. In addition to improving the meat's visual appeal, preserving the intended hue is essential for marketability and consumer acceptability.

Water-Holding Capacity and Juiciness: Water-holding capacity (WHC) is a fundamental physical property of meat that directly influences its juiciness and overall sensory experience. WHC is affected by several factors, including pH, muscle structure, and biochemical changes that occur postmortem. A lower WHC can lead to increased drip loss during cooking or storage, resulting in drier meat, while a higher WHC contributes to improved texture and juiciness. The pH of meat, which typically decreases after slaughter, plays a crucial role in WHC; optimal pH levels promote better protein-water interactions, enhancing moisture retention (Zhang et al., 2023a). Additionally, the structural integrity of muscle fibers and the presence of proteins that bind water can significantly affect WHC. Postmortem biochemical reactions, including the breakdown of proteins and muscle fibers, also influence moisture retention. Understanding these dynamics is crucial for meat processing and cooking, as maximizing WHC may result in better beef products that satisfy consumers' demands for flavor and juiciness.

WHC is largely determined by the structural properties of meat proteins, particularly myofibrillar proteins, which mixes with water molecules to bind them effectively. Several factors influence WHC, including pH, ionic strength, and temperature. A lower pH, which is often close to the isoelectric point of muscle proteins, can reduce water retention, resulting in drier and tougher meat. This phenomenon is particularly evident in post-mortem changes in meat, where the pH decreases due to lactic acid accumulation. Additionally, ionic strength can influence the solubility and interaction of proteins with water; higher salt concentrations can enhance WHC by promoting protein solubility and hydrophilicity. Temperature also plays a significant role; during cooking, proteins undergo denaturation, which can either enhance or diminish WHC depending upon the conditions. Understanding and optimizing WHC is important for meat processors and chefs alike, as it directly correlates with consumer satisfaction and the overall sensory experience of meat products.

Gelation and Emulsification Properties: Meat proteins, particularly myofibrillar proteins, play essential roles in gelation and emulsification, which are critical for the production of various processed meat products, such as sausages and patties. When meat is subjected to heat, these proteins undergo denaturation, allowing them to interact and form gels. This heat-induced gelation improves the processed beef products' structural integrity and adds to their consumer-pleasing textures (Khalid et al., 2023). Emulsification is another important property, particularly in products containing fat. The ability of meat proteins to stabilize emulsions prevents phase separation, ensuring that fat is evenly distributed throughout the product. The preservation of mouthfeel and texture depends on its stability. Furthermore, animal proteins' emulsifying qualities enable the inclusion of other substances, such flavorings and spices, without sacrificing quality. Understanding and optimizing these properties are crucial for meat processors seeking to create high-quality products that meet the demands of the market.

Chemical

Solubility and Heat Stability: A significant factor influencing the activity of animal proteins is their solubility, particularly in processed meat products. Proteins in meat can be broadly classified into sarcoplasmic and myofibrillar proteins, with different solubility characteristics. Sarcoplasmic proteins are water-soluble and play essential roles in metabolic processes, while myofibrillar proteins require the presence of salt to dissolve effectively. This salt-induced solubility is crucial during meat processing, as it allows for better binding of ingredients and improved texture. Heat treatment further complicates protein behavior, as it induces denaturation, leading to structure changes that impact both texture and moisture retention. Myofibrillar proteins upon heating, form heat-induced gels, which contribute to the firmness and chewiness of cooked meat (Xu and Xu, 2021). This gel formation is vital in the production of various meat products, like sausages and meatballs, where texture is a critical quality attribute. Additionally, the heat stability of these proteins affects their ability to retain moisture and maintain structural integrity during cooking, ultimately influencing consumer perception and product quality. Protein denaturation and gelation: Meat's texture and general quality are drastically changed by the process of protein denaturation, which is essential to cooking and processing. When meat is subjected to heat, proteins unfold and lose their native structures, leading to new interactions among the protein molecules. When proteins combine to produce gels, which give many meat products their desired textural properties, this denaturation is very important (Zhang et al., 2023b). Collagen, a key protein in connective tissue, undergoes denaturation when exposed to heat, converting into gelatin. This transformation is crucial for enhancing tenderness, as gelatin contributes to the moist mouthfeel and palatability of cooked meat. The gelation process is particularly essential for food processing applications like making sausages, where the hardness and slice the ability of the finished product may be determined by the proteins' capacity to gel (Hao et al., 2024). Understanding the mechanisms of protein denaturation and gelation allows meat processors to manipulate cooking conditions and formulations to achieve specific textural outcomes, ensuring that the final products meet consumer demands for taste and quality.

EFFECT OF PROCESSING ON, MEAT PROTEINS

Thermal Processing

Thermal processing is an essential aspect of meat preparation that significantly influences its texture, flavor, and overall quality. Cooking methods such as roasting, grilling, and boiling induce protein denaturation, a process where proteins unfold and lose their original structure due to exposure to heat (Bhat et al., 2021). This denaturation affects the meat's texture, often leading to enhanced tenderness when controlled heating methods are employed. Slow-cooking techniques break down tough connective tissues, resulting in a tenderer final product. However, excessive heat leads to undesirable outcomes, such as toughness and moisture loss, which detract from the eating experience. The Maillard reaction, which occur during high-heat cooking, enhance flavor by producing complex compounds that contribute to the savory taste and appealing color of cooked meat (Hossain et al., 2025). Understanding the dynamics of thermal processing is essential for chefs and food technologists, as optimal cooking techniques can maximize flavor and texture while minimizing negative effects associated with overcooking.

Curing and Fermentation

Curing and fermentation are traditional methods used to preserve meat while enhancing its flavor and texture. Curing involves the application of salt, nitrates, and other curing agents to meat, which alters the protein structure and provides preservation through moisture reduction and the inhibition of spoilage microorganisms (Dave & Ghaly, 2011). These process not only enhance shelf life but also contributes to the development of unique flavors and textures in cured products like ham and bacon. Fermentation, on the other hand, utilizes microbial activity to transform meat, as seen in products such as salami and pepperoni. During fermentation, beneficial bacteria break down proteins and fats, leading to the production of flavor compounds and improved digestibility (Sharma et al., 2020). The fermentation process also lowers the pH of the meat, enhancing its safety and stability. Together, curing and fermentation create distinctive taste profiles and textures that are highly valued in various cuisines worldwide, highlighting the importance of these methods in meat processing.

Freezing and Storage

Although freezing is a popular way to preserve meat, it has a big impact on the proteins and general quality of the meat. Ice crystals that develop inside muscle fibers when meat is frozen can alter the cellular structure and water-holding capacity (WHC). The juiciness and texture of the meat may be impacted by these disturbances as they may cause moisture loss during freezing. To mitigate these negative effects, proper freezing techniques are essential. Rapid freezing helps minimize ice crystal formation, reducing damage to muscle fibers and maintaining better quality. Additionally, controlled thawing methods, such as thawing in the refrigerator rather than at room temperature, can preserve the integrity of meat and prevent further moisture loss. Maintaining the quality and safety of frozen meat products is essential for giving consumers a satisfying eating experience after preparation. This is achieved by knowing how freezing affects meat proteins and using best practices for storage.

APPLICATIONS OF MEAT PROTEINS

Meat proteins are used in many different fields, including as nutrition, food production, and medicinal research. Their distinct functional qualities, such as emulsification, gelation, and solubility, make them indispensable in both industrial and culinary settings.

Food Processing and Product Development

Meat proteins are necessary for the production of processed meat products in the food business, such as patties, deli meats, and sausages. These proteins enhance overall quality, taste, and texture. Gel formation, which improves the structural integrity of products, depends on myofibrillar proteins. Emulsification properties allow for the incorporation of fats and other ingredients, creating stable mixtures that maintain desirable mouthfeel and appearance. Additionally, meat proteins can be modified through techniques like curing and fermentation to enhance flavor profiles and preservation, leading to products like salami and cured hams.

Nutritional Supplements

One important source of high-quality protein that offers the essential amino acids required for human health is meat. It is often used in meal replacement products, protein powders, and nutritional supplements targeted at athletes and fitness enthusiasts. These supplements promote muscle development, recuperation, and general nutritional balance by utilizing the whole amino acid profile of animal proteins. Meat proteins are a great option for anybody looking to effectively improve their protein consumption due to their bioavailability.

Culinary Applications

Meat proteins are essential for making a wide range of foods in kitchens at home and in restaurants. The culinary experience is improved by their capacity to absorb flavors and moisture. To create rich aromas and soft textures, methods like marinating, brining, and slow cooking take use of the qualities of beef proteins. Furthermore, because meat is so adaptable, it may be prepared using a variety of techniques, such as grilling, roasting, and braising, each of which highlights unique protein properties. Despite their convenience, processed meats sometimes provide lower-quality protein because of preservatives and chemicals. It is crucial for consumers to balance their intake of processed meats since excessive consumption has been linked to a number of health hazards.

Biomedical Research

Meat proteins are increasingly being investigated in biological research for purposes other than cooking. They can be used in tissue engineering and regenerative medicine because of their structural and functional characteristics. For instance, because of its biocompatibility and capacity to encourage cell attachment, collagen is being studied for its potential to aid in tissue development and repair. In order to increase the effectiveness of medicinal medicines, researchers are also looking at the possibility of using animal proteins as carriers for drug delivery systems.

Alternative Protein Sources

The extraction and modification of beef proteins for use in alternative protein products is gaining popularity as plant-based diets and sustainable food practices gain acceptance. This involves the creation of meat analogs, which use proteins obtained from plants to replicate the flavor and feel of meat. Food scientists may develop hybrid goods that satisfy the nutritional needs of a wider range of consumers while providing the sensory experiences of meat by comprehending the functional characteristics of meat proteins.

CONCLUSION

Meat proteins play a central role in human nutrition, offering high biological value through complete essential amino acid profiles and functional peptides. This chapter has systematically examined the composition, classification, structural organization, and functionality of meat proteins derived from various sources, including red meat, poultry, seafood, and processed products. Myofibrillar proteins, particularly myosin and actin, contribute significantly to muscle contraction, water-holding capacity, gelation, and emulsification are crucial properties for both raw and processed meat products. Sarcoplasmic proteins such as myoglobin dictate meat color and influence post-mortem pH, thereby affecting tenderness and flavor. Stromal proteins like collagen and elastin define meat texture and toughness, particularly in older animals or exercised muscles. Processing techniques, including thermal treatment, curing, fermentation, and freezing profoundly affect protein solubility, denaturation, and structural properties, ultimately shaping the sensory and nutritional qualities of meat. Meat proteins find extensive applications in food product development, nutritional supplementation, culinary preparation, and even biomedical research, such as collagen-based tissue scaffolds and drug delivery systems. This chapter underscores the multifaceted roles of meat proteins in ensuring meat quality, optimizing processing, enhancing health benefits, and expanding industrial applications.

REFERENCES

Ahmad, Saghir, and Abdol Ghafour Badpa. 2014. "Meat Products and Byproducts for Value Addition." In A. Malik, Z. Erginkaya, S. Ahmad, and H. Erten (Eds.), Food Processing*: Strategies for Quality Assessment* (pp. 125–153). Springer. https://doi.org/10.1007/978-1-4939-1378-7.

Alekseeva, E, and V Kolchina. 2019. "Amino Acid Composition of Beef Obtained from the Specialized Meat Cattle." In *Proceedings of the 1st International Conference on Life Sciences and Technology (IcoLiST)*, Malang, Indonesia. IOP Conference Series: Earth and Environmental Science, 341, 012136. IOP Publishing. https://doi.org/10.1088/1755-1315/341/1/012136.

Amirkhanov, Kumarbek, Aidyn Igenbayev, Almagul Nurgazezova, Eleonora Okuskhanova, Samat Kassymov, Nazerke Muslimova, and Zhanibek Yessimbekov. 2017. "Research Article Comparative Analysis of Red and White Turkey Meat Quality." *Pakistan Journal of Nutrition* 16: 412–16.

Asha, K, R Anandan, Suseela Mathew, and P T Lakshmanan. 2014. "Biochemical Profile of Oyster Crassostrea Madrasensis and Its Nutritional Attributes." *The Egyptian Journal of Aquatic Research* 40 (1): 35–41.

Ashaolu, Tolulope J, Ibrahim Khalifa, Matta A Mesak, Jose M Lorenzo, and Mohamed A Farag. 2023. "A Comprehensive Review of the Role of Microorganisms on Texture Change, Flavor and Biogenic Amines Formation in Fermented Meat with Their Action Mechanisms and Safety." *Critical Reviews in Food Science and Nutrition* 63 (19): 3538–55.

Ayaz, Qudsiya, Nadira Anjum, Haamiyah Sidiq, Zahida Naseem, Samira, Tajalli Asad, Barjees Ashaq, Naseh Nisar, Mansha Manzoor, and Iqra Bashir. 2024. "Structure, Composition, Nutritive Value and Biochemistry of Post-mortem Muscle." In S. A. Rather and F. A. Masoodi (Eds.), *Hand Book of Processed Functional Meat Products* (pp. 25–52). Springer. https://doi.org/10.1007/978-3-031-69868-2.

Bai, Yang, Feng Zhao, Tingting Wu, Fangchun Chen, and Xiaoxiao Pang. 2023. "Actin Polymerization and Depolymerization in Developing Vertebrates." *Frontiers in Physiology* 14: 1213668.

Baker, David H. 2009. "Advances in Protein–Amino Acid Nutrition of Poultry." *Amino Acids* 37: 29–41.

Bakhsh, Allah, Se-Jin Lee, Eun-Yeong Lee, Young-Hwa Hwang, and Seon-Tea Joo. 2021. "Characteristics of Beef Patties Substituted by Different Levels of Textured Vegetable Protein and Taste Traits Assessed by Electronic Tongue System." *Foods* 10 (11): 2811.

Barbut, Shai. 2024. "Measuring Water Holding Capacity in Poultry Meat." *Poultry Science* 103 (5): 103577.

Barone, Ada Maria, and Jessica Aschemann-Witzel. 2022. "Food Handling Practices and Expiration Dates: Consumers' Perception of Smart Labels." *Food Control* 133: 108615.

Barr, Benjamin, Danielle E Levitt, and Lauren Gollahon. 2025. "Red Meat Amino Acids for Beginners: A Narrative Review." *Nutrients* 17 (6): 939.

Bhat, Zuhaib F, James D Morton, Alaa El-Din A Bekhit, Sunil Kumar, and Hina F Bhat. 2021. "Thermal Processing Implications on the Digestibility of Meat, Fish and Seafood Proteins." *Comprehensive Reviews in Food Science and Food Safety* 20 (5): 4511–48.

Boada, Luis D, Luis Alberto Henríquez-Hernández, and O P Luzardo. 2016. "The Impact of Red and Processed Meat Consumption on Cancer and Other Health Outcomes: Epidemiological Evidences." *Food and Chemical Toxicology* 92: 236–44.

Boateng, Evans Frimpong, Mustapha Muhammad Nasiru, and Mary Agyemang. 2020. "Meat: Valuable Animal-Derived Nutritional Food. A Review." *Asian Food Sci. J.* 15: 9–19.

Bonilla, Diego A, Richard B Kreider, Jeffrey R Stout, Diego A Forero, Chad M Kerksick, Michael D Roberts, and Eric S Rawson. 2021. "Metabolic Basis of Creatine in Health and Disease: A Bioinformatics-Assisted Review." *Nutrients* 13 (4): 1238.

Bourouis, Imane, Zhihua Pang, and Xinqi Liu. 2023. "Recent Advances on Uses of Protein and/or Polysaccharide as Fat Replacers: Textural and Tribological Perspectives: A Review." *Journal of Agriculture and Food Research* 11: 100519.

Cao, Longxing, Brian Coventry, Inna Goreshnik, Buwei Huang, William Sheffler, Joon Sung Park, Kevin M Jude, Iva Marković, Rameshwar U Kadam, and Koen H G Verschueren. 2022. "Design of Protein-Binding Proteins from the Target Structure Alone." *Nature* 605 (7910): 551–60.

Chen, Bohan, Qi Yan, Dapeng Li, and Jing Xie. 2025. "Degradation Mechanism and Development of Detection Technologies of ATP-Related Compounds in Aquatic Products: Recent Advances and Remaining Challenges." *Critical Reviews in Food Science and Nutrition* 65 (1): 101–22.

Dave, D, and Abdel E Ghaly. 2011. "Meat Spoilage Mechanisms and Preservation Techniques: A Critical Review." *Am J Agric Biol Sci* 6 (4): 486–510.

Deshpande, Anagh. 2025. *Meat Technology and Processing*. Delhi: Educohack Press. ISBN 978-93-61526-50-3.

Dévi, G, D Gnimabou, K Christian, Kpatinvoh Brice, and B Farid. 2020. "Improvement in Sausage Manufacturing Process and Potential Use of Selected Aromatic Plants as Their Bio Preservatives in Benin." *GSC Biol. Pharma. Sci* 11 (3): 166–76.

Espe, Marit, Anne-Catrin Adam, Takaya Saito, and Kaja Helvik Skjærven. 2023. "Methionine: An Indispensable Amino Acid in Cellular Metabolism and Health of Atlantic Salmon." *Aquaculture Nutrition* 2023 (1): 5706177.

Faustman, Cameron, Surendranath P Suman, and Ranjith Ramanathan. 2023. "The Eating Quality of Meat: I Color." In F. Toldrá (Ed.), *Lawrie's Meat Science* (9th ed., pp. 363–392). Woodhead Publishing. https://doi.org/10.1016/B978-0-323-85408-5.00023-6.

Fu, Yu, Yuhao Zhang, Olugbenga P Soladoye, and Rotimi E Aluko. 2020. "Maillard Reaction Products Derived from Food Protein-Derived Peptides: Insights into Flavor and Bioactivity." *Critical Reviews in Food Science and Nutrition* 60 (20): 3429–42.

Gagaoua, Mohammed, Surendranath P Suman, Peter P Purslow, and Bénédicte Lebret. 2023. "The Color of Fresh Pork: Consumers Expectations, Underlying Farm-to-Fork Factors, Myoglobin Chemistry and Contribution of Proteomics to Decipher the Biochemical Mechanisms." *Meat Science* 206: 109340.

Gómez, Inmaculada, Rasmi Janardhanan, Francisco C Ibañez, and María José Beriain. 2020. "The Effects of Processing and Preservation Technologies on Meat Quality: Sensory and Nutritional Aspects." *Foods* 9 (10): 1416.

Gosline, John, Margo Lillie, Emily Carrington, Paul Guerette, Christine Ortlepp, and Ken Savage. 2002. "Elastic Proteins: Biological Roles and Mechanical Properties." *Philosophical Transactions of the Royal Society of London. Series B: Biological Sciences* 357 (1418): 121–32.

Guo, Jiaxin, Lujie Cui, and Zong Meng. 2023. "Oleogels/Emulsion Gels as Novel Saturated Fat Replacers in Meat Products: A Review." *Food Hydrocolloids* 137: 108313.

Hao, Shiqin, Min Qian, Yaru Wang, Kaiping Zhang, Jianjun Tian, and Xinliang Wang. 2024. "Research Progress on the Gel Properties of Fermented Sausage." *Food Materials Research* 4(1). https://doi.org/10.48130/fmr-0023-0042.

Hathwar, Swapna C, Amit Kumar Rai, Vinod Kumar Modi, and Bhaskar Narayan. 2012. "Characteristics and Consumer Acceptance of Healthier Meat and Meat Product Formulations—A Review." *Journal of Food Science and Technology* 49: 653–64.

Herz, Eva, Pascal Moll, Christophe Schmitt, and J Weiss. 2023. "Binders in Foods: Definition, Functionality, and Characterization." *Food Hydrocolloids* 145: 109077.

Hitchcock-DeGregori, Sarah E, and Bipasha Barua. 2017. "Tropomyosin Structure, Function, and Interactions: A Dynamic Regulator." In D. A. D. Parry and J. M. Squire (Eds.), *Fibrous Proteins: Structures and Mechanisms* (pp. 253–84). Springer. https://doi.org/10.1007/978-3-319-49674-0_9.

Hossain, Md Jakir, A M Nurul Alam, So-Hee Kim, Chan-Jin Kim, Seon-Tea Joo, and Young-Hwa Hwang. 2025. "Techniques and Emerging Trends in Flavor and Taste Development in Meat." *Food Science of Animal Resources* 45 (1): 266.

Hussain, Arif, Muhammad Hussain, Waqas Ashraf, Aiman Karim, Sahibzada Muhammad Aqeel, Adil Khan, Asif Hussain, Salman Khan, and Zhang Lianfu. 2024. "Preparation, Characterization and Functional Evaluation of Soy Protein Isolate-Peach Gum Conjugates Prepared by Wet Heating Maillard Reaction." *Food Research International* 192: 114681.

Iyiola, V O, and S U Bulus. 2024. "Effects of Slurries on the Quality of Kilishi." *Journal of Animal Science and Veterinary Medicine* 9 (5): 213–21.

Kadim, I T, M R Al-Ani, R S Al-Maqbaly, M H Mansour, O Mahgoub, and E H Johnson. 2011. "Proximate, Amino Acid, Fatty Acid and Mineral Composition of Raw and Cooked Camel (Camelus Dromedarius) Meat." *British Food Journal* 113 (4): 482–93.

Karimidastjerd, Atefeh, Zehra Gulsunoglu-Konuskan, Burcu Ersoy, and Zakir Showkat Khan. 2024. "Plant-Based Seafoods: A Sustainable and Nutritious Alternative." In F. Boukid, C. M. Rosell, and N. Gasparre (Eds.), *Handbook of Plant-Based Food and Drinks Design*. Elsevier.

Kathera, Chandrasekhar, and Jihoe Kim. 2024. "Serum Markers for Evaluating Beef Meat Quality: Potential Applications in Cell-Cultured Meat Production." (Version 1). Preprints. https://doi.org/10.20944/preprints202403.1223.v1.

Kemp, Caroline M, Paul L Sensky, Ronald G Bardsley, Peter J Buttery, and Tim Parr. 2010. "Tenderness–An Enzymatic View." *Meat Science* 84 (2): 248–56.

Khalid, Waseem, Aristide Maggiolino, Jasmeet Kour, Muhammad Sajid Arshad, Noman Aslam, Muhammad Faizan Afzal, Parkash Meghwar, Khair-ul-Wajeeha Zafar, Pasquale De Palo, and Sameh A Korma. 2023. "Dynamic Alterations in Protein, Sensory, Chemical, and Oxidative Properties Occurring in Meat during Thermal and Non-Thermal Processing Techniques: A Comprehensive Review." *Frontiers in Nutrition* 9: 1057457.

King, D Andy, M C Hunt, S Barbut, J R Claus, D P Cornforth, P Joseph, Y H Kim, G Lindahl, R A Mancini, M N Nair, K J Merok, A Milkowski, A Mohan, F Pohlman, R Ramanathan, C R Raines, M Seyfert, O Sørheim, S P Suman, and M Weber. 2023. "American Meat Science Association Guidelines for Meat Color Measurement." *Meat and Muscle Biology* 6 (4):1–81. https://doi.org/10.22175/mmb.12473.

Kumar, Rajiv Ranjan, Faslu Rahman, Bedika Bora, and Mohammed Shameeh. 2023. "Importance and Nutritive Value of Animal Proteins in Human Diet." In *Processing Technologies and Food Protein Digestion*, 1–25. Elsevier.

Liao, Jialong, Pengguang Zhang, Jingdong Yin, and Xin Zhang. 2024. "New Insights into the Effects of Dietary Amino Acid Composition on Meat Quality in Pigs: A Review." *Meat Science*, 109721.

Lim, Jae-Young, and Walter R Frontera. 2022. "Single Skeletal Muscle Fiber Mechanical Properties: A Muscle Quality Biomarker of Human Aging." *European Journal of Applied Physiology* 122 (6): 1383–95.

Linari, Marco, Gabriella Piazzesi, Irene Pertici, Jody A Dantzig, Yale E Goldman, and Vincenzo Lombardi. 2020. "Straightening out the Elasticity of Myosin Cross-Bridges." *Biophysical Journal* 118 (5): 994–1002.

Lombardi-Boccia, Lanzi, Lucarini, and Di Lullo. 2004. "Meat and Meat Products Consumption in Italy: Contribution to Trace Elements, Heme Iron and Selected B Vitamins Supply." *International Journal for Vitamin and Nutrition Research* 74 (4): 247–51.

Lonergan, Elisabeth Huff, Wangang Zhang, and Steven M Lonergan. 2010. "Biochemistry of Postmortem Muscle—Lessons on Mechanisms of Meat Tenderization." *Meat Science* 86 (1): 184–95.

López-Bote, Clemente, and Luis Calvo. 2023. "Chemical and Biochemical Constitution of Muscle." In F. Toldrá (Ed.), *Lawrie's Meat Science* (8th ed.), 105–57. Woodhead Publishing. https://doi.org/10.1016/B978-0-08-100694-8.00004-2.

Ma, Fang, Yang Li, Yuanlv Zhang, Qian Zhang, Xiaoxue Li, Qingqing Cao, Haiyang Ma, Delang Xie, Bingbing Zhang, and Jia Yu. 2024. "Effects of Umami Substances as Taste Enhancers on Salt Reduction in Meat Products: A Review." *Food Research International*, 185, 114248.

Maheswarappa, Naveena B, Muthukumar Muthupalani, Kiran Mohan, Rituparna Banerjee, Arup Ratan Sen, and Sukhdeo B Barbuddhe. 2022. "Chilling, Freezing, and Other Preservation Methods of Buffalo Meat." In *Asiatic Water Buffalo: A Sustainable and Healthy Red Meat Source* (1st ed.). Singapore: Springer Singapore. https://doi.org/10.1007/978-981-19-2619-8_9.

Maia, Rosana Cardoso, Rafael de Sousa Ferreira, Luiz Fernando Teixeira Albino, Arele Arlindo Calderano, Bruno Damaceno Faria, Diego Ladeira da Silva, Thaisa Pereira de França, and Horácio Santiago Rostagno. 2024. "Effect of Dietary Non-Essential Amino Acid Sources on Performance, Nitrogen Utilization and Blood Parameters for Broiler Chickens Fed a Low-Protein Diet." *Animal Feed Science and Technology* 315: 116010.

Manhiani, Paljinder Singh. 2010. Carnosine Content and Antioxidant Activity from Poultry Co-Products, Protein Meal and Stressed Poultry Tissues (Doctoral dissertation, Clemson University). *TigerPrints*. https://tigerprints.clemson.edu/all_dissertations/681.

Mao, Jingjing, Xinyi Wang, Hongfan Chen, Zhiping Zhao, Dayu Liu, Yin Zhang, and Xin Nie. 2024. "The Contribution of Microorganisms to the Quality and Flavor Formation of Chinese Traditional Fermented Meat and Fish Products." *Foods* 13 (4): 608.

Matarneh, Sulaiman K, Tracy L Scheffler, and David E Gerrard. 2023. "The Conversion of Muscle to Meat." In F. Toldrá (Ed.), *Lawrie's Meat Science* (8th ed.), 105–57. Woodhead Publishing.

Mathews, Martin B. 2012. *Connective Tissue: Macromolecular Structure and Evolution*. Springer Berlin Heidelberg. https://doi.org/10.1007/978-3-642-80904-0

Mohamed, Haidy E, Wafaa A Ibrahim, and Rehab E M Gaafar. 2022. "Impact of Moringa Olifera Leaves or Saccharomyces Supplementation on Carcass Quality, MRNA of Heat Shock Proteins and Antioxidants in Broilers Exposed to Heat Stress." *SVU-International Journal of Veterinary Sciences* 5 (4): 193–227.

Mohd Azmi, Syahira Izyana, Pavan Kumar, Neelesh Sharma, Awis Qurni Sazili, Sung-Jin Lee, and Mohammad Rashedi Ismail-Fitry. 2023. "Application of Plant Proteases in Meat Tenderization: Recent Trends and Future Prospects." *Foods* 12 (6): 1336.

Munekata, Paulo E S, Rubén Domínguez, Mirian Pateiro, Silvina Cecilia Andrés, Eva María Santos, Maria João Fraqueza, Paulo Cezar Bastianello Campagnol, and José Manuel Lorenzo. 2023. "Meat and Meat Products: Animal Species, Products, Processing, Quality, and Shelf Life." In H. L. Meiselman and J. M. Lorenzo (Eds.), *Meat and Meat Replacements: An Interdisciplinary Assessment of Current Status and Future Directions*, 45–76. Woodhead Publishing. https://doi.org/10.1016/C2020-0-02725-8.

Nam, Ki-Chang, Cheorun Jo, and Mooha Lee. 2010. "Meat Products and Consumption Culture in the East." *Meat Science* 86 (1): 95–102.

Nowacka, Malgorzata, Magdalena Trusinska, Paulina Chraniuk, Federico Drudi, Jakub Lukasiewicz, Nam Phuong Nguyen, Adrianna Przybyszewska, Katarzyna Pobiega, Silvia Tappi, and Urszula Tylewicz. 2023. "Developments in Plant Proteins Production for Meat and Fish Analogues." *Molecules* 28 (7): 2966.

Nurdiani, Rahmi, Asep Awaludin Prihanto, and Muhamad Firdaus. 2020. "Seafood as Source of Protein-based Functional Foods." In S.-K. Kim (Ed.), *Encyclopedia of Marine Biotechnology*, 2987–97. Wiley. https://doi.org/10.1002/.

Nusairat, Basheer, Guillermo Tellez-Isaias, and Rasha Qudsieh. 2022. *An Overview of Poultry Meat Quality and Myopathies*. In G. Téllez-Isaías, J. D. Latorre, and Y. Martínez-Aguilar (Eds.), *Broiler Industry*. IntechOpen. https://doi.org/10.5772/intechopen.104474.

Obernikhina, N, A Mykhailova, T Pradii, and T Sanzhur. 2020. "Biochemistry of Connective Tissue. *Guide for Practical Work on Biological Chemistry*. O. O. Department of Biological Chemistry, Bohomolets National Medical University.

Ohlau, Marlene, Achim Spiller, and Antje Risius. 2022. "Plant-Based Diets Are Not Enough? Understanding the Consumption of Plant-Based Meat Alternatives along Ultra-Processed Foods in Different Dietary Patterns in Germany." *Frontiers in Nutrition* 9: 852936.

Ouellet, Véronique, S John Weisnagel, Denis R Joanisse, Charles Lavigne, Junio Dort, André Marette, and Hélène Jacques. 2017. "Beneficial Impact of Cod Protein, L-Arginine, and Other Amino Acids on Insulin Sensitivity." In V. B. Patel, V. R. Preedy, and R. Rajendram (Eds.), *L-Arginine in Clinical Nutrition*, 433–47. Humana Press, Springer. https://doi.org/10.1007/978-3-319-26009-9.

Owusu-Apenten, Richard, and Ernest R Vieira. 2023. *Elementary Food Science* (5th ed.). Springer. https://doi.org/10.1007/978-3-030-65433-7.

Patted, Pradnya G, Rajashree S Masareddy, Archana S Patil, Ravikiran R Kanabargi, and Chetan T Bhat. 2024. "Omega-3 Fatty Acids: A Comprehensive Scientific Review of Their Sources, Functions and Health Benefits." *Future Journal of Pharmaceutical Sciences* 10 (1): 94.

Peng, Shiming, Chao Chen, Zhaohong Shi, and Lu Wang. 2013. "Amino Acid and Fatty Acid Composition of the Muscle Tissue of Yellowfin Tuna (Thunnus Albacares) and Bigeye Tuna (Thunnus Obesus)." *Journal of Food and Nutrition Research* 1 (4): 42–45.

Petrat-Melin, Bjørn, and Svend Dam. 2023. "Textural and Consumer-Aided Characterisation and Acceptability of a Hybrid Meat and Plant-Based Burger Patty." *Foods* 12 (11): 2246.

Pujol, Andrea, Juan Camilo Ospina -E, Hernan Alvarez, and Diego A Muñoz. 2023. "Myoglobin Content and Oxidative Status to Understand Meat Products' Color: Phenomenological Based Model." *Journal of Food Engineering* 348: 111439.

Ramesh, M N. 2020. "Canning and Sterilization of Foods." In M. S. Rahman (Ed.), *Handbook of Food Preservation* (2nd ed), 609–36. CRC.

Rasmussen, Monica, and Jian-Ping Jin. 2021. "Troponin Variants as Markers of Skeletal Muscle Health and Diseases." *Frontiers in Physiology* 12: 747214.

Ren, Qing-Shan, Kui Fang, Xin-Ting Yang, and Jia-Wei Han. 2022. "Ensuring the Quality of Meat in Cold Chain Logistics: A Comprehensive Review." *Trends in Food Science & Technology* 119: 133–51.

Ren, Yue, Yilin Wang, Yuxin Zhang, Zhongshuai Yang, Zhiming Ma, Jiaxin Chen, Xintong Chen, Zecheng Qiu, Jing Tian, and Aofei Pu. 2024. "Formation and Regulation Strategies for Volatile Off-Flavor Compounds in Livestock Meat, Poultry Meat, and Their Products: A Comprehensive Review." *Trends in Food Science & Technology*, 104689.

Safa, Hassan, Stéphane Portanguen, and Pierre-Sylvain Mirade. 2017. "Reducing the Levels of Sodium, Saturated Animal Fat, and Nitrite in Dry-Cured Pork Meat Products: A Major Challenge." *Food and Nutrition Sciences* 8: 419–43.

Salma, E O G, and M Nizar. 2015. "Atlantic Mackerel Amino Acids and Mineral Contents from the Tunisian Middle Eastern Coast." *International Journal of Agricultural Policy and Research*, 3 (2):77–83. https://doi.org/10.15739/IJAPR.029.

Di Santo, Morgan Vivianne Di. 2016. "Smokin'Out the Truth: An Early History of Southern Barbecue." United States, Bard College.

Shan, J G, Z L Tan, Z H Sun, J P Hu, S X Tang, H L Jiang, C S Zhou, M Wang, and G O Tayo. 2007. "Limiting Amino Acids for Growing Goats Fed a Corn Grain, Soybean Meal and Maize Stover Based Diet." *Animal Feed Science and Technology* 139 (3–4): 159–69.

Sharma, Ranjana, Prakrati Garg, Pradeep Kumar, Shashi Kant Bhatia, and Saurabh Kulshrestha. 2020. "Microbial Fermentation and Its Role in Quality Improvement of Fermented Foods." *Fermentation* 6 (4): 106.

Singh, Avtar, Ajay Mittal, and Soottawat Benjakul. 2022. "Full Utilization of Squid Meat and Its Processing By-Products: Revisit." *Food Reviews International* 38 (4): 455–79.

Singh, Awanish, and Nandan Sit. 2022. "Meat Analogues: Types, Methods of Production and Their Effect on Attributes of Developed Meat Analogues." *Food and Bioprocess Technology* 15 (12): 2664–82.

Squire, John. 2019. "The Actin-Myosin Interaction in Muscle: Background and Overview." *International Journal of Molecular Sciences* 20 (22): 5715.

Sultana, Khudsia, K Jayathilakan, and V A Sajeevkumar. 2022. "Chemistry of Animal Tissues." In O. P. Chauhan (Ed.), *Advances in Food Chemistry: Food Components, Processing and Preservation*, 385–437. Singapore: Springer Singapore. https://doi.org/10.1007/978-981-19-4796-4_11.

Sun, Peter D, Christine E Foster, and Jeffrey C Boyington. 2004. "Overview of Protein Structural and Functional Folds." *Current Protocols in Protein Science* 35 (1): 11–17.

Szikora, Szilárd, Péter Görög, and József Mihály. 2022. "The Mechanisms of Thin Filament Assembly and Length Regulation in Muscles." *International Journal of Molecular Sciences* 23 (10): 5306.

Teixeira, Alfredo, Severiano Silva, Cristina Guedes, and Sandra Rodrigues. 2020. "Sheep and Goat Meat Processed Products Quality: A Review." *Foods* 9 (7): 960.

Tobacman, Larry S, and Anthony Cammarato. 2021. "Cardiomyopathic Troponin Mutations Predominantly Occur at Its Interface with Actin and Tropomyosin." *Journal of General Physiology* 153 (3): e202012815.

Wang, Zhexin, and Stefan Raunser. 2023. "Structural Biochemistry of Muscle Contraction." *Annual Review of Biochemistry* 92 (1): 411–33.

Warriss, Paul D. 2010. "Meat Science: An Introductory Text. Cabi." *Meat Science* 56 (3): 319. https://doi.org/10.1016/s0309-1740(00)00053-x.

Williams, Peter. 2007. "Nutritional Composition of Red Meat." *Nutrition & Dietetics* 64:S113–19.

Wolk, Alicja. 2017. "Potential Health Hazards of Eating Red Meat." *Journal of Internal Medicine* 281 (2): 106–22.

Xiao-Hui, Gong, Wan Jing, Zhou Ye-Ling, Zhou Ying, Zhu Qiu-Jin, Liu Ling-Gao, Chen Dan, Huang Yan-Pei, Gu Sha, and Li Ming-Ming. 2023. "Mediated Curing Strategy: An Overview of Salt Reduction for Dry-Cured Meat Products." *Food Reviews International* 39 (7): 4565–80.

Xiong, Youling L, and William Benjy Mikel. 2001. "Meat and Meat Products." In Y. H. Hui, W.-K. Nip, and R. Rogers (Eds.), *Meat Science and Applications*. CRC Press. https://doi.org/10.1201/9780203908082.

Xu, Yujuan, and Xinglian Xu. 2021. "Modification of Myofibrillar Protein Functional Properties Prepared by Various Strategies: A Comprehensive Review." *Comprehensive Reviews in Food Science and Food Safety* 20 (1): 458–500.

Xu, Yujuan, Xinglian Xu, and Baocai Xu. 2024. "Glycosylation Modification: A Promising Strategy for Regulating the Functionalities of Myofibrillar Proteins." *Critical Reviews in Food Science and Nutrition* 64 (25): 8933–47.

Young, Owen A, and John West. 2001. "Meat Color." In Y. H. Hui, W.-K. Nip, and R. Rogers (Eds.), *Meat Science and Applications*. CRC Press. https://doi.org/10.1201/9780203908082.

Yuldashbaev, Yu. A, A M Abdulmuslimov, and I A Sazonova. 2021. "Amino-Acid Composition of Meat Produced by Ram Lambs of the Dagestan Mountain Breed and Their Crosses." *Russian Agricultural Sciences* 47 (6): 627–30. https://doi.org/10.3103/S1068367421060185.

Zając, Marzena, Paulina Pająk, and Gabriela Skowyra. 2021. "Characterization of Edible Collagen Casings in Comparison with the Ovine Casing and Their Effect on Sausage Quality." *Journal of the Science of Food and Agriculture* 101 (14): 6001–9.

Zduńczyk, Weronika, Katarzyna Tkacz, Renata Pietrzak-Fiećko, Benedetta Bottari, and Monika Modzelewska-Kapituła. 2024. "Pork as a Source of Nutrients in a Human Diet-Comparison of Meat Obtained from Conventional and Organic Systems Offered in the Polish Market." *NFS Journal* 37: 100199.

Zhang, Handan, Yungang Cao, Xinling Dong, Xia Li, and Chunhui Zhang. 2023a. "Effect of Different Postmortem Ageing Conditions on Physicochemical Properties, Structure and Water-Holding Capacity of Pork." *International Journal of Food Science and Technology* 58 (3): 1662–72.

Zhang, Haozhen, Weiyi Zhang, Xianming Zeng, Xue Zhao, and Xinglian Xu. 2022. "Recent Progress of Fat Reduction Strategies for Emulsion Type Meat Products." *Food Materials Research* 2 (1): 1–10.

Zhang, Yuemei, Genpeng Bai, Jinpeng Wang, Ying Wang, Guofeng Jin, Wendi Teng, Fang Geng, and Jinxuan Cao. 2023b. "Myofibrillar Protein Denaturation/Oxidation in Freezing-Thawing Impair the Heat-Induced Gelation: Mechanisms and Control Technologies." *Trends in Food Science & Technology* 138: 655–70.

8 Fish Proteins

Mumtahin Ul Kousar, Abida Jabeen, Aasima Rafiq, and Mudasir Ahmad Malik

FISH PROTEINS

Fish proteins are widely regarded as a high-quality, easily digestible protein source, offering a unique nutritional profile that includes essential amino acids (EAAs), vitamins, minerals, and bioactive peptides. Because of its high biological value (BV), which measures how well it supplies the EAAs needed by the human body, fish has become a much more important source of protein. Fish proteins have a better amino acid composition than those of terrestrial animals because they contain all nine EAAs, which makes them complete proteins. The soft texture and low connective tissue content of fish muscle fibres, which promote easier absorption in the gastrointestinal tract, are also responsible for the exceptionally high digestibility of fish protein. Fish protein is especially appropriate for consumption by vulnerable groups, including children, the elderly, and people with gastrointestinal disorders, due to its high digestibility (Zhou et al., 2017).

Essential sulphur-containing amino acids, like methionine and cysteine, which are necessary for a number of metabolic functions, such as protein synthesis, antioxidant defence, and detoxification, are frequently more abundant in fish proteins (Andersen et al., 2016). Additionally, these amino acids support overall cellular function by aiding in the synthesis of crucial peptides and enzymes. Furthermore, because fish proteins contain a high concentration of branched-chain amino acids (BCAAs) like leucine, isoleucine, and valine, their amino acid profile is especially advantageous for fostering muscle protein synthesis and recovery. Fish is a perfect protein source for athletes and people who engage in physically demanding activities because these amino acids are crucial for controlling muscle metabolism (Børsheim et al., 2002).

Fish is known for its abundance of bioactive peptides, which have both functional and therapeutic qualities, in addition to being a great source of high-quality protein. Protein fragments known as bioactive peptides have particular biological effects during digestion that frequently go beyond their basic nutritional purposes. These peptides, which are frequently obtained from the proteins of fish muscles, have been linked to a number of advantageous properties, such as immunomodulatory, antimicrobial, antihypertensive, and antioxidant effects (Korhonen & Pihlanto, 2006). For example, peptides made from fish collagen have demonstrated strong antioxidant qualities that help lower inflammation and oxidative stress, promoting cardiovascular health and avoiding skin damage brought on by ageing (Yuan et al., 2017). Enzymatic hydrolysis produces fish protein hydrolysates, which are especially rich in bioactive peptides and are being studied more and more for possible use in functional food products. These hydrolysates have shown health benefits, including

DOI: 10.1201/9781003613909-8

lowering blood pressure, promoting wound healing, supporting cognitive function, and improving the composition of the gut microbiota (Nasri, 2017).

Omega-3 polyunsaturated fatty acids (PUFAs), particularly eicosapentaenoic acid (EPA) and docosahexaenoic acid (DHA), which have been thoroughly researched for their potential health benefits, are also abundant in fish proteins. The cardioprotective benefits of omega-3 fatty acids, such as lowering blood pressure, triglyceride levels, and the risk of cardiovascular diseases, are well-established (Kris-Etherton et al., 2002). Additionally, omega-3 fatty acids have been associated with anti-inflammatory properties, which makes them useful in the treatment of inflammatory diseases like autoimmune disorders, asthma, and arthritis (Calder, 2006). Numerous studies indicate that regular consumption of EPA and DHA can improve cognitive function, lower the risk of neurodegenerative diseases like Alzheimer's and Parkinson's, and improve mental well-being. The benefits of omega-3 fatty acids for the brain are especially well-established (Freeman et al., 2006).

Fish is a great source of vitamins and minerals in addition to essential fatty acids, which balance out its protein content. For instance, fish is a great source of fat-soluble vitamins like vitamin D, which is essential for healthy bones, immune response, and calcium absorption (Calvo & Whiting, 2013). The nutritional profile of fish is further improved by the presence of minerals like iodine, selenium, zinc, and iron, which are crucial for immune system function, wound healing, thyroid function, and oxygen transport (Biesalski, 2016).

Fish proteins are being used more and more in a variety of food products, from conventional fish-based foods to cutting-edge creations like protein-enriched drinks, protein bars, and dietary supplements, as there is a growing global interest in sustainable and functional food sources. Fish proteins have special functional qualities that can enhance food products' texture, stability, and sensory qualities. These qualities include the capacity to form gels, emulsions, and foams (Kinsella, 1984). Because of these qualities, fish proteins are a desirable substitute for conventional animal proteins, especially for people who have dietary requirements or preferences. Particularly, there is growing interest in the use of fish protein hydrolysates and isolates in plant-based food formulations. This strategy might enhance the nutritional value of plant-based products, which frequently lack the bioactive substances and EAAs present in animal proteins. Fish protein can be added to plant-based formulations to produce nutritionally balanced goods with a more comprehensive amino acid profile and practical advantages (Shahidi & Ambigaipalan, 2018).

Fish proteins stand out as a valuable resource for creating new, health-promoting food products as the demand for functional foods keeps growing. Fish proteins contain bioactive peptides, omega-3 fatty acids, and micronutrients that have substantial therapeutic potential for managing and preventing chronic illnesses as well as improving overall health. They are a crucial part of contemporary diets because of their capacity to enhance cardiovascular health, promote cognitive function, and reduce inflammatory responses. Furthermore, because of their special qualities, fish proteins can be used as a variety of ingredients in food processing, especially when creating nutritionally enhanced goods that meet a range of consumer demands. The advantages of fish proteins for human nutrition and health will become even more apparent as more research is done on their bioactive components, which may improve gut health, regulate metabolism, and improve skin health. There is a lot of potential for satisfying the growing need for ingredients that promote health and promote optimal well-being across various demographics by incorporating fish proteins into functional foods and dietary supplements.

COMPOSITION OF DIFFERENT FISHES

Because of its high content of macronutrients, micronutrients, and bioactive compounds, fish is a highly prized food source. Depending on their diet, habitat, life stage, and whether they are farmed or wild-caught, fish species differ greatly in composition. To appreciate the variations among species and their corresponding health advantages, one can benefit from a general understanding of their nutritional makeup, which includes protein, lipids, water, vitamins, and minerals. With a focus on their macronutrient profiles and the particular health-promoting substances they contain, the discussion that follows gives a general overview of the composition of different fish species, including both lean and fatty ones.

Lean Fish

Because of their low fat content, lean fish like cod, haddock, and tilapia are a great source of high-quality protein at a comparatively low calorie count. Lean fish usually contain between 18% and 25% protein, with a good amino acid profile that includes all of the EAAs (Sikorski & Kolakowska, 2015). Because of their low connective tissue and soft muscle fibres, these fish are high in highly digestible proteins, which makes them appropriate for people with particular dietary requirements. Lean fish are still a great source of vital micronutrients even though they are low in fat. For example, cod contains a significant amount of vitamin B12, which is essential for the production of red blood cells and nerve function. These fish also supply minerals like iodine, phosphorus, and selenium; because of its marine habitat, cod is a great source of iodine (Biesalski, 2016). Strong antioxidant selenium is essential for lowering inflammation and shielding cells from oxidative damage. Furthermore, lean fish typically have low levels of omega-3 fatty acids, though species like cod liver still contain trace amounts of these fatty acids (Haug & Lückstädt, 2018).

Fatty Fish

High lipid content, which can make up as much as 20% of a species' weight, is a characteristic of fatty fish, including salmon, mackerel, sardines, and herring (Zhang et al., 2018). These fats are mostly omega-3 polyunsaturated fatty acids (PUFAs), such as docosahexaenoic acid (DHA) and eicosapentaenoic acid (EPA), which are well-known for their cardiovascular and anti-inflammatory qualities. The potential of these fatty acids to lower blood pressure, lower triglyceride levels, and stop blood clots from forming has been thoroughly investigated (Kris-Etherton et al., 2002). The presence of fat-soluble vitamins like A, D, and E is also linked to the fat content of fatty fish. These vitamins are necessary for healthy skin, bones, and the immune system, respectively. For instance, salmon is a great source of vitamin D, which is essential for bone mineralisation and calcium absorption. Vitamin E, a strong antioxidant that helps shield cells from oxidative damage, and vitamin A, which promotes immune system and vision health, are also abundant in fatty fish (Rønnestad et al., 2012). Fatty fish species typically contain high concentrations of iodine, selenium, and zinc, with mackerel and sardines offering the highest concentrations of these vital minerals. Another noteworthy feature of mackerel is its high iron content, which is necessary for the body to produce haemoglobin and carry oxygen (Sikorski & Kolakowska, 2015). These fish are rich in nutrients and good for the health because of their high fat content, which also helps make fat-soluble vitamins and antioxidants more bioavailable.

Shellfish and Other Marine Species

The composition of shellfish, such as prawns, crab, oysters, and mussels, varies from that of lean and fatty fish. Generally speaking, they are low in fat; prawns, for instance, have between 0.5% and 2% total fat. Though in smaller amounts than fatty fish, the fat content of shellfish is high in omega-3s and other polyunsaturated fatty acids (Haug & Lückstädt, 2018). With protein contents that are similar to those of lean fish, usually between 18% and 25%, shellfish are great sources of protein. Shellfish have high-quality protein that includes all of the EAAs required for human health. Minerals, especially zinc, copper, and selenium, are also abundant in shellfish. For instance, oysters are regarded as one of the best food sources of zinc, a mineral that is essential for wound healing, cellular metabolism, and immune system function (Biesalski, 2016). Because of their high iron content, mussels and oysters are also good for people who are iron deficient.

Aquaculture vs. Wild-Caught Fish

Whether a fish is farmed or wild-caught, it can affect its composition. The fish which is widely caught have typically a more varied diet which can cause variations in their nutritional profiles. For instance, compared to farmed salmon, wild salmon typically contain a higher percentage of omega-3 fatty acids. This could be because the fish's diet and difference in feeding environment (Sikorski & Kolakowska, 2015). On the other hand, fish which is farmed for food are frequently fed specially prepared diets that are meant to encourage growth. This may lead to a higher fat content, but in certain species, it may also result in a lower ratio of omega-3 to omega-6 fatty acids (Zhang et al., 2018). Additionally, fish raised for food may have higher concentrations of pollutants like pesticides and heavy metals, which can build up in the fish's fatty tissues (Haug & Lückstädt, 2018). The overall nutritional content and fatty acid profile of wild-caught fish are also influenced by environmental factors, such as water temperature and the availability of natural food sources.

Numerous nutritional advantages can be found in the various fish species. Low-fat diets can benefit from the high-quality protein and vital minerals that lean fish like cod and tilapia offer. Omega-3 fatty acids, fat-soluble vitamins, and vital minerals are abundant in fatty fish, such as salmon, mackerel, and sardines, and they support both cardiovascular and mental health. Despite having less fat, shellfish have a high protein content and a substantial mineral content, particularly iron and zinc.

PROTEIN CONTENT AND AMINO ACID PROFILE OF DIFFERENT FISHES

A premium protein source, fish is well-known for its advantageous amino acid composition and health advantages. Depending on the species, fish's protein content can vary from 15% to 25% of their total weight, with some variations depending on age, sex, diet, and environmental factors. Because of their high digestibility, nutrient bioavailability, and functional qualities, fish proteins are frequently regarded as superior to other animal-based proteins and are therefore an indispensable component of the human diet.

Protein Content in Fish

Depending on the species, fat content, and environmental circumstances, fish's protein content varies greatly. Lean and fatty fish are the two primary categories into which fish can be divided according to their fat content. Lean fish, like cod, haddock, and tilapia, are great sources of lean, easily digestible protein because they usually have lower fat content and higher protein content. According to Sikorski and Kolakowska (2015), the protein content of these species can range from 18% to 22% of their total weight. Conversely, fatty fish, like sardines, mackerel, and salmon, have a higher fat content but still offer substantial protein content, which can range from 15% to 20% of their weight (Zhang et al., 2018). Fatty fish provide extra nutritional advantages like omega-3 fatty acids, vitamins, and minerals, which add to their health-promoting qualities even though their protein content is marginally lower than that of lean fish.

Additionally, fish protein is very bioavailable and digestible, with digestibility rates frequently surpassing 90% (Sikorski & Kolakowska, 2015). This high digestibility is explained by the fish's muscle structure, which has a high concentration of myofibrillar proteins and fewer connective tissues, which facilitates the protein's breakdown and absorption by the human digestive system.

Amino Acid Profile of Fish

The nutritional value of fish protein is largely determined by its amino acid composition. All nine of the EAAs needed by the human body for healthy growth and maintenance are present in fish protein. The body is unable to produce these EAAs, which must be obtained through diet: histidine, isoleucine, leucine, lysine, methionine, phenylalanine, threonine, tryptophan, and valine. Since fish contains enough of each EAA to meet the body's needs, it is regarded as a complete protein source.

Essential amino acids, particularly lysine, methionine, and threonine, are comparatively abundant in fish protein (Sikorski & Kolakowska, 2015). Fish is a valuable dietary source of the EAA lysine, which is found in particularly high concentrations in fish and is crucial for protein synthesis, collagen formation, and calcium absorption (Haug & Lückstädt, 2018). Methionine, an EAA that contains sulphur, is essential for protein synthesis, methylation processes, and heavy metal detoxification (Colović et al., 2018). Fish protein is particularly rich in the sulphur amino acids cysteine and methionine, which support a number of bodily metabolic functions.

Because of variations in the composition of muscle fibres and the distribution of nutrients, fatty fish species typically have slightly higher levels of EAAs than lean fish (Zhang et al., 2018). The idea that these fish's protein is more nutritious is also influenced by their amino acid composition. For example, the amino acid makeup of salmon, sardines, and mackerel is ideal for promoting immune response, muscle growth, and metabolic functions.

Fish Protein Digestibility and Bioavailability

One of fish protein's distinguishing features is its high digestibility, which makes it a perfect protein source for people with weakened digestive systems, such as the elderly or those suffering from gastrointestinal disorders. Unlike beef and chicken, which have digestibilities of about 80%, fish protein can have a digestibility of over 90% (Sikorski & Kolakowska, 2015). The structure of fish muscle, which has less connective tissue than meat from terrestrial animals, is the main cause of this. Because fish muscle fibres are more delicate and shorter, the digestive tract can break them down and absorb them more quickly.

Additionally, fish protein has a high level of amino acid bioavailability, which is the body's capacity to absorb and use the amino acids in the protein. Glutamic acid, an amino acid that gives fish its umami flavour and functions as a key neurotransmitter in the brain, is abundant in fish protein. Furthermore, taurine, an amino acid that promotes bile salt production, heart health, and cellular function, is present in considerable amounts in fish protein (Haug & Lückstädt, 2018).

The presence of vitamins, minerals, and other bioactive substances that enhance nutrient absorption further increases the bioavailability of amino acids in fish. According to Rønnestad et al. (2012), the high vitamin D content of fatty fish like salmon, for instance, encourages calcium absorption, which can improve the use of amino acids for muscle protein synthesis. Zinc and selenium, which are essential for protein metabolism and amino acid utilisation, are also found in fish (Biesalski, 2016).

Amino Acid Profile in Specific Fish Species

Cod (*Gadus morhua*): The popular lean fish species known as cod (*Gadus morhua*) has a protein content of 18% to 22%. Lysine, methionine, and leucine are abundant in its amino acid profile, which offers a well-balanced range of vital amino acids. Because of its high concentration of EAAs and low fat content, cod is a valuable protein source for both muscle growth and repair (Sikorski & Kolakowska, 2015).

Salmon (*Salmo salar*): Known for its high omega-3 fatty acid content, salmon is a fatty fish that also offers high-quality protein. Salmon's amino acid profile contains high levels of lysine, methionine, threonine, and valine, all of which promote metabolism, immune response, and general body growth. Salmon's antioxidant qualities are also influenced by its high methionine and cysteine content (Zhang et al., 2018).

Mackerel (*Scomber scombrus*): Another fatty fish with a substantial protein content of 18% to 20% is mackerel (*Scomber scombrus*). With a high concentration of vital amino acids like leucine, lysine, and methionine, mackerel shares a similar amino acid profile to salmon. The high concentration of histidine, an amino acid crucial for growth, tissue repair, and metabolic regulation, in mackerel is especially noteworthy (Haug & Lückstädt, 2018).

Tilapia (*Oreochromis niloticus*): The lean fish known as tilapia (*Oreochromis niloticus*) has a protein content of 20% to 24%. It has a balanced amino acid profile with notable levels of phenylalanine, methionine, and lysine. Tilapia is a good source of protein for a variety of people, including athletes and those with high protein needs, because it still contains a sufficient amount of EAAs despite its relatively low fat content (Biesalski, 2016).

Sardines (*Sardina pilchardus*): Small fatty fish like sardines have about 20% protein and are especially rich in vital amino acids like valine, lysine, and leucine. Additionally, they are a good source of omega-3 fatty acids, which support the health of the heart and brain. According to Sikorski and Kolakowska (2015), sardines are a complete protein and micronutrient source that includes calcium, which is critical for healthy bones. Table 8.1 highlights the nutritional value of various fish species by summarising their protein content and amino acid profile:

Fish proteins provide important health benefits and are a high-quality, easily digested source of EAAs (Table 8.2). Key amino acids like lysine, methionine, threonine, and

TABLE 8.1
Amino Acid Composition and Protein Content of Various Fish

Fish Species	Protein Content (%)	Key Amino Acids	Additional Nutrients
Cod (*Gadus morhua*)	18–22	Lysine, methionine, leucine, threonine	Low fat, Vitamin B12, Selenium, Iodine
Salmon (*Salmo salar*)	18–20	Lysine, methionine, threonine, valine	Omega-3 fatty acids (EPA, DHA), Vitamin D, Selenium
Mackerel (*Scomber scombrus*)	18–20	Lysine, leucine, methionine, histidine	Omega-3 fatty acids (EPA, DHA), Iron, Selenium
Tilapia (*Oreochromis niloticus*)	20–24	Lysine, methionine, phenylalanine, threonine	Low fat, Vitamin B12, Phosphorus
Sardines (*Sardina pilchardus*)	20	Lysine, leucine, valine, methionine	Omega-3 fatty acids (EPA, DHA), Calcium, Vitamin D
Haddock (*Melanogrammus aeglefinus*)	18–22	Lysine, methionine, threonine, leucine	Low fat, Vitamin B12, Selenium
Tuna (*Thunnus* spp.)	22–25	Lysine, methionine, histidine, leucine	Omega-3 fatty acids, Vitamin D, Selenium

Source: Sikorski and Kolakowska (2015).

TABLE 8.2
Essential Amino Acid Composition of Selected Fish Species (g/100 g Edible Portion)

Amino Acid	Cod (*Gadus morhua*)	Salmon (*Salmo salar*)	Mackerel (*Scomber scombrus*)	Tilapia (*Oreochromis niloticus*)	Sardines (*Sardina pilchardus*)	Tuna (*Thunnus spp.*)	Reference
Histidine	0.45	0.65	0.70	0.58	0.62	0.69	Haug and Luckstad (2018) Biesalski (2016)
Isoleucine	0.95	1.05	1.10	1.02	1.08	1.10	Sikorski and Kolakowska (2015) Biesalski (2016)
Leucine	1.60	1.78	1.85	1.65	1.80	1.90	Haug and Luckstad (2018) Biesalski (2016)
Lysine	1.90	2.10	2.15	2.05	2.20	2.30	Sikorski and Kolakowska (2015) Biesalski (2016)
Methionine	0.60	0.72	0.75	0.66	0.70	0.73	Haug and Luckstad, (2018) Biesalski (2016)
Phenylalanine	0.85	1.00	1.02	0.92	1.00	1.05	Sikorski and Kolakowska (2015) Biesalski (2016)
Threonine	0.90	1.00	1.05	0.95	0.98	1.00	Haug and Luckstad, (2018) Biesalski (2016)
Tryptophan	0.21	0.26	0.27	0.23	0.25	0.28	Sikorski and Kolakowska (2015) Biesalski (2016)
Valine	1.00	1.12	1.18	1.05	1.10	1.15	Sikorski and Kolakowska (2015) Biesalski (2016)

leucine are abundant in the amino acid profiles of different fish species and are crucial for immune system function, muscle growth, and tissue repair. For people with particular dietary requirements, fish protein is the perfect source of protein because it is readily digested and bioavailable, with digestibility rates frequently surpassing 90%. Lean fish, such as cod and tilapia, offer a concentrated source of protein, while fatty fish, such as salmon, mackerel, and sardines, offer additional health benefits because of their high omega-3 fatty acid content. Utilising fish as a valuable component in human nutrition requires an understanding of the protein content and amino acid profile of various fish species.

COMPOSITION OF DIFFERENT FISH-BASED FOOD PRODUCTS

Fish-based food products are consumed by people all over the world due to their remarkable nutritional value, convenience, and adaptability. These products go through a variety of processing techniques that affect their composition and are made from whole fish, fish muscle, or by-products. Fish species, processing methods, and additional ingredients all influence the makeup of fish-based products. These goods are frequently utilised in both industrial and dietary settings and are abundant in proteins, EAAs, omega-3 fatty acids, vitamins, and minerals (Table 8.3).

Fresh Fish Fillets

Because of their high nutritional content, low processing requirements, and simple preparation techniques, fresh fish fillets are frequently regarded as the best fish-based food product. Fresh fish can vary in composition depending on the species, but generally speaking, it contains:

- **Water Content:** 70% to 80% of fish fillets are made of water, which is one of the elements that gives fish their moist texture and perishable quality (Oluwaniyi et al., 2010). Fish is low in calories due to its high water content.
- **Protein:** Fish fillets normally have a protein content of 15–25%. Because of their exceptional amino acid profile, which includes all of the EAAs needed for human nutrition, fish proteins have a high biological value (FAO, 2020). Fish protein is on par with meat and dairy in terms of quality (Costa et al., 2017).
- **Lipids:** Although the amount of fat in fish varies by species, fillets typically contain between 2% and 15% fat (Pyz-Łukasik et al., 2020). Polyunsaturated fatty acids (PUFAs), especially omega-3 fatty acids like eicosapentaenoic acid (EPA) and docosahexaenoic acid (DHA), are abundant in fish oils and are essential for brain function, inflammation reduction, and cardiovascular health (Calder, 2015).
- **Vitamins:** A variety of vitamins, including fat-soluble vitamins A, D, and E, are found in fish fillets. Oily fish are especially high in vitamin D, which is necessary for bone health and calcium metabolism (Lu et al., 2027).

TABLE 8.3
Composition of Various Food Products Made from Fish

Fish-Based Food Product	Key Components	Composition	Comments
Fresh fish fillets	Water	70%–80%	High water content, contributes to low calorie content.
	Protein	15%–25%	High biological value with all essential amino acids.
	Lipids	2%–15%	Rich in omega-3 fatty acids (EPA, DHA).
	Vitamins	A, D, E	Rich in fat-soluble vitamins, particularly in oily fish.
	Minerals	Iodine, selenium, zinc, phosphorus	Important for thyroid function, immunity, and bone health.
Canned fish	Water and brine	60%–70%	Maintains moisture; sodium from brine.
	Protein	20%–25%	Retained after canning, slight loss of amino acids.
	Lipids	10%–20%	Higher fat in oil-packed fish, omega-3 rich.
	Sodium	High	Brine contributes to high sodium content.
	Vitamins & minerals	A, D, B-complex, iodine, Selenium	Vitamin losses occur during heat processing.
Fish sauces	Amino acids	High (e.g., glutamate)	Contributes to umami flavour, rich in free amino acids.
	Salt	15%–25%	High salt content for preservation and flavour.
	Vitamins & Minerals	B vitamins, iodine	Vitamin content may decrease due to fermentation.
	Antioxidants	Bioactive peptides	Derived from fermentation, may offer health benefits.
Fish powders	Protein	40%–80%	High protein content, concentrated after drying.
	Fat	5%–15%	Omega-3 fatty acids retained in powder form.
	Minerals	Calcium, phosphorus, iron, magnesium	Retained during drying, beneficial for bone health.
	Vitamins	A, D	Some vitamins may be lost during the drying process.
Surimi	Protein	8%–15%	Extracted through washing, concentrated protein paste.
	Carbohydrates	5%–15%	Added starches for texture improvement.
	Fat	<5%	Low fat, with some fat removed during processing.
	Additives	Preservatives, flavour enhancers, colorants	Often included to mimic fish texture and appearance.
Fish-based snacks	Protein	20%–30%	Rich in protein, depending on the fish species.
	Fat	10%–30%	Higher in fat if fried, includes omega-3 fatty acids.
	Carbohydrates	10%–20%	Added starches/flours to improve texture and flavour.
	Preservatives	Antioxidants, flavour enhancers	Used to improve shelf life and prevent oxidation.

- **Minerals:** Essential minerals like iodine, selenium, zinc, and phosphorus are abundant in fish. Selenium has antioxidant properties (Hurst et al., 2019), and iodine, in particular, is critical for thyroid function (Larsen et al., 2016).

Canned Fish

Because they are convenient, have a long shelf life, and retain many of their nutritional value, canned fish products like sardines, mackerel, and tuna are very popular. The processing technique, which frequently includes heat treatment and the addition of brine or oil, affects the composition of canned fish. The general composition includes:

- **Water and Brine:** A large portion of the moisture in canned fish is retained, and the brine solution preserves the fish's flavour and texture. The water content is usually between 60% and 70%, just like fresh fish.
- **Protein:** Depending on the species and preparation, canned fish contains 20% to 25% protein. After canning, the quality of the protein is maintained, but the heat processing may cause a minor decrease in the amount of amino acids (McDonald et al., 2019).
- **Lipids:** The type of fish and the packing medium (water, oil, or brine) affect how much fat is in canned fish. Oil-packed fish typically contains 10%–20% more fat, including omega-3 fatty acids. Oil raises the product's total fat content while also enhancing its sensory qualities (Caponio et al., 2010).
- **Sodium:** The sodium content of canned fish is one of the major issues, mostly because brine is used for preservation. Consuming a lot of sodium can increase the risk of cardiovascular disease and hypertension (He et al., 2017). There are many lower sodium substitutes on the market.
- **Minerals and Vitamins:** Canned fish is still a good source of vitamins and minerals, just like fresh fish. But heat processing can cause vitamin losses, particularly for heat-sensitive nutrients like vitamin C and certain B vitamins (Ahmed et al., 2016).

Fish Sauces

A mainstay of Southeast Asian cooking, fish sauce is made by typically fermenting fish with salt. The main ingredient in fish sauce is liquid, and the type of fish used, the amount of salt, and the length of fermentation all affect the sauce's flavour.

- **Amino Acids:** Glutamate and other free amino acids are abundant in fish sauce and give it its umami flavour. Its savoury profile is enhanced by the fermentation process, which breaks down proteins into peptides and amino acids (Mouritsen et al., 2017).
- **Salt:** With 15%–25% salt by weight, fish sauce has a high sodium content. In order to preserve the product and enhance its flavour, the high salt content is essential (Li et al., 2025).

- **Vitamins and Minerals:** Iodine and B vitamins are among the trace amounts of vitamins and minerals found in fish sauce. However, the vitamin content might be lower than in the original fish because of the fermentation process and the concentration of salt (Blankenship & Kimbrell, 2018).

Fish Powders

Fish powders are dehydrated fish products that are frequently added to food as ingredients in processed foods, nutritional supplements, or flavour enhancers. The final product's nutritional profile is impacted by the drying and grinding procedures.

- **Protein:** Depending on the species of fish and the drying method, fish powder typically has a protein content of 40% to 80%, making it a great source of protein. Protein concentration is aided by the drying process (Gulzar et al., 2020).
- **Fat:** A portion of the original fish's fat, including healthy omega-3 fatty acids, is preserved in fish powder. Depending on the species and preparation, the fat content can vary from 5% to 15% (Byrd et al., 2021).
- **Minerals:** Minerals like calcium, phosphorus, iron, and magnesium are abundant in fish powder. Fish powder is a great supplement for bone and immune health because these minerals are retained even after drying (Sroy et al., 2023).
- **Vitamins:** Water-soluble vitamins, like vitamin C, may be diminished during the drying process, even though fish powder keeps a lot of fat-soluble vitamins, like A and D (FAO, 2020).

Surimi

Made from fish meat, usually from species like pollock or whiting, surimi is a processed fish product. It is an essential component of many goods, such as fish cakes and imitation crab meat.

- **Protein:** Protein content in surimi typically ranges from 8% to 15%, depending on the type of fish used. A concentrated protein paste is left behind after the protein is extracted using a washing procedure that eliminates fats and contaminants (Cheng et al., 2020).
- **Carbohydrates:** Carbohydrates like starch are frequently added to surimi to improve texture and water retention. Usually, 5%–15% of the product is made up of starch (Cheng et al., 2020).
- **Fat:** Since a large portion of the fat is eliminated during processing, surimi usually has a low fat content of less than 5%. But for better mouthfeel and flavour, some products are made with extra fats (Yuan et al., 2017).
- **Additives:** In order to replicate the look and flavour of whole fish or shellfish, surimi products frequently include additional preservatives, flavour enhancers, and colourants (Matsumoto et al., 2021).

Fish-Based Snacks

As healthier snack options, fish-based snacks like fish jerky and chips are becoming more and more well-liked. These products go through drying, baking, or frying processes after combining fish with different ingredients like seasonings, starches, and oils.

- **Protein:** Depending on the kind of fish and the processing technique, fish-based snacks usually have 20%–30% protein, with fully complete amino acid profiles and high protein quality (Kumoro et al., 2022).
- **Fat:** Fish-based snacks are high in fat, particularly if they're fried. Oils add flavour and texture, and the fat content can range from 10% to 30% (Zhou et al., 2017).
- **Carbohydrates:** In order to enhance texture and provide energy, a lot of fish-based snacks include additional carbohydrates, mostly from starches and flours (Cheng et al., 2020).
- **Preservatives:** Preservatives, antioxidants, and other additives may be added to fish-based snacks to prolong their shelf life. Antioxidants such as vitamin E or rosemary extract can be added to help stop fats from oxidatively deteriorating (Sánchez et al., 2020).

CLASSIFICATION OF FISH PROTEINS

Among the highest quality proteins in the human diet, fish proteins provide a full range of EAAs, vitamins, minerals, and health-promoting bioactive compounds. Fish muscle, the main source of fish protein, contains many different protein fractions with a variety of functions that contribute to its nutritional, sensory, and functional qualities. Fish proteins can be categorised according to their structure, solubility, functionality, and biological role in humans and fish, which helps to better understand how various protein fractions behave and contribute to food systems, including those used in fish-based products. The nutritional significance of fish proteins, in particular, is crucial in their classification.

General Classification of Fish Proteins

Myofibrillar proteins, sarcoplasmic proteins, and stromal (connective tissue) proteins are the three general categories into which fish proteins fall. Each fraction has distinct functional characteristics, such as gel formation, water binding, and emulsification, and these classes are determined by their solubility and role within muscle tissue.

Myofibrillar Proteins: Fish muscle is primarily composed of myofibrillar proteins, which give muscle fibres their ability to contract. They are essential to the functionality and texture of processed foods like surimi and fish fillets.

These proteins are water-soluble and form the structural framework of muscle tissue. Fish's primary myofibrillar proteins are as follows:

- **Actin:** One of the main contractile proteins found in fish muscle is actin. It works with myosin to promote muscle contraction and creates thin filaments in muscle fibres. Although the amount of actin in fish varies by species, it typically makes up 5%–15% of the total protein in the muscle (Rajan et al., 2023).
- **Myosin:** The fundamental structure of the contractile filaments is formed by the motor protein myosin and actin. Because myosin in fish affects texture and gel formation, particularly in products like surimi, it has been the subject of much research. Fish muscle can contain anywhere between 10% and 20% myosin (Jiang et al., 2020). Myosin's ability to form gels in processed fish products is dependent on its level of denaturation as well as the presence of additional ingredients like salt or starch.
- **Tropomyosin:** Another crucial myofibrillar protein found in fish muscle is tropomyosin. By blocking binding sites when calcium ions are not present, it controls the interaction between actin and myosin. Fish have a major allergy to tropomyosin, which is responsible for muscle contraction (Kim et al., 2021). Although it varies by species, it is usually found in considerable amounts in fish muscles.
- **Troponin:** The protein complex known as troponin, which includes troponin T, I, and C, functions in tandem with tropomyosin to control muscle contraction in response to calcium ions. Troponin T is specifically involved in binding to tropomyosin, whereas troponin I inhibits the actin-myosin interaction when calcium is not present. Troponin is particularly significant when examining the functionality of fish proteins in products such as surimi, where the protein structure is altered to achieve the desired texture and gel strength (Boris et al., 2021).

Sarcoplasmic Proteins: Soluble proteins called sarcoplasmic proteins are present in the cytoplasm of muscle cells and play a role in the tissue's metabolic and regulatory processes. Although these proteins are crucial for the general health and functionality of muscles, they do not directly affect the contractile characteristics of muscle fibres. Enzymes, structural proteins, and metabolic proteins are examples of sarcoplasmic proteins that are involved in energy production, regulation, and muscle contraction. Important sarcoplasmic proteins consist of:

- **Myoglobin:** Fish muscle cells contain a haem-containing protein called myoglobin, which binds and stores oxygen in muscle fibres. It is essential for enabling the movement of oxygen when muscles contract (Zhao et al., 2019). Generally speaking, fish muscle has less myoglobin than that of terrestrial animals, though this varies by species and the oxidative capacity of the muscle.
- **Enzymes:** Fish muscle tissue contains enzymes like lactate dehydrogenase (LDH) and citrate synthase, which are involved in the citric acid

cycle and glycolysis, which produce energy. According to Zhou et al. (2017), these enzymes are essential for controlling metabolic processes and preserving muscle function.
- **Other Regulatory Proteins:** Protein kinases and other proteins that control metabolism and muscle contraction are examples of sarcoplasmic proteins. These proteins affect how fish muscle adjusts to various physiological and environmental circumstances (Liu et al., 2020).

Stromal (Connective Tissue) Proteins: The extracellular matrix (ECM) of fish muscle tissue contains stromal proteins, which support the structural integrity of muscle fibres. They control the general structure of the tissue and support the muscle cells. In fish muscle, collagen, elastin, and other glycoproteins are the primary stromal proteins. The texture of fish products is greatly influenced by these proteins, especially in processed foods like fish gels and surimi.

- **Collagen:** The most prevalent structural protein in fish connective tissues, collagen is essential for preserving the integrity of muscle tissue. The network of collagen fibres gives muscle tissues their strength and suppleness. Compared to mammalian collagen, fish collagen has distinct properties such as a higher degree of solubility and a lower cross-linking density (Rezaei et al., 2020). Because of its superior gelling and emulsifying qualities, fish collagen is used in a variety of food and pharmaceutical products.
- **Elastin:** Fish muscle tissues can stretch and regain their original shape thanks to a protein called elastin, which is less common than collagen but is crucial for preserving the resilience of fish muscles (Kong et al., 2021).
- **Glycoproteins:** Glycoproteins, which are also found in fish muscle, play a role in the ECM's structural organisation and cellular communication. Particularly in products like surimi and fish-based gels, these proteins help give processed fish products their texture and gel-forming qualities (Matsumoto et al., 2021).

Classification Based on Amino Acid Composition

The amino acid profile of fish proteins is very favourable for human consumption since they contain all EAAs in sufficient amounts. Fish proteins are categorised according to their amino acid composition, which is a crucial factor in determining their nutritional quality.

- **Essential Amino Acids:** Fish is a complete protein source because its proteins are high in EAAs like lysine, leucine, phenylalanine, and methionine. The overall quality of fish protein is influenced by the highly bioavailable balance of EAAs in fish (Zhou et al., 2017).

- **Non-Essential Amino Acids:** Non-EAAs like glycine, alanine, and proline, which are critical for collagen synthesis and other physiological processes, are also abundant in fish proteins (Nielsen et al., 2020).
- **Bioactive Peptides:** Bioactive peptides, which are protein fragments with particular health-promoting properties, can be found in some fish proteins. These peptides contribute to the functional qualities of foods based on fish by having anti-inflammatory, antihypertensive, and antioxidant properties (Kong et al., 2021).

STRUCTURE OF FISH PROTEINS

Functional characteristics like solubility, gelling, emulsifying, and foaming all of which are critical for food processing and the quality of fish-based products are largely determined by the structure of proteins. The linear chains of amino acids that make up fish proteins and those of other vertebrates fold into particular three-dimensional shapes. Developing novel processing methods, enhancing the texture and flavour of fish products, and optimising their nutritional value all depend on an understanding of the structure of fish proteins. The four levels of fish protein structure are primary, secondary, tertiary, and quaternary. The stability and functionality of the protein depend heavily on these interconnected structural levels.

Primary Structure

The particular arrangement of amino acids connected by peptide bonds is known as a protein's primary structure. The protein's overall three-dimensional structure and functionality are determined by this sequence. The amino acid makeup of fish proteins is what distinguishes them from those of other animals. The high nutritional value of fish proteins is attributed to their primary structure, which is chains of amino acids that include both essential and non-essential amino acids.

- **Amino Acid Composition:** A variety of amino acids found in fish proteins are essential for human health. Essential amino acids such as lysine, methionine, leucine, and valine are especially abundant in fish proteins. Protein synthesis, muscle repair, and immunological function all depend on these amino acids (Boris et al., 2021). Furthermore, glycine, proline, and hydroxyproline—all of which are critical for collagen formation and preserving the health of the skin and joints—are abundant in fish proteins (Zhou et al., 2017).
- **Peptide Bonds:** Polypeptides are made up of individual amino acids joined by peptide bonds. The folding and general stability of the protein will be influenced by the type and arrangement of the amino acids involved. Fish proteins differ from those of mammals in their sequences, in part because of the species-specific adaptations, environmental factors, and fish species (Boris et al., 2021).

Secondary Structure

Hydrogen bonds between the backbone atoms of the amino acid residues stabilise the local spatial arrangement of polypeptide chains, which is known as a protein's secondary structure. Alpha-helix and beta-sheet secondary structures are the two most prevalent forms present in fish proteins.

- **Alpha-Helix:** Hydrogen bonds between the backbones of amino acids stabilise the coiled structure known as the alpha-helix. Myofibrillar proteins, such as myosin, a key contractile protein in fish muscle, frequently contain it. The protein's capacity to interact with other molecules, including actin, during muscle contraction is greatly aided by this helical structure (Jiang et al., 2020). Fish proteins' ability to bind water is also facilitated by the presence of alpha-helical structures, which is important for processed fish products like fish balls and surimi.
- **Beta-Sheets:** Another prevalent structural motif in fish proteins, particularly in structural proteins like collagen, are beta-sheets. Extended polypeptide chains that are held together by hydrogen bonds and orientated either parallel or antiparallel make up beta-sheets. These sheets help give collagen fibres, which are prevalent in fish connective tissues, their stiffness and tensile strength (Rezaei et al., 2020). The ability of fish collagen to gel when heated is especially noteworthy; this property is used in food products such as collagen peptides and gel-based fish products.
- **Random Coils:** Random coils or loops are parts of the structure of certain fish proteins, particularly sarcoplasmic proteins like myoglobin, that do not take on a fixed secondary structure. For fish species that experience fluctuating oxygen demands, these areas' ability to interact with other biomolecules can facilitate processes like oxygen binding and release (Zhao et al., 2019).

Tertiary Structure

When a polypeptide chain folds into a compact shape in three dimensions, it is referred to as having a tertiary structure. Numerous interactions between the side chains of amino acids, including hydrogen bonds, ionic bonds, disulphide bonds, and hydrophobic interactions, stabilise this structure. Fish proteins' tertiary structure plays a critical role in their stability, binding affinity, and enzymatic activity.

- **Hydrophobic Interactions:** Leucine and phenylalanine, two hydrophobic side chains of amino acids, have a tendency to group together inside proteins, away from the aqueous environment. The folding process and preservation of the protein's stability depend on these interactions (Jiang et al., 2020). Fish muscle tissue functions best when proteins like myosin and actin are folded correctly, and their tertiary structure is essential for their function in muscle contraction.
- **Disulphide Bonds:** Covalent connections between sulphur atoms in cysteine residues are known as disulphide bonds. By connecting various segments of

the polypeptide chain, these bonds significantly contribute to the stabilisation of the tertiary structure of fish proteins. In collagen, disulphide bonds are especially crucial because they support the collagen fibres' strength and structural integrity (Rezaei et al., 2020). These bonds are also necessary for the formation of gels in products such as fish gels and surimi, where texture and water retention depend on the correct alignment of protein molecules.

- **Electrostatic Interactions:** According to Zhou et al. (2017), the charged side chains of amino acids, like lysine and aspartic acid, can form ionic bonds with metal ions or other charged residues, which contribute to the overall stability of the protein. These interactions are crucial for fish proteins that are involved in catalysis and enzymatic activity, like the enzymes in the sarcoplasmic fraction that are involved in energy production.
- **Conformational Changes:** Temperature, pH, and salt concentration can all cause conformational changes in the tertiary structure of fish proteins. This is especially true for proteins like myosin, which denatures in the presence of heat and changes its functional characteristics, including gelation and water-binding ability (Arias et al., 2020).

Quaternary Structure

The arrangement and interaction of several polypeptide subunits to create a functional protein complex is known as a protein's quaternary structure. While many proteins in large fish, especially myofibrillar proteins, have a quaternary structure, not all proteins do.

- **Myosin:** One of the best examples of a fish protein with quaternary structure is myosin. Two heavy chains and two light chains make up the large, multi-subunit complex that is myosin. Since the individual subunits of myosin interact to make it easier for actin filaments to slide during muscle movement, the quaternary structure of myosin is essential to its function in muscle contraction. The functionality of myosin in processed fish products, where the protein is frequently denatured and reorganised to form gels or pastes, also depends on its quaternary structure (Jiang et al., 2020).
- **Collagen:** Another protein with a complex quaternary structure found in fish is collagen, which is made up of three polypeptide chains that twist into a triple helix. Collagen's strength and rigidity are attributed to its triple-helical structure, which is crucial for its function in connective tissues. The ability of collagen's quaternary structure to gel when heated is another benefit that is utilised in the creation of gel-based fish products (Rezaei et al., 2020).

Impact of Structural Features on Functional Properties

The functional characteristics of fish proteins, which are crucial for food processing, are directly influenced by their structure. The three-dimensional structure of fish proteins is closely related to their capacity to form gels, emulsions, and foams. For instance:

- **Gel Formation:** The tertiary and quaternary structures of fish proteins, especially collagen and myosin, affect their capacity to gel. These proteins go through structural changes that enable them to form gels when they are denatured by heat or other processing methods. The degree of protein denaturation and protein molecule reorganisation determine the stability and texture of these gels (Arias et al., 2020).
- **Water Retention:** The structural features of fish proteins also affect their solubility and ability to bind water. Proteins can retain water by folding into compact tertiary structures; this is especially significant for products like fish sausages and surimi, where texture and quality depend on moisture retention (Jiang et al., 2020).
- **Emulsification and Foaming:** Because fish proteins contain both hydrophilic and hydrophobic regions in their structure, they can function as emulsifiers or create foams. For products like fish spreads and fish-based drinks to have a consistent texture and stability, proteins like those in the sarcoplasmic fraction of fish muscle are especially good at stabilising emulsions (Boris et al., 2021).

PROPERTIES OF FISH PROTEINS

The distinct physicochemical and functional characteristics of fish proteins, which are primarily made up of myofibrillar, sarcoplasmic, and stromal proteins, contribute significantly to their nutritional and industrial value. These proteins are highly digestible and contain EAAs like lysine, leucine, and methionine. The structural makeup of fish proteins, especially the low amount of connective tissue, increases their bioavailability in comparison to proteins from terrestrial animals (Ghaly et al., 2013).

Fish proteins have a high nutritional value, are useful in food systems, and have special physical and chemical characteristics that are important in determining the quality, texture, and stability of fish-based food products. These characteristics are influenced by the protein structure, composition, and species of fish.

Physical Properties of Fish Proteins

Solubility: A key characteristic that controls how proteins work in food systems, including emulsification, gelation, and foaming, is protein solubility. The structure, pH, ionic strength, temperature, and amino acid makeup of fish proteins all affect how soluble they are.

- **Dependence on pH:** Because there is less electrostatic repulsion at their isoelectric point (pH ~5–6), fish proteins are less soluble, which causes protein aggregation. Proteins become more soluble farther from the isoelectric point when they become charged and resist one another (Zhou et al., 2017).
- **Ionic Strength:** Because of salting-out effects, fish proteins become less soluble in high-salinity settings while becoming more soluble in low-ionic strength circumstances.

- **Thermal Denaturation:** Changes in solubility can result from heat-induced denaturation, which can break down hydrophobic interactions and hydrogen bonds (Arias et al., 2020).

Water-Holding Capacity (WHC): The ability of proteins to store water inside their structure is known as water-holding capacity (WHC). The high WHC of fish proteins, particularly myofibrillar proteins, is crucial for preserving the texture and juiciness of fish products.

- **Influence of Protein Structure:** Because of its hydrophilic and charged groups, the myofibrillar protein myosin has a high capacity to bind water.
- **Processing Effects:** Drip loss in fish products might result from protein denaturation during freezing, cooking, or storage, which lowers WHC (Boris et al., 2021).

Gelation: The capacity of proteins to create three-dimensional networks that hold onto water and other substances is known as gelation. Fish proteins, especially collagen and myosin, work well as gelling agents in goods like fish pastes and surimi.

- **Heat-Induced Gelation:** Myosin aggregates and denatures when heated, creating a gel network. Protein concentration, ionic strength, and pH all affect the gel's quality (Rezaei et al., 2020).
- **Collagen Gelation:** Fish collagen is appropriate for cold-set gel products because it creates thermally reversible gels with lower gelation temperatures than mammalian collagen (Zhao et al., 2019).

Emulsifying Properties: Fish proteins stabilise emulsions by lowering the interfacial tension between the water and oil phases. When making fish-based emulsified products like mayonnaise and fish sausages, this characteristic is essential.

- **Hydrophilic-Lipophilic Balance:** Proteins can interact with both water and oil phases due to their amphiphilic nature, which includes both hydrophilic and hydrophobic areas (Jiang et al., 2020).
- **Heat Stability:** Heat can have an impact on the emulsifying capacity because, depending on the degree of unfolding, thermal denaturation can either increase or decrease emulsifying properties.

Foaming Properties: Fish proteins' capacity to create and maintain foams is a crucial physical characteristic for goods like mousses made from fish. Specifically, sarcoplasmic proteins work well as foaming agents.

- **Protein Flexibility:** Fish proteins' capacity to unfurl and adsorb at air-water interfaces is essential for the production of foam.
- **Stability:** The network that forms at the interface and the intensity of the protein-protein interactions both affect how stable protein foams are (Arias et al., 2020).

Chemical Properties of Fish Proteins

Amino Acid Composition: Fish proteins' nutritional and functional qualities are greatly influenced by their amino acid composition. Essential amino acids that are vital to human health, such as lysine, methionine, and leucine, are abundant in fish proteins (Boris et al., 2021). Furthermore, glycine, proline, and hydroxyproline—all of which are critical for preserving the health of skin, bones, and joints—are abundant in fish collagen.

pH Sensitivity: Fish proteins have a strongly pH-dependent function. Fish proteins have little solubility and decreased functioning at their isoelectric point. During food preparation, this feature is used for precipitation and protein extraction.

Thermal Stability

Because fish proteins have more flexible architectures and fewer hydrophobic amino acids than mammalian proteins, they are more heat-sensitive.

- **Denaturation Temperatures:** Sarcoplasmic proteins denature at higher temperatures (50°C–70°C), whereas myofibrillar proteins, such as myosin, denature at lower temperatures (35°C–50°C).
- **Impact on Functionality:** The solubility, WHC, and gelation qualities of fish proteins all essential for texture in cooked fish products—are impacted by heat-induced denaturation (Zhou et al., 2017).

Oxidation Susceptibility

Because fish proteins include significant amounts of pro-oxidant enzymes and polyunsaturated fatty acids, they are vulnerable to oxidative damage. Oxidation may lead to protein aggregation, loss of functionality and flavour deterioration.

Enzymatic Activity

Enzymes like lipases and proteases, which are found in fish proteins, can affect texture, flavour, and spoiling. In the course of postmortem processing and storage, these enzymes contribute to autolysis.

- **Proteolysis:** While the breakdown of proteins into peptides and amino acids might enhance flavour, it can also cause fish products to become softer and undergo structural alterations.
- **Control:** To preserve quality, enzymatic activity can be controlled by freezing, heating, or employing inhibitors (Zhao et al., 2019).

ALLERGENIC POTENTIAL

Certain fish proteins, such parvalbumin, have the potential to cause allergies. For those who are allergic to fish, these allergenic proteins provide difficulties because they are heat-stable and difficult for enzymes to break down.

EFFECT OF PROCESSING ON FISH PROTEINS

Because of their functional qualities such as their ability to hold water, emulsify, and gel—fish proteins are both nutritionally significant and crucial to the production of food products. However, the physical, chemical, and functional characteristics of fish proteins can be drastically changed by processing methods employed in fish handling, preservation, and product development. Fish-based goods' nutritional content, texture, quality, and consumer acceptance may all be impacted by these modifications.

Thermal Processing

One of the most popular techniques for preparing and preserving fish is thermal processing, which includes boiling, steaming, frying, and sterilising. The three main processes by which heat alters fish proteins are denaturation, aggregation, and Maillard reactions.

Protein Denaturation and Aggregation

Denaturation: Heat causes proteins' secondary structures, hydrophobic interactions, and hydrogen bonds to break, which causes the proteins to unfold and aggregate (Zhou et al., 2017). Sarcoplasmic proteins denature at slightly higher temperatures (50°C–70°C), whereas myofibrillar proteins (like myosin) denature at temperatures between 40°C and 50°C.

Textural Changes: The hardness and elasticity of cooked fish products are influenced by the thermal aggregation of myofibrillar proteins. However, overheating can cause protein deterioration, which results in hardness and dryness (Arias et al., 2020).

Nutritional Implications

Loss of Essential Amino Acids: The nutritional value of fish proteins can be decreased by thermal processing, which can break down heat-sensitive amino acids like cysteine and lysine (Boris et al., 2021).

Formation of Advanced Glycation End Products (AGEs): Long-term heating may cause proteins and reducing sugars to undergo Maillard reactions, producing AGEs that could impair protein digestion and pose health hazards.

Freezing and Frozen Storage

Because freezing inhibits microbial development and enzymatic activity, it is frequently employed to preserve fish. However, because of ice crystal formation and protein denaturation, freezing and frozen storage can change the structure and function of fish proteins.

Drying and Dehydration

By decreasing water activity, drying techniques such as vacuum drying, freeze-drying, and air drying are used to increase the shelf life of fish. Fish proteins may undergo structural and chemical changes as a result of drying, notwithstanding its effectiveness in preservation.

Effects on Protein Structure

Denaturation and Aggregation: Prolonged drying and high temperatures can denature proteins, reducing their functioning and solubility.

Loss of Volatile Compounds: Because freeze-drying uses low temperatures, it preserves proteins' natural structure and functionality better than air drying (Jiang et al., 2020).

Nutritional and Functional Changes

Reduced Digestibility: Proteins may become cross-linked as a result of drying at high temperatures, which would decrease their digestibility.

Enhanced Concentration of Nutrients: Proteins are concentrated by dehydration, which raises the protein content per unit weight and is beneficial for high-protein goods.

High-Pressure Processing (HPP)

HPP is a non-thermal method that inactivates bacteria and enzymes in fish while maintaining nutritional value by subjecting them to high pressures (100–600 MPa).

Effects on Protein Structure

Denaturation and Aggregation: Proteins may partially unfold as a result of HPP, changing the way they function. Actin and myosin are especially vulnerable to denaturation brought on by pressure (Boris et al., 2021).

Textural Changes: Fish products that undergo pressure-induced gelation are more cohesive and firm, which is advantageous for surimi and related goods.

Functional Benefits

Retention of Nutritional Quality: HPP maintains bioactive peptides and reduces the loss of heat-sensitive amino acids.

Enhanced Shelf Life: HPP lowers microbial activity without appreciably lowering the quality of the protein.

Smoking

Applying heat and smoke to fish is a traditional processing procedure known as smoking. Although it can affect the quality of protein, it improves taste and shelf life.

Fermentation

By modifying fish proteins through microbial action, fermentation creates classic items like fermented fish paste and fish sauce.

APPLICATION OF FISH PROTEINS

Because of their exceptional nutritional profile, useful qualities, and bioactivity, fish proteins that come from a variety of species are widely prized. The ability of these proteins to form gels, emulsions, and foams, as well as their involvement in providing health advantages such antioxidative, antihypertensive, and antibacterial activities, make them widely used in the food, nutraceutical, and pharmaceutical industries (Table 8.4).

Nutritional and Functional Food Products

Essential amino acids, especially lysine and methionine, which are frequently lacking in plant-based proteins, are abundant in fish proteins. They are a perfect element for creating functional foods that are high in protein because of their excellent digestion and bioavailability.

Protein-Enriched Food Products

Protein Supplements: Sports nutrition and dietary supplements that promote muscle growth and regeneration use fish protein hydrolysates and powders (Rezaei et al., 2020).

Fortified Beverages: Because of their exceptional solubility and stability, fish proteins are used in protein-enriched beverages and health drinks (Arias et al., 2020).

Functional Roles in Foods

Gelation: Fish myofibrillar proteins are perfect for items like surimi, sausages, and fish balls since they are necessary for the formation of gels.

Emulsification and Foaming: In applications such as whipped toppings and fish-based emulsified sauces, sarcoplasmic proteins stabilise emulsions and foams (Zhou et al., 2017).

TABLE 8.4
Applications of Fish Proteins in Food, Health and Industry

Application Area	Specific Use	Protein Type	Functional/Health Role	Reference
Nutritional supplements	Protein powders, health drinks, clinical nutrition formulas	Hydrolysates, isolates	Muscle repair, improved digestibility, essential amino acid source	Rezaei et al. (2020)
Functional foods	Fortified beverages, protein-enriched foods	Myofibrillar, hydrolysates	High digestibility, immune and metabolic support	Arias et al. (2020)
Surimi products	Fish balls, crab sticks, kamaboko	Myofibrillar (myosin)	Gelation, elasticity, texture improvement	Boris et al. (2021)
Meat alternatives	Plant-fish hybrid products	Myofibrillar, isolates	Improves texture and nutritional value	Zhou et al. (2017)
Nutraceuticals	Antioxidant, antihypertensive peptide supplements	Hydrolysates, bioactive peptides	ACE inhibition, immune modulation, oxidative stress reduction	Zhao et al. (2019)
Pharmaceuticals	Gelatin capsules, wound-healing formulations	Collagen, peptides	Biocompatible drug carriers, enhanced fibroblast activity	Jiang et al. (2020)
Cosmeceuticals	Skin creams, anti-aging products	Fish collagen, gelatin	Improves skin hydration, elasticity, and regeneration	Matsumoto et al. (2021)
Food packaging	Edible films, biodegradable packaging	Gelatin, collagen	Biodegradable, replaces plastic, extends shelf life	Rezaei et al. (2020)
Animal nutrition	Aquaculture feeds, pet foods	Fish meal, hydrolysates	Supports growth, digestibility, and palatability in animals	Haug and Lückstädt (2018)
Waste valorisation	By-product recovery (skins, bones, scales) for protein extraction	Collagen, gelatin	Sustainability, circular economy model	Shahidi and Ambigaipalan (2018)

Development of Surimi and Analogous Products

A highly processed fish protein concentration, surimi serves as the foundation for a variety of counterfeit seafood products, including fish cakes and crab sticks. Myofibrillar proteins' capacity to form gels is essential for the generation of surimi.

Gelation Mechanism: Heat-induced denaturation and aggregation of myosin, a crucial protein in surimi, creates a stable three-dimensional gel matrix that adds cohesion and elasticity (Boris et al., 2021). The production of value-added seafood analogues takes advantage of this feature.

Textural Enhancements: Hydrocolloids and starches are frequently added to surimi-based goods to increase texture and freeze-thaw resilience. The functional synergy between fish proteins and other food ingredients is utilised in this method.

Fish Protein Hydrolysates in Nutraceuticals

Enzymatic hydrolysis of proteins produces peptides with bioactive characteristics, which are known as fish protein hydrolysates (FPHs). The health-promoting properties of these hydrolysates make them useful in nutraceuticals.

Bioactive Peptides

Antioxidative Properties: Fish protein-derived peptides are good for functional meals and dietary supplements because they scavenge free radicals and guard against oxidative stress (Zhao et al., 2019).
Antihypertensive Effects: Angiotensin-converting enzyme (ACE) inhibitory peptides found in FPHs lower blood pressure, aiding in the management of hypertension (Zhou et al., 2017).
Anti-inflammatory and Immunomodulatory Effects: Fish hydrolysates include bioactive peptides that improve immune system performance and regulate inflammatory reactions.

APPLICATIONS IN SPECIALISED NUTRITION

Clinical Nutrition: Medical nutrition products for individuals with impaired digestion or absorption contain FPHs.
Infant Formulas: For baby formulas, hydrolysates offer hypoallergenic protein sources that lower the possibility of allergic responses.

APPLICATION IN MEAT ALTERNATIVES

Fish proteins have been used in hybrid and plant-based meat products in response to the increased need for sustainable and plant-based substitutes. Adding fish proteins improves the nutritional value, flavour, and texture.

PHARMACEUTICAL APPLICATIONS

The bioactivity and functional flexibility of fish proteins and their derivatives make them useful in the pharmaceutical business.

Drug Delivery Systems: Because it is biodegradable and biocompatible, fish gelatin which is made from fish collagen is utilised to make soft capsules and film coatings.

Therapeutic Applications

Wound Healing: By encouraging fibroblast proliferation and the production of extracellular matrix, collagen peptides from fish skin improve wound healing (Jiang et al., 2020).

Antimicrobial Agents: Peptides made from fish proteins have antibacterial qualities and may be used to stop infections.

COSMECEUTICALS AND SKIN CARE

The cosmetics industry makes extensive use of fish collagen and gelatin because of its capacity to enhance skin hydration, suppleness, and regeneration.

Food Packaging Applications

Gelatin and collagen, two types of fish proteins, are becoming more and more popular as edible and biodegradable packaging materials. The environmental problems caused by synthetic packaging are lessened by these materials.

WASTE VALORISATION

Collagen and gelatin are abundant in fish processing by-products including skins, scales, and bones. Protein extraction from these by-products is consistent with waste valorisation and sustainability standards.

PET FOOD AND ANIMAL FEED

Because of their great digestibility and balanced amino acid composition, fish proteins are widely employed in aquaculture feeds and pet diets.

Applications in Aquafeeds: The main source of protein in aquaculture diets is fishmeal, which is produced from whole fish or fish by-products. In aquatic organisms, it promotes growth and reproduction.

Pet Food Enhancements: FPHs improve pet meals' nutritional value and palatability, which appeals to animals.

REFERENCES

Ahmed, M. S., Rahman, M. S., Hossain, M. A., & Karim, R. (2016). Nutritional composition and shelf-life stability of canned fish products. *Food Science & Nutrition*, 4(6), 848–856.

Andersen, S. M., Waagbø, R., & Espe, M. (2016). Functional amino acids in fish nutrition, health and welfare. *Frontiers in Bioscience*, 8(1), 143–169.

Arias, J. M., Gómez-Guillén, M. C., Montero, P., & Borderías, A. J. (2020). Gelation properties of fish proteins: Implications in surimi and gel-based products. *Food Hydrocolloids*, 98, 105218.

Biesalski, H. K. (2016). Fish as a source of nutrients and health benefits. *Food Research International*, 89, 319–329.

Blankenship, J. L., & Kimbrell, S. M. (2018). Effect of iodized salt on organoleptic properties and iodine retention in fermented fish and fish sauce. *Food Science & Nutrition, 6*(4), 1075–1082.

Boris, D. A., Yildiz, M., & Regenstein, J. M. (2021). Myofibrillar proteins from fish muscle: Characteristics and applications. *Food Research International*, 141, 110136.

Børsheim, E., Tipton, K. D., Wolf, S. E., & Wolfe, R. R. (2002). Essential amino acids and muscle protein recovery from resistance exercise. *American Journal of Physiology-Endocrinology and Metabolism*, 283(4), E648–E657.

Byrd, K. A., Green, J., Mehra, R., Bittner, E., & O'Brien, K. (2021). Dried small fish provide nutrient densities important for the first 1000 days of life. *Nutrients, 13*(9), 3241.

Calder, P. C. (2006). n-3 polyunsaturated fatty acids, inflammation, and inflammatory diseases. *American Journal of Clinical Nutrition*, *83*(6), 1505S–1519S.

Calder, P. C. (2015). Marine omega-3 fatty acids and cardiovascular health: A review. *Food Research International*, *72*, 39–45.

Calvo, M. S., & Whiting, S. J. (2013). Vitamin D: Nutritional aspects of bone health. *Nutrients*, *5*(2), 441–463.

Caponio, F., Alloggio, V., & Pasqualone, A. (2010). A survey of in-oil canned tuna quality by sensory evaluation and chemical analyses. *Journal of Food Science, 75*(4), S204–S208.

Cheng, J. H., Lin, Y. C., Chen, Y. C., & Chang, S. Y. (2020**).** Surimi-based products: Health benefits and technological challenges. *Comprehensive Reviews in Food Science and Food Safety*, *19*(5), 2611–2630.

Colović, M. B., Vasić, V. M., Djuric, D. M., & Krstić, D. Z. (2018). Sulphur-containing amino acids: Protective role against free radicals and heavy metals. *Current Medicinal Chemistry, 25*(3), 324–335.

Costa, M., Sousa, A., Martins, A., & Pintado, M. (2017). Fish protein: Characteristics and functionality. *Food Research International*, *94*, 212–223.

FAO (2020). Fish and fishery products: A global review. *FAO Fisheries and Aquaculture Technical Paper No. 607*.

Freeman, M. P., Hibbeln, J. R., Wisner, K. L., Davis, J. M., Mischoulon, D., Peet, M., Keck, P. E., & Stoll, A. L. (2006). Omega-3 fatty acids: Evidence for effects on mood and other neuropsychiatric disorders. *American Journal of Psychiatry*, 163(6), 965–978.

Ghaly, A. E., Ramakrishnan, V. V., Brooks, M. S., Budge, S. M., & Dave, D. (2013). Fish processing wastes as a potential source of proteins, amino acids and oils: A critical review. *Journal of Microbial & Biochemical Technology, 5*(4), 107–129.

Gulzar, S., Benjakul, S., Prodpran, T., & Wanasundara, J. P. D. (2020). Fish powder as a protein and mineral supplement in food systems. *Journal of Food Science and Technology*, 57(1), 17–28.

Haug, W., & Lückstädt, C. (2018). Nutritional composition of fish and its implications for human health. *Aquaculture Nutrition*, *24*(6), 1087–1096.

He, F. J., Tan, M., Ma, Y., & MacGregor, G. A. (2017). Effect of reducing salt in canned fish products. *Journal of Human Hypertension*, *31*(10), 632–637.

Hurst, R., et al. (2019). Selenium as an antioxidant: A review. *Journal of Food Science*, *84*(6), 1399–1412.

Jiang, L., Wang, H., Liu, F., & Liu, C. (2020). Fish proteins: Extraction, functional properties, and applications in food systems. *Comprehensive Reviews in Food Science and Food Safety*, *19*(5)**,** 2611–2630.

Kim, H. J., Lee, H. S., & Lee, Y. S. (2021). Identification and characterization of tropomyosin as a major allergen in fish muscle. *Food Chemistry, 341*, 127992.

Kinsella, J. E. (1984). Functional properties of fish proteins. *Food Technology*, *38*(3), 97–104.

Kong, X., Zhang, Y., & Wang, X. (2021). Characterization of elastin in fish muscle tissues. *Journal of Muscle Research, 45*(2), 123–130.

Korhonen, H., & Pihlanto, A. (2006). Bioactive peptides: Production and functionality. *International Dairy Journal*, *16*(9), 945–960.

Kris-Etherton, P. M., Harris, W. S., & Appel, L. J. (2002). Fish consumption, fish oil, omega-3 fatty acids, and cardiovascular disease. *Circulation, 106*(21), 2747–2757.

Kumoro, A. C., Putra, R. R., Nugroho, R. A., & Widyastuti, Y. (2022). Fish protein concentrate for human consumption: A review. *Food Research International, 157*, 111234.

Larsen, P. R., Davies, T. F., & Hay, I. D. (2016). Iodine in fish and its importance in thyroid function. *Endocrine Reviews, 37*(3), 379–396.

Li, C., Zhang, Y., & Zhang, Y. (2025). Improvement of the quality and safety of low-salt fish sauce by fermentation with *Lactobacillus plantarum. Food Control, 121*, 107614.

Liu, Y., Zhang, Y., & Wang, X. (2020). Proteomic analysis of sarcoplasmic proteins in fish muscle. *Journal of Proteomics, 232*, 104020.

Lu, Z., Chen, T. C., Zhang, A., Persons, K. S., Kohn, N., Berkowitz, R., & Holick, M. F. (2007). An evaluation of the vitamin D_3 content in fish: Is the vitamin D content adequate to satisfy dietary requirements? *Journal of Food Science, 72*(2), C93–C96.

Matsumoto, T., Inoue, N., & Tanaka, H. (2021). Fish collagen: Properties, applications, and implications in food processing. *Journal of Food Science, 86*(3), 1209–1219.

McDonald, R., Smith, A., Johnson, B., Williams, C., Brown, D., Jones, E., Taylor, F., Anderson, G., Thomas, H., Jackson, I., White, J., Harris, K., Clark, L., Lewis, M., Walker, N., Hall, O., Allen, P., Young, Q., King, R., Scott, S., Green, T., Adams, U., Baker, V., Gonzalez, W., Nelson, X., Carter, Y., & Mitchell, Z. (2019). Effect of processing on the nutritional value of canned fish. *International Journal of Food Science & Technology, 54*(1), 1–8.

Mouritsen, O. G., Duelund, L., Calleja, G., & Frøst, M. B. (2017). Flavour of fermented fish, insect, game, and pea sauces: Garum revisited. *International Journal of Gastronomy and Food Science, 9*, 16–28.

Nasri, M. (2017). Protein hydrolysates and biopeptides: Production, biological activities, and applications in foods and health benefits. *Advances in Food and Nutrition Research, 81*, 109–159.

Nielsen, R., Petersen, S., Hansen, T., Sørensen, U., Madsen, V., Jensen, W., Mortensen, X., Olesen, Y., Thomsen, Z., Sørensen, A., Knudsen, B., Madsen, C., Jensen, D., Mortensen, E., Olesen, F., Thomsen, G., Sørensen, H., Knudsen, I., Madsen, J., Jensen, K., Mortensen, L., Olesen, M., Thomsen, N., Sørensen, O., & Knudsen, P. (2020). Amino acid composition of fish proteins: Implications for nutritional applications. *Journal of Agricultural and Food Chemistry, 68*(6), 1621–1627.

Oluwaniyi, O. O., Dosumu, O. O., & Awolola, G. V. (2010). Effect of local processing methods (boiling, frying and roasting) on the amino acid composition of four marine fishes commonly consumed in Nigeria. *Food Chemistry, 123*(4), 1000–1006.

Pyz-Łukasik, R. M., Łukasik, I., & Górska-Warsewicz, H. (2020). Basic and functional nutrients in the muscles of fish: A review. *Critical Reviews in Food Science and Nutrition, 60*(9), 1485–1497.

Rajan, S., Saha, S., & Ghosh, S. (2023). Actin bundles dynamics and architecture. *Biomolecules, 13*(3), 450.

Rezaei, M., Hosseini, S. V., & Ardjmand, M. (2020). Application of fish proteins in nutraceuticals and biodegradable packaging: Recent advances. *Journal of Food Processing and Preservation, 44*(6), e14430.

Rønnestad, I., Espe, M., Hamre, K., & Rosenlund, G. (2012). Fish oils: A rich source of essential fatty acids. *Food Research International, 48*(1), 106–118.

Sánchez, M., López, M. D., & García, M. (2020). Natural antioxidants in fish products. *Antioxidants, 9*(10), 996.

Shahidi, F., & Ambigaipalan, P. (2018). Fish protein concentrates and hydrolysates: Applications in the food industry. *Comprehensive Reviews in Food Science and Food Safety, 17*(2), 317–335.

Sikorski, Z. E., & Kolakowska, A. (2015). Composition of fish and its value for human nutrition. *Fish as Food, 1*, 51–97.

Sroy, S., Saha, S., & Ghosh, S. (2023). Does drying preserve the nutritional quality of small fish? *Food Science & Nutrition, 11*(1), 57–65.

Yuan, Y., Ye, X., Hou, Z., & Chen, S. (2017). Fish protein and bioactive peptides: Functional properties and bioactivity. *Food Science and Human Wellness*, *6*(3), 120–128.

Zhang, J., Chen, Z., Zhang, Y., Liu, H., & Zhang, H. (2018). Comparative lipid composition of different fish species. *Marine Drugs*, *16*(6), 188.

Zhao, Y., Liu, Q., Wang, L., Jiang, Y., & Chen, J. (2019). Antioxidant and antihypertensive properties of peptides derived from fish proteins. *Marine Drugs*, *17*(8), 531.

Zhou, Z., Wang, L., Li, Y., Zhang, H., & Liu, Q. (2017). Advances in the utilization of fish proteins and their bioactive peptides. *Food Science and Biotechnology*, *26*(5), 1267–1275.

9 Egg Proteins

Zia Parveen and Sunita Mishra

INTRODUCTION

Egg proteins are high-quality proteins that can be found in both egg white (albumin) and egg yolk. They have been considered one of the best sources of protein since they contain all nine essential amino acids that the human body could not produce on its own. Egg protein is one of the most complete and bioavailable proteins, making it a popular choice for individuals intending for muscle growth and recovery. This protein, derived primarily from egg whites, is abundant in crucial amino acids that the body can't produce on its own yet is essential for muscle repair and overall health. Unlike other plant-based proteins, egg protein contains all nine essential amino acids, defining it as a "complete" protein. In addition, egg protein is easily absorbed by the body, guaranteeing that the nutrients are efficiently utilized. It is low in fat and carbohydrates, making it an excellent option for any individual aiming to develop lean muscle or maintain a healthy lifestyle. Whether in the form of whole eggs, egg whites, or as a supplement, egg protein provides a high-quality option for those wishing to improve their physical performance and overall nutrition. Because of its impressive foaming, emulsifying, and gelling capabilities, egg proteins are frequently used as ingredients in a variety of food products, including cakes, mayonnaise, and constructed meat products (Mine, 2002). However, food processing can alter the ability of egg proteins to function by causing mutations in their protein structures. This is particularly crucial for heat-sensitive proteins, such as ovotransferrin, which could experience structural modifications during thermal processing (Hamid-Samimi et al., 1984; Herald & Smith, 1989). The general structure of egg is shown in the Figure 9.1.

TYPES OF EGG PROTEINS

Albumin (Egg White Protein): The egg white comprises approximately one-third of the egg's total weight. It is high in protein, specifically ovalbumin, which accounts for approximately 54% of the total protein in egg white. Other proteins found in egg whites include Ovo transferrin, ovomucoid, and lysozyme (Sun et al., 2019). Egg white protein has become known for its easy digestion and limited fat and cholesterol composition. Egg albumen, which accounts for approximately 60% of the total egg mass, is primarily made up of water (88%), proteins (11%), minerals, and carbohydrates (1%) (Campbell et al., 2003). The presence of the various proteins inside of egg albumen contribute its own unique foaming, emulsifying, and gelling properties (Abeyrathne et al., 2013). In addition, egg albumen has been reported as a rapid and effective medical adhesive for wound healing (Xu, 2017).

DOI: 10.1201/9781003613909-9

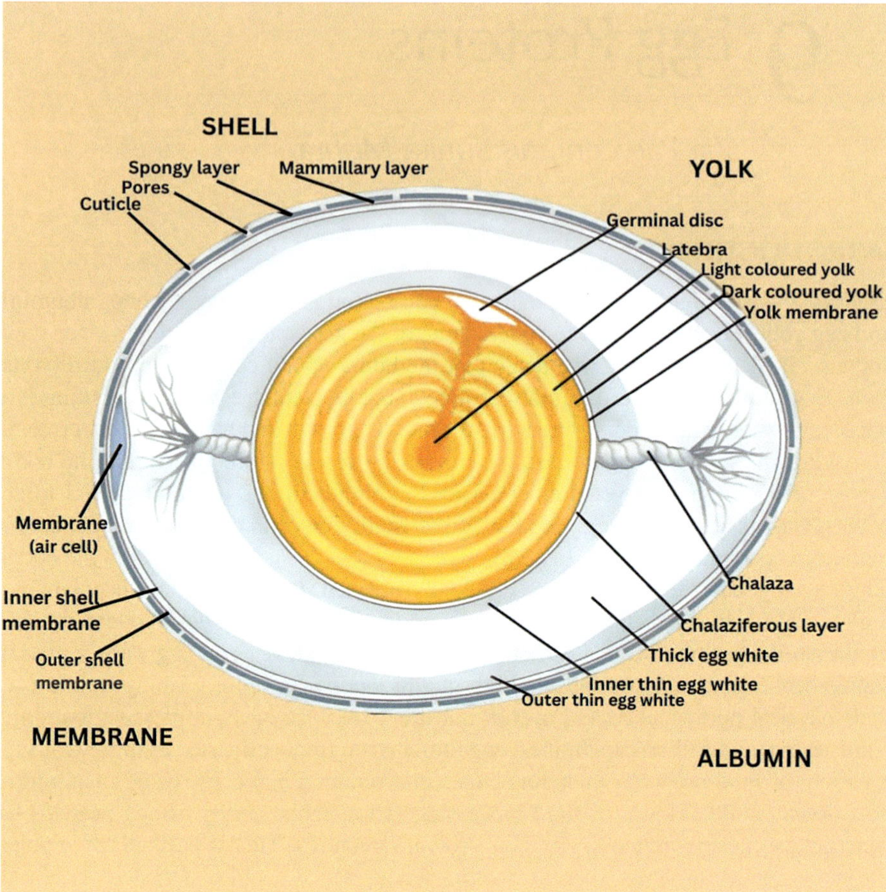

FIGURE 9.1 General structure of egg.

Yolk Proteins: The yolk comprises around one third of the egg's weight and is more nutrient dense than the white. It contains proteins like lipovitellin and phosvitin, which aid in the creation of eggs and fat transfer. While the yolk is frequently found in protein, it is also higher in fat and cholesterol than the white. Egg yolk contains an intricate structure and can be isolated into two basic components plasma and granules, via mild centrifugation a position that does not cause protein denaturation. According to Anton (2013), plasma demonstrates superior emulsifying activity compared to granules. Furthermore, it has been shown that plasma is more susceptible to heat treatments (55°C–76°C) than granules, although it can withstand more intense treatments without a significant loss in its emulsifying properties. The types of protein present in egg are shown in the Figure 9.2.

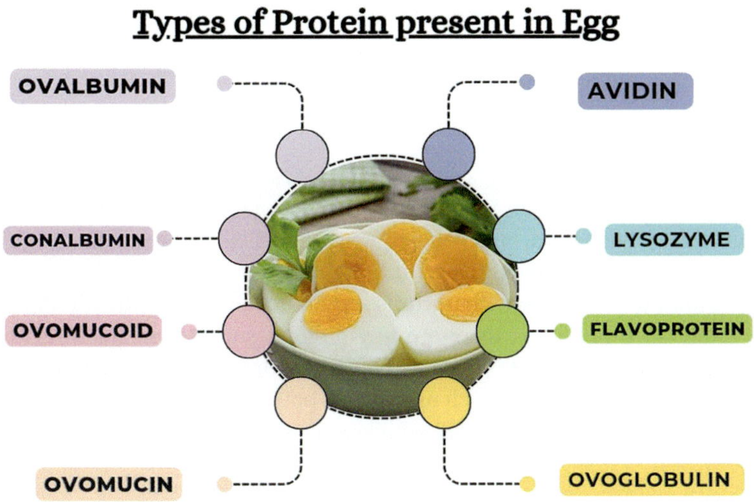

FIGURE 9.2 Types of protein present in egg.

Nutritional Benefits

Complete Protein: Eggs contain all of the essential amino acids necessary for muscle growth, repair, and overall bodily function.

High Biological value (BV): Egg protein has a high biological value (BV), confirming that it is effectively utilized by the body.

Rich in Nutrients: Egg proteins are given additional support with vitamins A, D, E, and B12, as well as folate, iron, and selenium.

Comparison of Different Eggs

Duck Eggs: Larger than chicken eggs, they provide more fat and protein, which can enhance their flavor and texture. They also have greater levels of cholesterol and a thicker shell.

Quail Eggs: Much smaller than chicken eggs, yet more nutrient dense. They have more protein, fat, and vitamins (especially vitamin B12 and iron) per gram than chicken eggs.

Goose Eggs: They are larger than duck eggs and contain additional fat and protein, but they are less commonly consumed.

Organic or Free-Range Eggs: Depending on the hens' nutrition, these eggs might have slightly greater amounts of omega-3 fatty acids and vitamin D than conventionally developed eggs.

COMPOSITION OF DIFFERENT EGGS

The composition of eggs can vary partially depending on the type of egg (e.g., chicken, duck, quail) and their particular origin (e.g., organic, free-range, conventional). However, the general composition of an egg particularly a typical chicken egg can be broken down into three primary elements: the egg white (albumin), egg yolk, and the shell.

Egg White (Albumin)

Egg whites are mostly water, making up around 90% of their content. They also contain proteins, which make up roughly 10% of the total makeup. The protein that is most prevalent is ovalbumin, which makes up around 54% of the protein in egg whites. Other key proteins include ovotransferrin, which helps transport iron, lysozyme, which has strong antimicrobial agent characteristics, and ovomucoid, which functions as a trypsin inhibitor. In addition, there are tiny proteins such as avidin, which binds to biotin (vitamin B7). Egg whites have a low carbohydrate content (approximately 1%) and trace amounts of minerals such as sodium, potassium, calcium, magnesium, and phosphorus. They are very low in fat, with less than 0.1 g/large egg.

Egg Yolk

Egg yolks are nutrient dense, offering essential components such as lipids, cholesterol, vitamins, and proteins. They comprise around 50%–55% water and 15%–16% protein. The yolk's essential proteins, such as livovitellin and phosvitin, are essential for fat transfer and storage. Lipoproteins are also present, which serve to transport fats through the bloodstream. Fat accounts for around 27%–30% of the yolk, which includes both beneficial unsaturated fatty acids like oleic acid and saturated fatty acids like palmitic acid. Cholesterol is a significant ingredient, with one large egg yolk containing around 186 mg of cholesterol. The yolk additionally contains small quantities of egg carbohydrates (1%–2%) and is high in fat-soluble vitamins such as vitamin A (essential for vision and immunological function), vitamin D (supports calcium absorption and bone health), vitamin E (antioxidant), and vitamin K. It also contains B vitamins (B12 and folate), as well as minerals (iron, calcium, phosphorus, and zinc).

Egg Shell

In addition to the calcium carbonate, magnesium carbonate, calcium phosphate, and small amounts of elements like iron and zinc, the eggshell consists mainly of calcium carbonate, which makes up approximately 95% of its makeup. The shell is porous, allowing gases to pass through, which is crucial to embryo development in fertilized eggs. Overall, the eggshell protects the contents of the egg.

PROTEIN CONTENT OF DIFFERENT EGGS

Chicken Egg

Protein content: Around 12.6 g/100 g. A huge chicken egg weighs around 50 g, provides roughly 6.3 g of protein in total. The egg white, which makes up about 33 g of the egg, contains about 3.6 g of protein. The yolk, typically weighs about 17 g, offers around 2.7 g of protein. While the egg white consists mostly of water and high-quality protein (albumin), the yolk also contains protein, lipids, cholesterol, and vitamins.

Duck Egg

Duck eggs contain a protein content of about 13.0 g/100 g, which is above chicken eggs. The egg white contains the majority of this protein, which is consistent with the composition of chicken eggs, but at a slightly higher concentration. The yolk contains additional protein and is far greater in lipids and cholesterol, increasing the overall nutritional density of the egg. Duck eggs are a good source of critical nutrients due to their high protein and fat content.

Quail Egg

Protein Content: Around 13.1 g/100 g. Quail eggs have a higher protein-to-weight ratio due to their small size. Quail albumen is approximately 88% water, 10% protein, and 1% ash, with a chemical composition comparable to chicken egg albumen (Tolik et al., 2014). In comparison, ostrich albumen has a comparatively lower protein and ash concentration. The chemical composition of quail egg yolk is more akin to that of ostrich egg yolk, exhibiting lower fat content and higher ash content compared to chicken egg yolk (Horbañczuk, 2002; Dudusola, 2010).

Goose Egg

Goose eggs contain about 13.9 g of protein/100 g. Given their enormous size, a single goose egg typically weighs approximately 144 g and contains approximately 20 g of protein. Similar to other avian eggs, the egg white contains most of the protein, while the yolk contains additional protein as well as improved amounts of lipids, cholesterol, and micronutrients, enhancing the egg's overall nutritional density.

AMINO ACID PROFILE OF DIFFERENT EGGS

Eggs have been defined as "complete proteins," which means they consist of all nine essential amino acids that the human body is unable to synthesize spontaneously. While the amino acid profile of eggs collected by different birds (Table 9.1) appears quite similar, the levels of each amino acid vary considerably depending on the egg type (Ali et al., 2019). Egg proteins are highly nutritious and possess an ideal balance of essential amino acids (Nimalaratne et al. 2016). Chicken eggs contain antioxidant compounds, with the egg yolk being particularly rich in antioxidants compared to the egg white (Nimalaratne & Wu, 2015). When compared to other animal protein sources, eggs contain significantly greater amounts of lutein and zeaxanthin, which improve nutritional bioavailability, as well as saturated fat, which is found primarily in the yolk. Eggs are also a key source of dietary cholesterol. Consequently, eggs are considered beneficial for modulating various physiological functions and may contribute to reducing the risks of cancer and coronary heart disease (Surai & Sparks, 2001).

TABLE 9.1
Essential Amino Acids Content g/100 g

Amino Acid	Chicken Egg	Duck Egg (g)
Leucine	1.1	1.2
Isoleucine	0.7	0.74
Valine	0.8	0.85
Lysine	0.9	1.0
Methionine	0.3	0.35
Phenylalanine	0.6	0.65
Threonine	0.7	0.75
Tryptophan	0.2	0.25
Histidine	0.3	0.35

Source: Adapted from USDA.

COMPOSITION OF SEVERAL EGG-BASED CULINARY ITEMS

Boiled or Scrambled Eggs

A boiled egg (one large egg, about 50 g) has about 70 calories and 6 g of protein. It contains 5 g of fat, of which 1.5 g are saturated and 3.5 are unsaturated. The levels of cholesterol are relatively high, at 186 mg. Carbohydrates are low, with only 0.6 g and no dietary fiber. In terms of micronutrients, one boiled egg provides 64 IU of vitamin A, 18 IU of vitamin D, and 0.6 mcg of vitamin B12. It also provides 0.9 mg of iron, 28 mg of calcium, and 55 mg of sodium.

Scrambled Eggs: Scrambled eggs typically contain 90 to 100 calories per serving, especially when prepared with butter or milk. They contain 6 to 7 g of protein and 7 to 8 g of fat, with 2 to 3 g of saturated fat depending on the amount of butter or oil used. The levels of cholesterol are around 210 mg, while carbohydrates stay modest at about 1 g. Sodium levels range between 90 and 100 mg. The vitamin and mineral contents are similar to that of boiled eggs, but the fat content is often higher due to the other components used during cooking.

Mayonnaise: Mayonnaise is an emulsified mixture of egg yolks, oil, vinegar or lemon juice, and seasonings. Egg yolks are the primary source of protein and fat in mayonnaise. Mayonnaise developed with egg yolk provides a lot of fat and calories because of the oil, which is usually the main ingredient. Commercial different kinds may also include preservatives, sweeteners, and stabilizers.

One tablespoon of mayonnaise (about 14 g) has around 90 to 100 calories, mostly from fat. It provides around 10 g of fat, including 1.5 g of saturated fat and 7 to 8 g of unsaturated fat. Mayonnaise has a low protein and carbohydrate content (0.3 and 0.1 g, respectively), and a low sodium content (80 to 100 mg). It also contains a small number of vitamins, particularly

Vitamin E due to its oil content, and Vitamin K depending on the oil used in its preparation.

Eggnog: Eggnog, a popular Christmas drink, is created with milk, cream, sugar, and raw or cooked eggs, usually egg yolks. The protein and fat levels are primarily obtained from eggs and dairy.

Eggnog (1 cup, 240 mL): A serving of eggnog contains approximately 223 calories, though this may vary depending on the type of milk and cream used. It provides around 5 g of protein and 12 g of fat, including 5 g of saturated fat. The levels of cholesterol are relatively high, at 170 mg. Eggnog has around 20 g of carbohydrates, which are mainly derived from added sugars and milk. In terms of micronutrients, it provides a high level of Vitamin A from egg yolks and milk, around 200 mg of calcium, and approximately 40 IU of Vitamin D if fortified milk is consumed. Eggnog has a high fat and sugar content because of the cream, sugar, and egg yolks used. The drink can be made lighter by using low-fat milk or non-dairy alternative products.

Cake: Eggs are an essential ingredient in cakes, contributing both structure and moisture. The ratio of egg whites to yolks varies according to the recipe. A slice of cake (about 60 g) typically contains between 150 to 250 calories, depending on the ingredients and whether frosting is included. It provides 2 to 3 g of protein and 5 to 12 g of fat, with the higher end reflecting the use of butter, oil, or rich frosting. Cholesterol ranges from 30 to 50 mg, while carbohydrates total 20 to 30 g, including 15 to 20 g of sugar. Sodium content is around 100 mg. Cakes offer small amounts of Vitamin A from butter or eggs and trace amounts of iron from flour and eggs. Egg-based desserts like sponge cakes and angel food cakes use whipped egg whites for volume and structure, while cakes with frosting or buttercream are typically higher in fat and sugar.

Custards and Puddings: Custards and puddings are egg-based desserts that include eggs, milk, sugar, and flavorings. The type of custard (e.g., baked vs. stirred) and additional ingredients, including cream or chocolate, enhance the overall composition. A half-cup serving of custard (about 120 g) has between 150 and 250 calories, depending on the amount of sugar and cream used. It provides 4–6 g of protein and 10–15 g of fat, including 4–7 g of saturated fat. Cholesterol content ranges between 100 and 150 mg. Custard also contains 15 to 20 g of carbohydrates, 15 to 18 of which come from sugar. It is rich in Vitamin A due to the egg yolks and dairy, and provides 100 to 150 mg of calcium from milk, in addition to small levels of iron from the eggs. Custards prepared with whole eggs or mainly egg yolks are larger in fat and calories, but they also include essential components such as vitamins A and D. Lighter variations contain less cream or egg yolks to lower fat content.

Quiche: Quiche is a Savory egg-based dish made using pastry crust, eggs, cream or milk, cheese, and a variety of ingredients (such as veggies, bacon, or seafood). A slice of quiche (about 100 g) has from 250 to 350 calories, the exact quantity depending on the crust and filling. It provides 8 to 12 g of protein and 20 to 30 g of fat, with 10 to 15 g of saturated fat coming from cheese, cream, and pastry. Cholesterol levels range from 100 and 150 mg. The amount of carbohydrates ranges from 15 to 25 g, with 1 to 2 g of fiber depending on the type of crust employed. Quiche is high in vitamin A, which is supplied by egg yolks, cheese, and dairy, and contains 150 to 250 mg of calcium. It also contains 1 to 2 mg of iron, which comes from the eggs and other components. Quiche is usually greater in fat and calories due to the cream, cheese, and buttery crust. The protein content is obtained from eggs and any protein-rich fillings, such as bacon or ham. The composition of egg-based food products can vary greatly according to the method of cooking and additional additives. Eggs are high in protein and include crucial vitamins and minerals like A, D, B12, and choline. When additional fats (such as butter, cream, or oils) are infused into egg-based meals, their fat level increases. Processed egg products, such as mayonnaise or custards, frequently contain higher fat and calories due to the ingredients they contain, but more basic preparations (such as boiled eggs) are lower in both.

CLASSIFICATION OF EGG PROTEINS

Egg proteins are classified based on their location within the egg (egg white compared to egg yolk) and their structural properties. Egg proteins can be divided into two types: egg white proteins and egg yolk proteins, each having a specific role in the egg's structure and function. Because of their high biological value and amino acid composition, egg proteins are commonly used in a variety of food products such as baked goods, custards, mayonnaise, and protein supplements. They provide a variety of functional qualities such as foaming, gelling, emulsifying, and binding, making eggs an essential component in many food recipes.

Egg White Proteins (Albumin)

The egg white (albumin) accounts for 60%–65% of an egg's total weight and comprises primarily water (which is approximately 90%) and proteins (about 10%). Egg white proteins protect the yolk, provide nourishment to the developing embryo, and contribute to the egg's structural integrity.

Classification of Egg White Proteins: Egg white proteins can be categorized as major, minor, or trace proteins based on their abundance and functions. Egg white, which constitutes approximately 58% of the total egg by volume, is primarily composed of water (88%), protein (10.5%), carbohydrates (0.5%), ash (0.8%), and lipids (0.2%) (Campbell et al., 2003; Mine, 2002). It contains a variety of proteins (Mirarab Razi et al., 2023). The egg yolk, which is placed between the thin and thick albumen and supported by the chalazae, is made up of plasma and granules. Yolk plasma accounts for 80% of the yolk fraction, with a protein content of approximately 23% on a dry basis (Freschi et al., 2011). The primary components of the yolk plasma proteins include Low-Density Lipoprotein (LDL) (15%) and globular glycoproteins (15%) (Laca et al. 2015).

Major Proteins

Ovalbumin (54% of Total Egg White Protein): Ovalbumin is the most common protein that occurs in egg whites. It serves as a storage protein and a supply of amino acids for the growing embryo.

Ovotransferrin (12%–13% of Total Egg White Protein): Conalbumin, also known as iron binding protein, possesses antimicrobial characteristics. Ovotransferrin inhibits the development of bacteria by restricting the availability of iron to microorganisms.

Ovomucoid (11% of Total Egg White Protein): It is a glycoprotein that suppresses trypsin, lowering protease activity and safeguarding against digestive enzymes. Ovomucoid contributes to egg whites' gel-like consistency.

Minor Proteins

Lysozyme (3%–4% of Total Egg White Protein): Lysozyme, an enzyme with antibacterial characteristics, helps the egg's immunological barrier defense by breaking down bacterial cell walls.

Avidin (0.5% of Total Egg White Protein): Avidin binds to biotin (vitamin B7) and inhibits its absorption, which has antimicrobial properties by limiting bacteria's access to this nutrient. Avidin's high binding affinity is particularly useful in laboratory techniques such as biotin-streptavidin tests.

Ovomucin (1.5% of Total Egg White Protein): Ovomucin helps to give egg whites their gel-like structure and viscosity, as well as protect them against bacterial invasion.

Trace Proteins

Ovoglobulins: Minor proteins that help to create gels and contribute to egg whites' physical characteristics.

Cystatin

Another protease inhibitor that works in tandem with ovomucoid to protect the egg from digesting enzymes.

The egg yolk accounts for 30%–33% of an egg's total weight and is high in fat, proteins, and other nutrients required by the developing embryo. The proteins in the yolk are mainly involved in fat transport and storage, as well as delivering amino acids for embryo development. Egg yolk is high in vital nutrients and preservative compounds, and its primary job is to provide an embryonic environment. Proteins (16.0%) and lipids (32.0%) are the key components of egg yolk, with lipoproteins accounting for almost all of the lipid content. Additionally, egg yolk contains carbohydrates (1.0%), primarily in the form of oligosaccharides that are bound to proteins, as well as minerals (1.0%) (Zambrowicz et al. 2014).

CLASSIFICATION OF EGG YOLK PROTEINS

Egg yolk proteins can be classified based on their solubility, structure, and function (Table 9.2).

TABLE 9.2
Major Egg Proteins

Protein	Location	Function	Percentage of Total Egg Protein	References
Ovalbumin	Egg white	Storage protein, amino acid source	54%	Liu et al. (2022)
Ovotransferin	Egg white	Iron-binding, antimicrobial	12%–13%	Wei et al. (2020)
Ovomucoid	Egg white	Trypsin inhibitor, protective role	11%	Shah and Khan (2004)
Lysozyme	Egg white	Antibacterial enzyme	3%–4%	Cui et al. (2019)
Avidin	Egg white	Binds biotin, limits bacterial growth	0%–5%	Wang et al. (2020)
Lipovitellin	Egg yolk	Fat storage, transport	60%–70%	Deseta et al. (2021)
Phosvitin	Egg yolk	Iron-binding, antioxidant	10%–20%	Tang et al. (2022)
Vitellogenin	Egg yolk	Precursor to lipovitellin, involved in yolk formation	Precursor protein, not in mature eggs	Gu et al. (2017)

Major Proteins

Lipovitellin (60%–70% of Yolk Protein): The most abundant yolk protein, which is essential for fat storage as well as movement. Lipovitellin is a component of the lipoproteins in the yolk that transfer lipids to the embryo.

Phosvitin (10%–20% of Yolk Protein): A phosphoprotein responsible for binding and storing iron in the yolk. Phosvitin has a role in iron metabolism and has potential antioxidant effects.

Vitellogenin (Precursor of Lipovitellin): Vitellogenin is a precursor protein produced in the liver of the laying hen and transported to the ovary, where it is incorporated into the yolk. It is not typically found in mature eggs but is important in the early stages of egg formation.

Lipoproteins: Lipoproteins (protein-lipid complexes) transport lipids in the yolk, allowing the embryo to develop properly by delivering needed fatty acids.

Minor Proteins

Ovoinhibitor: A minor protease inhibitor in the yolk that helps protect the egg from digestive enzyme.

Granulins: Small proteins in thc yolk that may play a role in cellular activities during embryogenesis.

Trace Proteins

Yolk Sac Proteins: These proteins play a key part in the early stages of embryonic development.

Ovalbumin-Like Proteins: These proteins have a similar structure as ovalbumin, however are found in significantly smaller concentrations in yolk.

FUNCTION-BASED CLASSIFICATION OF EGG PROTEINS

Egg proteins can be classified not only by their location, but also by their functions:

Structural Proteins

Ovalbumin: Conservation of amino acids.
Ovomucin: Provides Egg whites with a gel-like consistency while preserving structural integrity.

Protective/Antimicrobial Proteins

Lysozyme: It breaks down bacterial cell walls.
Avidin: Binds biotin, preventing bacterial growth.
Ovotransferrin: Binds iron, limiting microbial growth.

Transport and Storage Proteins

Lipovitellin: Stores lipids in the yolk.
Phosvitin: Stores iron and may also have antioxidant functions.

Enzyme Inhibitors

Ovomucoid: Inhibits proteases.
Ovoinhibitor: Protects the egg from protease degradation.

Immune Proteins

Lysozyme and Avidin also function as immunological barrier defense proteins, defending the egg against infections caused by bacteria.

STRUCTURE OF EGG PROTEINS

The structure of egg proteins is crucial to their function within the egg. Egg proteins are divided into two types: egg white proteins and egg yolk proteins, each with unique structural characteristics that contribute to their roles in egg protection, embryo development, and physical properties (Table 9.3). These proteins' structures can be studied at multiple levels, including primary, secondary, tertiary, and quaternary structure. The following is a detailed discussion of the structure of important egg proteins. Egg proteins' various architectures allow them to perform specific activities such as antimicrobial defense, nutrition transport, fat storage, enzyme inhibition, and the provision of necessary amino acids for embryogenesis. The capacity of these proteins to adopt many structural configurations contributes significantly to eggs' diversity and nutritional worth (Figure 9.3).

TABLE 9.3
Summary of Protein Structures

Protein	Location	Structure	Function
Ovalbumin	Egg white	Globular, α-helix & β-sheet	Storage protein, amino acid source
Ovotransferin	Egg white	Bilobal, β-sheet	Iron-binding, antimicrobial
Ovomucoid	Egg white	Globular, β-sheet	Trypsin inhibitor, protective
Lysozyme	Egg white	Globular, α-helix	Antibacterial enzyme
Avidin	Egg white	Globular, β-sheet	Binds biotin, limits bacterial growth
Lipovitellin	Egg yolk	Globular, α-helix & β-sheet	Fat storage, transport
Phosvitin	Egg yolk	β-sheet, unstructured	Iron-binding, antioxidant
Vitellogenin	Egg yolk	Globular, α-helix & β-sheet	Lipid and protein transport

Source: USDA.

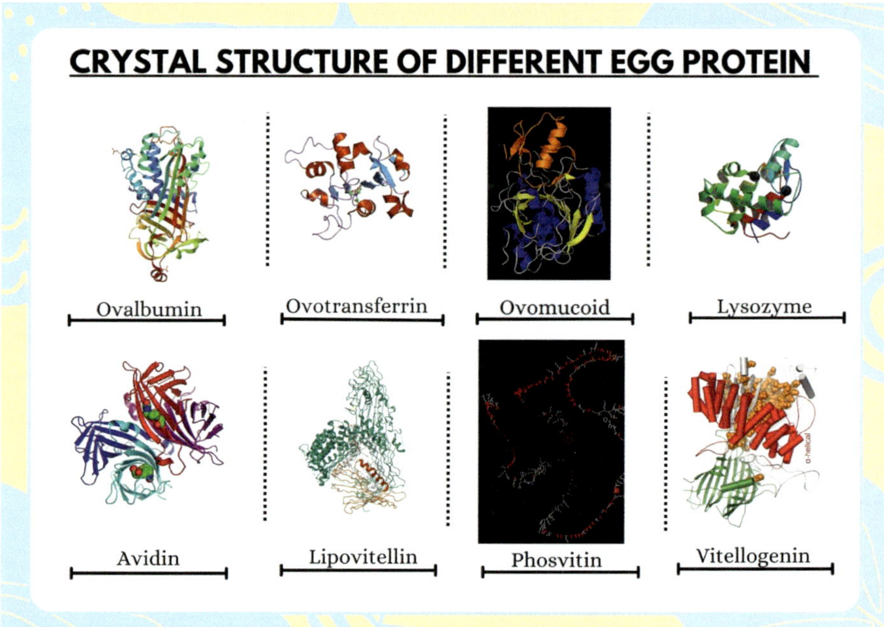

FIGURE 9.3 Crystal Structure of Egg Protein.

PROPERTIES OF EGG PROTEINS

Egg proteins have a number of distinctive features that contribute to their functional roles in both the egg and dietary supplements. These qualities are regulated by structural characteristics, amino acid content, and the presence of certain functional groups. The following is a summary of egg proteins' key qualities, which include solubility, thermal stability, emulsifying properties, gelation ability, foam generation, and nutritional value. Egg proteins are crucial to the structural integrity and useful characteristics of eggs in culinary applications. Their versatility in emulsification, foaming, gelation, and binding makes them indispensable in a wide range of food products, while their nutritional value provides significant dietary benefits.

Physical Properties

The physical properties of egg proteins are critical for both their utility in food preparation and their structural significance in the egg (Table 9.4). These properties are influenced by temperature, pH, ionic strength, and mechanical processing. Egg proteins' main physical qualities are viscosity, gelation, foam formation, denaturation, solubility, emulsification, and thermal stability. Here is a full explanation of each of these physical properties:

Viscosity: Viscosity, or a liquid's resistance to flow, is a key feature for many egg-based goods.

Egg White Proteins: The viscosity of egg whites rises with protein concentration and denaturation. When ovalbumin and other proteins in egg whites are denatured (e.g., by heat or mechanical agitation), they create a gel-like network, increasing viscosity.

Egg Yolk Proteins: Egg yolk's low viscosity is due to the presence of lipovitellin and phosvitin, which are absent in egg whites. However, when egg yolk is emulsified (as in mayonnaise or hollandaise sauce), the viscosity rises due to the creation of stable emulsion.

TABLE 9.4
Physical Properties of Egg Proteins

Property	Egg White Proteins	Egg Yolk Proteins
Viscosity	Increase with concentration and denaturation	Relatively low, increases with emulsification
Gelation	Forms firm with gels when heated	Softer gels, more viscous especially when emulsified
Foam formation	High foaming capacity	Limited foaming capacity when whipped
Denaturation & thermal stability	Denatures and coagulates around 80⁰C	More heat-stable, denatures at high temperature
Solubility	Soluble in water at neutral pH, less when heated	Insoluble in water, soluble in lipids
Emulsification	Minor emulsifying role	Lecithin
Color	Colorless/opaque when cooked	Yellow orange due to carotenoids
Protein-protein interactions	Forms stable networks	Emulsion
Elasticity & firmness	Firm and rigid when cooked	Elastic and creamy, especially when emulsified

Gelation: Gelation is the process by which proteins form a gel-like structure after heating and cooling. This characteristic can be particularly significant in many egg-based dishes.

Egg White Proteins: When egg whites are cooked, ovalbumin and ovomucoid desaturated and agglomerate, forming a network. This causes gel formation, which affects the texture of cooked egg whites (e.g., in boiled eggs or custards). The degree of gelation is determined by temperature and protein content.

Egg Yolk Proteins: The egg yolk contains lipovitellin, which contributes significantly to its gelation. When heated, yolk proteins desaturated and create gels, although the yolk gels are softer than egg white gels. This feature is vital for producing smooth-textured items such as custards and puddings.

Foam Formation: Egg whites are well-known for having the ability to produce stable foams, which is essential in baking, sweets, and gourmet preparations.

Foam Stability: When egg white proteins are whipped, they generate foams, particularly ovalbumin, ovotransferrin, and ovomucoid. These proteins unfold when mechanically agitated, trapping air and producing a stable network of protein molecules that help to stabilize the air bubbles.

Denaturation and Thermal Stability: Denaturation is an important physical property of egg proteins that refers to the unfolding or alteration of a protein's natural three-dimensional structure as a result of conditions like as heat, pH, or chemicals. This characteristic has an important significance in egg cooking.

Egg White Proteins: The thermal stability of egg white proteins, such as ovalbumin, is temperature dependent. At approximately 80°C (176°F), ovalbumin begins to denature and coagulate, transforming the egg white from a clear, viscous liquid to an opaque, solid content. This is an essential phase in boiling or frying eggs.

Egg Yolk Proteins: Egg yolk proteins, particularly lipovitellin, are more heat stable than egg whites. They start to denature at greater temperatures, but the yolk is more viscous and does not coagulate as quickly as the egg white.

Solubility: Solubility is the ability of egg proteins to dissolve in water or other solvents. The solubility of egg proteins can be determined by the protein type, solvent, and environmental factors such as pH and ionic strength.

Egg White Proteins: Most egg white proteins, including ovalbumin, are soluble in water at a neutral pH (about 7). However, ovalbumin loses solubility when heated (due to denaturation) or subjected to high pH levels. Ovotransferrin is somewhat more soluble than ovalbumin and can bind iron in aqueous solutions.

Egg Yolk Proteins: Lipovitellin in egg yolk is insoluble in water but soluble in lipid environments due to its hydrophobic properties. Phosvitin, a phosphoprotein, is soluble in water and helps to bind minerals like iron. Emulsification

Emulsification: Emulsification is the mixing of oil and water to create a stable dispersion. Egg yolk is a powerful emulsifier due to the presence of lecithin, a phospholipid, along with proteins like lipovitellin and ovovitellin (Tian et al., 2024). The excellent emulsifying properties of egg proteins have positioned them as a focal point of research in the development of food-grade Pickering emulsions (Ribeiro et al., 2021; Shi et al., 2020). Additionally, egg proteins have potential applications in the delivery of bioactive substances (Liu et al., 2022; Zhang et al., 2023). Lecithin: The most important emulsifier in egg yolk, lecithin, can decrease the surface tension of oil and water, allowing them to form a stable emulsion. This feature is essential for manufacturing dishes such as mayonnaise, salad dressings, and hollandaise sauce.

Egg White Proteins: While egg white proteins do not contain lecithin, they help with emulsification in some products by stabilizing the emulsion created by other components.

Color: The color of egg proteins, particularly egg yolk, plays an important role in food preparation.

Egg Yolk: The yellow to orange color of egg yolk is mainly caused by the presence of carotenoids such as lutein and zeaxanthin. These carotenoids are lipid-soluble and are present in lipovitellin and other yolk proteins.

Egg White: Egg whites usually appear colorless or slightly opaque, but when boiled, proteins such as ovalbumin denaturate, causing them to become opaque white. The texture and appearance of cooked egg whites are mostly due to protein aggregation and water retention.

Protein-Protein Interactions: Egg proteins, particularly those found in egg white, can form complex networks via protein-protein interactions, especially when denatured by heat or mechanical stress.

Egg Whites: When egg whites are whipped, proteins such as ovalbumin and ovotransferrin come together to create a network that stabilizes the foam. The hydrophobic and hydrophilic proteins contribute to the formation of this network by defining links and trapping air.

Elasticity and Firmness: The protein network derived after heating influences critical texture characteristics like as elasticity and stiffness.

Egg Whites: The hardness of cooked egg whites rises as ovalbumin and ovomucoid coagulate. When eggs are boiled or fried, the proteins cluster, resulting in a solid, firm consistency.

Egg Yolk: Because the egg yolk has a less rigid protein network than egg whites, it retains a more elastic and creamy quality even after heating. The presence of lipids also adds to the texture.

Egg proteins have a wide range of physical qualities that make them appropriate for a number of culinary uses. From foam production in baking to emulsification in sauces, egg proteins' physical qualities are critical for achieving texture, consistency, and stability in a variety of products.

Chemical Properties of Egg Proteins

The chemical characteristics of egg proteins play an important role in defining their functioning and behavior in a variety of applications, including cooking, food

TABLE 9.5
Chemical Properties of Egg Proteins

Property	Description	Key Proteins Involved
Amino acid composition	Complete protein with all essential amino acids	Ovalbumin, Ovo transferrin, Lipovitellin
Protein folding & denaturation	Heat pH, or mechanical forces can cause proteins to unfold and coagulate	Ovalbumin, Ovomucoid
pH sensitivity	Proteins aggregate or precipitate at their isoelectric point	Ovalbumin, Lipovitellin
Protein-lipid interactions	Form stable emulsions and interact with lipids	Lecithin, Lipovitellin
Oxidation stability	Egg protein and lipids are susceptible to oxidation	Ovalbumin, Lecithin, Lipovitellin
Binding of metals	Proteins like ovotransferrin bind metals like iron	Ovotransferrin, Phosvitin
Millard reaction	Reaction between amino acid and sugars contribute to browning and flavor	Ovalbumin, Cysteine containing proteins
Disulfide bond formation	Covalent bonds that stabilize proteins structures	Ovalbumin, Lipovitellin

processing, and nutritional value (Table 9.5). These chemical characteristics include amino acid composition, protein folding and denaturation, pH sensitivity, protein-lipid interactions, oxidative stability, metal binding, and disulfide bond formation. Here's a comprehensive description of egg protein's chemical features. The chemical features of egg proteins lead to their varied use in food science, nutrition, and culinary. They regulate how egg proteins interact with other ingredients, their stability under varied conditions, and their capacity to function in a variety of food items, including emulsions, foams, gels, and coatings. These characteristics additionally affect the texture, flavor, and appearance of cooked eggs and egg-based products.

Amino Acid Composition: Egg proteins contain a variety of amino acids, which are the building blocks of proteins. The amino acid profile determines the protein's structure, function, and nutritional value.

Complete Protein: Egg proteins are complete proteins, which means they include all nine essential amino acids that the human body cannot produce. These amino acids are: lysine, tryptophan, methionine, phenylalanine, valine, leucine, isoleucine, threonine, and histidine.

High Biological Value: Egg proteins have an adequate amino acid content, making them suitable for promoting tissue growth and repair. This is one reason why eggs are often referred to as having the highest biological value (BV) for protein.

Important Amino Acids:

Methionine: A sulfur-containing amino acid that plays an important role in protein synthesis.

Cysteine: Contains a thiol group (–SH) that can create disulfide bonds, which are critical for protein structure.

Glutamine: The most abundant amino acid in egg whites, and it is crucial for cellular metabolism.

Protein Folding and Denaturation: Proteins are composed of lengthy amino acid chains that fold into certain shapes with three dimensions known as native conformations. This folding is essential to the protein's functional properties.

Denaturation: When egg proteins become exposed to heat, acid, or mechanical agitation, they denaturate, or unfold from their original structure. Denaturation decreases proteins' biological activity and solubility, and in the case of egg whites, it induces coagulation (solidification).

For example, ovalbumin, the most prevalent protein in egg whites, denatures at approximately 80°C (176°F), transforming it from a liquid to a solid state.

Disulfide Bonds: The structure of egg proteins, such as ovalbumin and ovotransferrin, is based on disulfide bonds. These connections assist to maintain protein stability and solubility in their original state. When egg proteins are denatured, the disulfide connections break, and the proteins lose their natural structure.

pH SENSITIVITY

Egg proteins are sensitive to pH fluctuations. pH influences protein solubility, shape, and behavior.

Protein-Lipid Interactions: Egg proteins, particularly those found in egg yolks, interact with lipids (fats) in a variety of ways, which is essential for their emulsifying properties.

Emulsification: When egg yolk proteins, particularly lecithin (a phospholipid), combine with oil and water, they form stable emulsions. Lecithin contains both hydrophobic (water-repelling) and hydrophilic (water-attracting) regions, allowing it to stabilize the oil-water contact. This trait explains why egg yolks are utilized in mayonnaise, hollandaise sauce, and salad dressings.

Lipid Binding: Lipovitellin, the primary yolk protein, which is capable of binding lipids, plays a function in lipid transit and storage within the egg. Lipid-protein interactions also influence the texture and reliability of egg-based food products.

OXIDATION AND STABILITY

Egg proteins are vulnerable to oxidation, particularly in the presence of oxygen, light, and heat. Oxidation can impair the functional characteristics and quality of egg proteins.

Lipid Oxidation in Egg Yolk: The lipid content of egg yolk is susceptible to oxidation, which can result in off-odors and a decrease in the quality of egg-based food products. Oxidation is a key concern in the storage of eggs and egg products because it can cause rancidity.

Protein Oxidation: Proteins, particularly ovalbumin in egg white, can oxidize sulfhydryl groups (-SH), modifying their functional characteristics. Oxidation of proteins can cause disulfide bonds to form, protein chains to cross-link, and solubility to decrease.

Binding of Metals: Certain egg proteins, particularly ovotransferrin and phosvitin, have a tendency to bind metal ions, influencing their nutritional value and stability.

Iron Binding: Ovotransferrin Egg whites include an iron-binding protein that may preserve iron and hinder microorganism development by making iron less accessible. This is one of the explanations behind ovotransferrin's antimicrobial properties. capabilities.

Calcium and Phosphorus Binding: Phosvitin, a phosphoprotein found in egg yolk, can bind to calcium and phosphorus, making them available for absorption in the human body. This improves bone health and other metabolic processes.

Maillard Reaction: The Maillard reaction is a chemical reaction between amino acids (particularly lysine and cysteine) and reducing carbohydrates (such as glucose and fructose). This process occurs during cooking and helps generate the color and flavor of egg-based food items.

Browning: The Maillard process is responsible for the browning of eggs when cooked at high temperatures (for example, frying). It also produces pleasant chemicals that improve the flavor and scent of cooked eggs.

EFFECT OF PROCESSING ON EGG PROTEINS

Egg processing can have a significant influence on egg protein attributes such as functionality, texture, nutritional value, and bioavailability. The effect of processing on egg proteins is primarily determined by the type of processing (e.g., heating, freezing, drying), the processing conditions (e.g., temperature, duration, pH), and the specific proteins in the egg (egg white vs. yolk) (Table 9.6). The following are the main effects of various types of processing on egg proteins. Processing egg proteins can have a significant effect on their functional properties, including foaming, gelation, emulsification, and solubility. Understanding the effect of different processing processes is important for optimizing egg consumption in food manufacturing and assuring the correct texture, stability, and nutritional content in the final outcome.

APPLICATION OF EGG PROTEINS

Egg proteins have been employed in a variety of sectors due to their varied functional capabilities, which include emulsification, foaming, gelation, thickening, and binding. These qualities make egg proteins extremely valuable in food processing,

TABLE 9.6
Effects of Processing on Egg Proteins

Processing Method	Effect on Egg White Proteins	Effect on Egg Yolk Proteins	References
Boiling/Frying	Denaturation and coagulation, loss of foaming ability	Firming, but retains smooth texture and elasticity	Van der Plancken et al. (2006)
Pasteurization	Partial denaturation, minimal textural changes	Minimal effect, retains texture and nutritional value	Van der Plancken et al. (2005)
Freezing	Reduced foaming ability, watery texture when thawed	Gel- like or granular texture capacity, retains nutritional value	Zhao et al. (2014)
Spray Drying	Partial denaturation, reduced foaming ability	Reduced emulsifying capacity, retains nutritional value	Singh and Ramaswamy (2013)
Salt Addition	Increased solubility and foaming ability	Increased solubility, enhanced emulsifying properties	Chen et al. (2015)
Acid Addition	Denaturation, reduced solubility	Denaturation, reduced solubility	Liu et al. (2017)
Alkali Addition	Enhanced foaming and viscosity	Potential for increased emulsifying ability	Chen et al. (2015)

TABLE 9.7
Applications of Egg Proteins

Industry	Application
Food industry	Emulsification (mayonnaise, dressings), foaming (meringues, cake), binding (meat products), texture (ice cream)
Pharmaceuticals	Vaccine production, protein supplements, wound healing
Cosmetics	Skin care, emulsifiers (creams, lotions)
Biotechnology	Protein research, diagnostic tests, antimicrobial properties
Agriculture	Biodegradable films, animal feed additives, fertilizers
Art and craft	Egg tempera painting, craft adhesives

baking, medicines, cosmetics, and biotechnology (Table 9.7). The following are the most common uses of egg proteins in various industries. Egg proteins' distinctive abilities such as emulsification, foaming, gelling, and binding, make them vital in a wide range of applications across industries. Their nutritional value, safety, and variety of functions ensure their continued application in food production, pharmaceuticals, personal care products, and other applications.

Food Industry

Egg proteins, particularly those present in egg whites and yolks, are employed in a range of culinary products because they enhance texture, flavor, and nutritional value. They are noted for their capacity to generate stable emulsions, foams, and gels, making them essential ingredients in many recipes (Abeyrathne et al., 2014). Lysozyme is a significant bacteriolytic protein found in egg white that inhibits foodborne pathogens such as Listeria monocytogenes and Clostridium botulinum (Cegielska-Radziejewska et al., 2008).

Baking

Egg Whites: The foamy characteristics of egg whites make them indispensable in the baking business. When whipped, the proteins in egg whites (mainly ovalbumin and ovomucoid) trap air, resulting in light and airy textures in goods such as meringues, angel food cakes, and soufflés.

Egg Yolk: The emulsifying qualities of egg yolk are important in baking, especially for biscuits, cakes, and pastries. The lecithin in egg yolk helps to combine fat and water, resulting in a smooth batter or dough.

EMULSIONS AND SAUCES

Mayonnaise and Dressings: Although egg yolk contains lecithin, a natural emulsifier, it is a key component in the preparation of mayonnaise, hollandaise sauce, and other emulsified sauces. Egg proteins help to stabilize the oil-water combination by preventing it from separating.

ICE CREAM AND FROZEN DESSERTS

Egg Yolks: The proteins found in egg yolk, specifically lipovitellin, contribute to the texture and creaminess of ice cream and custard-based frozen treats. Egg yolk proteins also aid in emulsification, resulting in a smooth, homogenous product with adequate air retention.

FOAMS AND WHIPPED PRODUCTS

Egg Whites: Egg whites are used to make stable foams in confections such as marshmallows, cakes, whipped toppings, and meringues. The ovalbumin protein in egg whites produces a network that traps air, leading to the ideal texture.

MEAT AND FISH PRODUCTS

Egg Proteins: Egg proteins serve as binders in goods such as sausages, meatballs, and fish cakes, helping to keep the contents together and improving texture. They also aid in moisture retention during cooking, resulting in a juicier final product.

NUTRITIONAL FOODS

Egg Proteins: Because eggs contain complete proteins, they are frequently found in nutritional goods such as protein bars, meal replacements, and infant feeds. Egg protein isolates (such as egg white protein powder) have become popular as a supplement in sports nutrition.

Pharmaceutical and Medical Application

Egg proteins are widely used in the pharmaceutical business, owing to their biological qualities and high nutritional value. Ovotransferrin has the ability to bind iron and rapidly release the bound iron at pH values less than 4.5 (Ko & Ahn, 2008), making it an attractive prospective source of iron supplementation for humans (Abdallah & Chahine, 1999). Ibrahim and Kiyono (2009) demonstrated that ovotransferrin undergoes thiol-linked auto-cleavage following reduction, resulting in partially hydrolyzed products with potent anticancer activity against colon and breast cancer cells. Additionally, lysozyme functions as an immune-modulating agent and has the ability to suppress tumor cell growth (Kovacs-Nolan et al., 2005), positioning it as a potential anticancer agent.

VACCINE PRODUCTION

Ovalbumin and other egg proteins are employed in the creation of vaccines, particularly those for influenza. Chicken eggs are used to cultivate viruses, particularly the influenza virus, for the manufacturing of inactivated vaccines. The viral antigens are extracted from the eggs and employed in vaccine formulations.

PROTEIN SUPPLEMENTS

Egg White Proteins: Egg white protein powders are frequently used in nutritional supplements to provide high-quality protein. They have become highly popular in sports nutrition to aid in muscle repair and growth. Egg whites are a good source of vital amino acids for athletes due to their high biological value (BV) and protein content.

WOUND HEALING AND THERAPEUTICS

Because of their antimicrobial properties and anti-inflammatory qualities, egg-derived proteins could be used to treat wounds. Some egg proteins, such as ovotransferrin, have antibacterial characteristics that can be used as topical wound and burn treatments.

COSMETIC AND PERSONAL CARE PRODUCTS

Egg proteins are utilized in beauty products because they have been shown to improve skin texture, suppleness, and moisture.

SKIN CARE

Egg White Proteins: Egg white proteins, particularly albumin, are employed in face masks and skin treatments to tighten and firm. They are said to decrease pore size and improve skin texture by temporarily tightening the skin.

HAIR CARE

Egg Proteins: Shampoos and conditioners include egg proteins, particularly ovalbumin. These proteins help to strengthen and repair damaged hair by adhering to the hair shaft and giving moisture and nutrients. They also enhance shine and aid in detangling.

EMULSIFIERS IN CREAMS AND LOTIONS

Lecithin in egg yolk is a natural emulsifier that can be used in creams, lotions, and moisturizers to assist blend oil and water while also improving the product's smooth texture.

Biotechnology and Research

Egg proteins are frequently used in biotechnology and research, especially in investigations on protein chemistry and immunology.

Protein Research

Egg proteins including ovalbumin, ovotransferrin, and phosvitin are used as model proteins in structural biology and biochemistry studies. Researchers often utilize these proteins to investigate protein folding, denaturation, and protein-protein interactions.

Diagnostic Tests

Some diagnostic tools use egg proteins to identify antibodies or antigens in medical diagnoses. For example, ovalbumin can be utilized in enzyme-linked immunosorbent tests (ELISA) and other immunoassays.

Antimicrobial and Antiviral Research

Ovotransferrin, an egg white protein, has been investigated for antimicrobial properties and antiviral activities. It binds metal ions, notably iron, rendering them inaccessible to microbes and has direct antimicrobial properties. This has prompted investigation into its possible applications in food safety and healthcare.

Industrial and Agricultural Application

Egg proteins are also used in various kinds of industrial and agricultural areas due to their unique features.

Biodegradable Films

Egg Proteins: In the packaging industry, egg proteins, particularly those from egg whites, have been explored as biodegradable alternatives to plastic films. These films are eco-friendly and can be used for food packaging or as protective coatings for agricultural products.

Fertilizer and Feed Additives

Egg Proteins: Some egg proteins, such as egg albumin, are utilized as animal feed additives due to their high nutritional value. They are also utilized as fertilizers to provide protein-based organic matter, which may assist in enhancing soil quality.

Art and Craft: Egg proteins have been used in various kinds of artistic applications, particularly traditional painting and crafts.

EGG TEMPERA PAINT

Egg Yolk: Historically, egg tempera was used as a painting medium, combining egg yolk with pigments to produce a stable, long-lasting paint. The proteins in egg yolk aid in bonding the pigments, enhancing the adherence and endurance of paint on canvas or wood.

FUTURE PERSPECTIVE OF EGG PROTEINS

Egg proteins have a promising future in a variety of industries due to their unique functional characteristics and nutritional benefits. As the demand for sustainable, healthful, and diverse ingredients grows, egg proteins are projected to play a more significant role in a variety of developing trends and technologies. The following are some key areas where egg proteins are expected to undergo significant development in the coming years:

SUSTAINABLE AND ALTERNATIVE PROTEINS

With a global push for sustainability as well as lowering the environmental impact of food production, egg proteins are becoming more popular as a protein source with a lower carbon footprint than traditional animal proteins such as beef, pork, and chicken.

Eco-friendly Production: Egg protein synthesis facilities require fewer resources than many other animal-based protein sources. Eggs require less land, water, and energy than other meat sources, making them a sustainable choice for food manufacturers seeking high-quality proteins with less impact on the environment.

Plant-Based and Hybrid Products: As part of the movement toward plant-based and hybrid meals, eggs are expected to be mixed with plant proteins to improve the flavor, texture, and nutritional profile of plant-based goods. Companies are already combining egg proteins with plant-based components to make products that have the flavor and feel of meat or dairy.

HEALTH AND NUTRITION

Egg proteins are termed complete proteins because they contain all nine essential amino acids in sufficient quantities. As health-conscious consumers seek protein-rich diets that provide additional nutritional benefits, egg proteins will remain a key element in functional foods and dietary supplements.

Protein Supplements: The requirement for high-quality protein sources, particularly among athletes and active consumers, is growing. Egg white protein powders are projected to gain popularity due to their low fat and high protein content. Protein bars, meal replacements, and other sports nutrition products may gradually include easily digestible egg protein isolates and hydrolyzed proteins.

Bioactive Peptides: Research into bioactive peptides produced by egg proteins is progressing. These peptides have been demonstrated to have antioxidant, antibacterial, antihypertensive, and immunomodulatory properties, and they could be turned into functional foods or nutraceuticals to support overall health and wellness.

FOOD INNOVATION AND NOVEL APPLICATIONS

The future of egg proteins in food products also depends on inventive uses in new food categories, which are driven by consumer preferences for convenience, texture, and nutritional benefits.

Functional Foods and Beverages: Egg proteins, specifically egg white protein, are projected to be used in a broader range of functional beverages and snacks, particularly protein-rich drinks, smoothies, and granola bars. These items are aimed at consumers looking for high-protein snacks that can be ingested on the go and help them maintain an active lifestyle.

Clean Label Products: As consumers become more concerned with clean labels and natural ingredients, egg proteins may become a popular choice in products that eschew synthetic additives and preservatives. Egg proteins' natural emulsifying and stabilizing qualities might be used frequently in products that value plant-based ingredients and sustainability, such as dairy alternatives and gluten-free baking.

Plant-Egg Hybrids: Advances in food technology may result in the development of hybrid food products mixing egg proteins with plant-based or microbial proteins. These hybrid goods may have the same flavor and feel as traditional egg-based foods, appealing to people looking to reduce their animal-based product consumption.

Enzyme-assisted Extraction: Emerging biotechnology technologies, such as enzyme-assisted extraction, have the potential to produce high-quality egg proteins that have greater functional properties. This may lead to new uses in food products where the performance of egg proteins is improved.

Non-thermal Processing: High-pressure processing (HPP) and pulsed electric fields (PEF) can be utilized to preserve the nutritional value and functional characteristics of eggs while avoiding the damage caused by typical thermal processing. These improvements may improve egg protein stability while preserving its inherent characteristics in packaged and processed foods.

3D Food Printing: As 3D food printing improves, egg proteins may become an essential element in creating customized food textures and shapes in printed dishes. Egg proteins are excellent for food printing due to their viscosity and ability to form solid shapes.

BIOTECHNOLOGY AND MEDICAL USES

Egg proteins may see increased use in biotechnology, notably for medicinal applications and biopharmaceutical research.

Vaccine Production: Egg-based platforms for vaccine production are already in use for influenza and other vaccines, and research into egg-based production systems is expanding. Egg-derived proteins may also be further developed for use in gene therapy or cell culture technologies.

Immunotherapies and Drug Delivery: Egg proteins, particularly ovotransferrin, are being examined for potential applications in drug delivery systems. Egg proteins' capacity to bind to certain molecules or pathogens may allow them to be used to target specific parts of the body, leading to more effective medication delivery and immunotherapy.

Biodegradable Packaging: Research into the production of biodegradable egg protein films may lead to their use in medical packaging or pharmaceutical coatings. In the medical field, these films could be utilized for packing sterile drugs or as wound dressings.

GLOBAL TRADE AND MARKET EXPANSION

Egg proteins, particularly egg powder and egg protein isolates, are projected to witness increased demand in international markets, particularly in developing countries where people have become more aware of the benefits of high-quality protein.

Export Opportunities: Egg protein exports are expected to increase as globally demand for protein-rich meals rises, particularly in population-growing regions (such as Asia, Africa, and Latin America). Egg protein isolates and concentrates may become increasingly prevalent components in global food supply connections.

ETHICAL AND CONSUMER AWARENESS

As people become more concerned about responsible consumption and animal welfare, the future of egg proteins may see advancements in ethical egg production methods.

Animal Welfare: The development of free-range or pasture-raised eggs, which prioritize animal welfare standards, may result in a market demand for higher-quality, ethically produced egg proteins. Consumers who shun animal-based products for ethical reasons may benefit from innovations in lab-grown eggs or egg protein substitutes.

CONCLUSION

Egg proteins, which are primarily generated from egg whites and yolks, are some of the most nutritious, functional, and versatile proteins known. They include a distinct blend of high-quality amino acids, including all nine essential amino acids, making them a complete protein source. This, combined with their high bioavailability, makes egg proteins a good dietary choice for promoting muscular power, immune system function, and general wellness. Egg proteins' functional qualities, such as emulsification, foaming, gelation, and binding, make them vital in a variety of food products, including baked goods and confections, dairy substitutes, and meat substitutes. Egg proteins serve as essential elements in the manufacturing of stable emulsions, aerated foams, and smooth textures, which are required in the formulation of many processed foods. Egg proteins are also being investigated in biotechnology, particularly for the creation of vaccines, drug delivery systems, and biopharmaceuticals, as well as in medicinal applications such as wound healing and immunotherapies. They have incredible potential for individualized nutrition, as their high protein content and digestibility may be adapted to specific health demands. Furthermore, egg proteins show great promise in the goal of sustainable food production. As worldwide demand for sustainable and ethically sourced protein grows, egg proteins offer a more environmentally friendly option than other animal proteins, with a lower carbon footprint and less use of resources. With the rise of plant-based foods and the growing demand for functional and clean-label products, egg proteins are projected to remain a key ingredient in novel food compositions. They are predicted to play a significant part in the development of novel food technologies, such as 3D food printing and alternative protein blends that combine the health advantages of both animal and plant proteins.

Finally, egg proteins will remain a significant, flexible, and sustainable element in a variety of industries, including food production, health supplements, biotechnology, and environmental applications. As consumer demand for health-conscious, sustainable, and high-quality proteins rises, egg proteins will remain at the forefront of meeting these demands, making them a key part of both today's and future food systems.

REFERENCES

Abdallah, F. B., & Chahine, J. M. E. H. (1999). Transferrins, the mechanism of iron release by ovotransferrin. *European Journal of Biochemistry*, 263, 912–920.

Abeyrathne, E. D. N. S., Lee, H. Y., Jo, C., Nam, K. C., & Ahn, D. U. (2014). Enzymatic hydrolysis of ovalbumin and the functional properties of the hydrolysates. *Poultry Science*, 93(10), 2678–2686.

Abeyrathne, E. N. S., Lee, H. Y., & Ahn, D. U. (2013). Egg white proteins and their potential use in food processing or as nutraceutical and pharmaceutical agents—A review. *Poultry Science*, 92(12), 3292–3299.

Ali, A. A. M., Noor, H. S. M., Chong, P. K., Babji, A. S., & Lim, S. J. (2019). Comparison of amino acids profile and antioxidant activities between edible bird nest and chicken egg. *Malaysian Applied Biology*, 48(2), 63–69.

Anton, M. (2013). Egg yolk: Structures, functionalities and processes. *Journal of the Science of Food and Agriculture*, 93(12), 2871–2880.

Campbell, L., Raikos, V., & Euston, S. R. (2003). Modification of functional properties of egg-white proteins. *Food*, 47(6), 369–376.

Cegielska-Radziejewska, R., Leśnierowski, G., & Kijowski, J. (2008). Properties and application of egg white lysozyme and its modified preparations–a review. *Polish Journal of Food and Nutrition Sciences*, 58, 5–10.

Chen, Z., Li, J., Tu, Y., Zhao, Y., Luo, X., Wang, J., & Wang, M. (2015). Changes in gel characteristics of egg white under strong alkali treatment. *Food Hydrocolloids*, 45, 1–8.

Cui, J., Zhou, J., Huang, L., Jing, J., Wang, N., & Wang, L. (2019). Curcumin encapsulation and protection based on lysozyme nanoparticles. *Food Science & Nutrition*, 7(8), 2702–2707. doi: 10.1002/fsn3.1129.

Deseta, M. L., Sponton, O. E., Erben, M., Osella, C. A., Frisón, L. N., Fenoglio, C., … & Perez, A. A. (2021). Nanocomplexes based on egg white protein nanoparticles and bioactive compounds as antifungal edible coatings to extend bread shelf life. *Food Research International*, 148, 110597. doi: 10.1016/j.foodres.2021.110597.

Dudusola, I. O. (2010). Comparative evaluation of internal and external qualities of eggs from quail and guinea fowl. *International Research Journal of Plant Science*, 1, 112–115.

Freschi, J., Razafindralambo, H., Danthine, S., & Blecker, C. (2011). Effect of ageing on different egg yolk fractions on surface properties at the air–water interface. *International Journal of Food Science and Technology*, 46, 1716–1723.

Gu, L., Su, Y., Zhang, M., Chang, C., Li, J., McClements, D. J., & Yang, Y. (2017). Protection of β-carotene from chemical degradation in emulsion-based delivery systems using antioxidant interfacial complexes: Catechin-egg white protein conjugates. *Food Research International*, 46(8), 1716–1723. doi: 10.1016/j.foodres.2017.03.015.

Hamid-Samimi, M., Swartzel, K. R., & Ball, H. R. (1984). Flow behavior of liquid whole egg during thermal treatments. *Journal of Food Science*, 49(1), 132–136.

Herald, T. J., & Smith, D. M. (1989). Functional properties and composition of liquid whole egg proteins as influenced by pasteurization and frozen storage. *Poultry Science*, 68(11), 1461–1469.

Horbañczuk, J. O. (2002). *The Ostrich*. European Ostrich Group, Ribe, Denmark.

Ibrahim, H. R., & Kiyono, T. (2009). Novel anticancer activity of the autocleaved ovotransferrin against human colon and breast cancer cells. *Journal of Agricultural and Food Chemistry*, 57, 11383–11390.

Ko, K. Y., & Ahn, D. U. (2008). An economic and simple purification procedure for the large-scale production of ovotransferrin from egg white. *Poultry Science*, 87, 1441–1450.

Kovacs-Nolan, J., Phillips, M., & Mine, Y. (2005). Advances in the value of eggs and egg components for human health. *Journal of Agricultural and Food Chemistry*, 53, 8421–8431.

Laca, A., Paredes, B., Rendueles, M., & Díaz, M. (2015). Egg yolk plasma: Separation, characteristics and future prospects. *LWT-Food Science and Technology*, 62, 7–10.

Liu, J., Chai, J., Yuan, Y., Zhang, T., Saini, R. K., Yang, M., & Shang, X. (2022). Dextran sulfate facilitates egg white protein to form transparent hydrogel at neutral pH: Structural, functional, and degradation properties. *Food Hydrocolloids*, 122, 107094. doi: 10.1016/j.foodhyd.2021.107094.

Liu, Y.-F., Oey, I., Bremer, P., Silcock, P., & Carne, A. (2017). In vitro peptic digestion of ovomucin-depleted egg white affected by pH, temperature and pulsed electric fields. *Food Chemistry*, 231, 165–174.

Mine, Y. (2002). Recent advances in egg protein functionality in the food system. *World's Poultry Science Journal*, 58(01), 31–39.

Nimalaratne, C. & Wu, J. (2015). Hen egg as an antioxidant food commodity: A review. *Nutrients,* 7: 8274–8293.

Nimalaratne, C., Schieber, A. & Wu, J. (2016). Effect of storage and cooking on antioxidant capacity of laying hen egg. *Food Chemistry*, 194, 111–116.

Razi, S. M., Fahim, H., Amirabadi, S., & Rashidinejad, A. (2023). An overview of the functional properties of egg white proteins and their application in the food industry. *Food Hydrocolloids*, 135, 108183.

Ribeiro, E. F., Morell, P., Nicoletti, V. R., Quiles, A., & Hernando, I. (2021). Protein- and polysaccharide-based particles used for Pickering emulsion stabilisation. *Food Hydrocolloids*, 119, 106839.

Shah, R. B., & Khan, M. A. (2004). Protection of salmon calcitonin breakdown with serine proteases by various ovomucoid species for oral drug delivery. *Journal of Pharmaceutical Sciences*, 93(2), 392–406. doi: 10.1002/jps.10557.

Shi, A., Feng, X., Wang, Q., & Adhikari, B. (2020). Pickering and high internal phase Pickering emulsions stabilized by protein-based particles: A review of synthesis, application and prospective. *Food Hydrocolloids*, 109, 106117.

Singh, A., & Ramaswamy, H. S. (2013). Effect of high-pressure processing on color and textural properties of eggs. *Journal of Food Research*, 2(4), 11.

Sun, C., Liu, J., Yang, N., & Xu, G. (2019). Egg quality and egg albumen property of domestic chicken, duck, goose, turkey, quail, and pigeon. *Poultry Science*, 98(10), 4516–4521.

Surai, P. & Sparks, N. (2001). Designer egg: Form improvement of egg composition functional food. *Food Science & Technology*, 12, 7–16.

Tang, S., Y. Jiang, T. Tang, H. Du, Y. Tu, and M. Xu. (2022). Effects of grafting degree on the physicochemical properties of egg white protein-sodium carboxymethylcellulose conjugates and their aerogels. *Applied Sciences*, 12(4):2017. doi: 10.3390/app12042017

Tian, Y., Lv, X., Oh, D. H., Kassem, J. M., Salama, M., & Fu, X. (2024). Emulsifying properties of egg proteins: Influencing factors, modification techniques, and applications. *Comprehensive Reviews in Food Science and Food Safety*, 23(5), e70004.

Tolik, D., Poawska, E., Charuta, A., Nowaczewski, S., & Cooper, R. (2014). Characteristics of egg parts, chemical composition and nutritive value of Japanese quail eggs–a review. *Folia Biologica (Kraków)*, 62(4), 287–292.

Van der Plancken, I., Van Loey, A., & Hendrickx, M. E. (2005). Changes in sulfhydryl content of egg white proteins due to heat and pressure treatment. *Journal of Agriculture and Food Chemistry*, 53, 5726–5733.

Van der Plancken, I., Van Loey, A., & Hendrickx, M. E. (2006). Effect of heat-treatment on the physico-chemical properties of egg white proteins: A kinetic study. *Journal of Food Engineering*, 75(3), 316–326.

Wang, Y., Zhang, L., Wang, P., Xu, X., & Zhou, G. (2020). pH-shifting encapsulation of curcumin in egg white protein isolate for improved dispersity, antioxidant capacity and thermal stability. *Food Research International*, 137, 109366.

Wei, Z., Chen, Y., Wijaya, W., Cheng, Y., Xiao, J., & Huang, Q. (2020). Hydrogels assembled from ovotransferrin fibrils and xanthan gum as dihydromyricetin delivery vehicles. *Food & Function*, 11(2), 1478–1488. doi: 10.1039/c9fo02564b.

Xu, K. (2017). Egg albumen as a fast and strong medical adhesive glue. *Advanced Healthcare Materials*, 6, 1700132.

Zambrowicz, A., Dąbrowska, A., Bobak, Ł., & Szołtysik, M. (2014). Egg yolk proteins and peptides with biological activity. *Postepy Higieny i Medycyny Doswiadczalnej*, 68, 1524–1529.

Zhang, Y., Guo, Y., Liu, F., & Luo, Y. (2023). Recent development of egg protein fractions and individual proteins as encapsulant materials for delivery of bioactives. *Food Chemistry*, 403, 134353.

Zhao, Y., Tu, Y., Xu, M., Li, J., & Du, H. (2014). Physicochemical and nutritional characteristics of preserved duck egg white. *Poultry Science*, 93(12), 3130–3137.

Index

Note: **Bold** page numbers refer to tables and *italic* page numbers refer to figures.

U

V

W

Z

For Product Safety Concerns and Information please contact our
EU representative GPSR@taylorandfrancis.com Taylor & Francis
Verlag GmbH, Kaufingerstraße 24, 80331 München, Germany